CURRENT TOPICS IN PHOTOSYNTHESIS

Current topics in photosynthesis

Dedicated to Professor L.N.M. Duysens on the occasion of his retirement

edited by

J. AMESZ, A.J. HOFF and H.J. VAN GORKUM

Department of Biophysics
State University of Leiden
Leiden, The Netherlands

Reprinted from Photosynthesis Research, Volume 9, Numbers 1–2

1986 **MARTINUS NIJHOFF PUBLISHERS**
a member of the KLUWER ACADEMIC PUBLISHERS GROUP
DORDRECHT / BOSTON / LANCASTER

Distributors

for the United States and Canada: Kluwer Academic Publishers, 190 Old Derby Street, Hingham, MA 02043, USA
for the UK and Ireland: Kluwer Academic Publishers, MTP Press Limited, Falcon House, Queen Square, Lancaster LA1 1RN, UK
for all other countries: Kluwer Academic Publishers Group, Distribution Center, P.O. Box 322, 3300 AH Dordrecht, The Netherlands

ISBN-13: 978-94-010-8463-5   e-ISBN:978-94-009-4412-1
DOI: 10.1007/978-94-009-4412-1

Copyright

Table of contents

Foreword

Four decades ago, when Lou Duysens was about to start his work on fluorescence and energy transfer in photosynthesis that would lead to his thesis [1], very little was known about the molecular mechanisms of photosynthesis, certainly from our present-day point of view. However, this state of affairs would rapidly change in the ensuing years by the introduction of modern physical and biochemical techniques. Especially the field of optical spectroscopy, on which the work of Duysens had such a significant impact, has proved to be one of the most fruitful techniques in the study of primary processes and electron transfer reactions in photosynthesis.

Duysens' thesis established the role of energy transfer in photosynthesis and also showed for the first time the existence in photosynthetic bacteria of light-induced absorbance changes of what is now known as the primary electron donor P-870. Subsequent studies by the same method demonstrated the photo-oxidation of cytochromes, both in bacteria [2] and in algae [3, 4] and of the absorbance changes [3] that were later found to be due to electrochromic band shifts of antenna pigments. Measurements of cytochrome kinetics in light of various wavelengths led to the concept of two photosystems in green plant photosynthesis [5], whereas a study of the factors affecting the fluorescence yield of chlorophyll gave the first information on the electron acceptor Q of photosystem II [6].

Research in many laboratories all over the world has led to an understanding of primary and secondary processes in photosynthesis and of the structure of the photosynthetic membrane that is vastly more detailed than could have been envisaged 40 years ago. This is well illustrated by this special issue on the occasion of Lou Duysens' retirement, which we hope will give an impression of the present status of research.

As editors of this issue of Photosynthesis Research we are very grateful to our colleagues — many of which at one time or another worked at the Leiden laboratory for some months or years — for their contributions. We are much indebted to the Editors of Photosynthesis Research, Prof. Govindjee and Dr. R. Marcelle who endorsed our plans, and to Ir. A.C. Plaizier of Martinus Nijhoff Publishers for his kind help.

The Editors,
J. Amesz
A.J. Hoff
H.J. van Gorkom

References

1. Duysens LNM (1952) Doctoral thesis, University of Utrecht

2. Duysens LNM (1954) Nature 173:692
3. Duysens LNM (1954) Science 120:353–354
4. Duysens LNM (1955) Science 121:210–211
5. Duysens LNM, Amesz J and Kamp BM (1961) Nature 190:510–511
6. Duysens LNM and Sweers HE (1963) In 'Studies on Microalgae and Photosynthetic Bacteria', Special Issue Plant Cell Physiol, pp 353–372. Tokyo, University of Tokyo Press

Introduction

A major event in the history of photosynthesis research was the establishment of the Biophysical Research Group in Utrecht with support of the Rockefeller Foundation about 1935. This group was initiated by Prof. A.J. Kluyver, the Delft microbiologist, and Prof. Ornstein of the Physics Department at Utrecht to combine expertise in both biology and physics. This group, directed by Prof. Wassink and later by Prof. Thomas, produced an abundance of top quality research and also trained very distinguished scientists. One of these, Prof. Louis N.M. Duysens, whose impending retirement is the occasion for this volume, has indeed more than fulfilled the hopes of the Utrecht Group's founders by his distinguished research career and productive leadership of the Biophysical laboratory at Leiden.

Because Lou Duysens had been thoroughly trained in basic physics he was able to think clearly in mathematical terms about biological problems even as a graduate student. Starting with studies of energy transfer between photosynthetic pigments, Duysens' 1952 Utrecht thesis clearly presented the basic optics of cell suspensions. The concluding part of that thesis reported the ingenious measurement of very small changes of absorption induced by light or by chemical treatment in photosynthetic bacteria. Such measurements of absorption changes have become widely used in many laboratories and have made possible the chemical identification of many intermediate steps in the process of photosynthesis. One of his discoveries when he came as a postdoctoral fellow to the Carnegie Institution was to find an absorption change at 515 nm in *Chlorella*. The challenge to explain this electrochromic change and its usefulness in studies of photosynthesis is attested by the numerous papers that have appeared on this subject since then. In the red alga *Porphyridium cruentum* Duysens and Amesz and Kamp found the 420 nm absorption change of cytochrome-f to indicate its oxidation by illumination at 680 nm, light absorbed by chlorophyll, and its reduction by 562 nm, light absorbed by phycoerythrin. They proposed the names system 1 and system 2 for these two separately acting groups of pigments which has become universally used. Studies of chlorophyll fluorescence, of absorption changes, and of energy transfer in photosynthetic organisms has been a continuing activity in Duysens laboratory. Photosynthetic bacteria and red or blue green algae, each with their own pigment systems, have been used as well as green algae and leaf chloroplasts throughout his career. The general principles of photochemical energy conversion have been elucidated by selection of the most suitable organism for each investigation. As this volume will show Duysens' efforts have been outstandingly successful in this competitive field of worldwide interest.

Department of Plant Biology   C. Stacy French
Carnegie Institution of Washington
Stanford, CA 94305, USA

Photosynthesis Research 9, 3–12 (1986)
© *1986 Martinus Nijhoff/Dr. W. Junk Publishers, Dordrecht.*

Molecular arrangement of pigment-protein complex of photosystem 1

V.V. SHUBIN, N.V. KARAPETYAN and A.A. KRASNOVSKY

A.N. Bakh Institute of Biochemistry, USSR Academy of Sciences, Moscow 117071, USSR

(*Received 25 June 1985*)

Key words: light-harvesting complex (pea); cPII, optical absorption, linear dichroism, chroism, protein secondary structure

Abstract. The circular dichroism (CD) method was applied to study the molecular organization of P700, antenna chlorophyll and protein of photosystem 1 complexes (CP1), isolated from chloroplasts under mild treatment with Triton X-100. Analysis of CD spectra and protein:chlorophyll:P700 ratios for CP1 complexes that were different in their chlorophyll content indicate that CP1 preparations can be considered as a mixture of CP1-RC, containing P700 (10–20%), and CP1-LH without P700 (80–90%). Both types of complexes contain approximately 25 chlorophyll molecules, and the destruction of their spatial organization with detergents represents a cooperative transition. The rate of chlorophyll destruction in CP1-LH is much higher than that in CP1-RC. In both complexes a 65 kDa polypeptide predominates, whose secondary structure (typical for α/β proteins) is stable to Triton X-100 and does not depends on the chlorophyll content. Chlorophyll seems to be grouped in clusters (5–7 molecules) in the hydrophobic cores of 2–3 parallel α/β domains of the 65 kDa protein. Only one of the clusters in CP1-RC includes P700; on P700 photooxidation the change of its interaction with the nearest pigment environment results in a complicated shape of the light-induced CD spectra.

Abbreviations

PS1, photosystem 1; CP1 pigment-protein complex of PS1; Chl, chlorophyll *a*; CP1-140, CP1 with ratio Chl:P700 140; RC, reaction center; LH, light-harvesting pigment; CP1-RC, CP1, containing P700; CP1-LH, CP1 without P700 (containing LH); CD, circular dichroism; SDS, sodium dodecyl sulfate.

Introduction

It is generally accepted that the primary photosynthetic reactions take place in the reaction centers (RC) of photosystems of green plants and photosynthetic bacteria. The suggestion [5, 10] that only a part of Chl is photochemically active anticipated the discovery of the RC. Bacterial RC can be isolated in purified state and crystallized, whereas the RC's of green plants contain a large amount of antenna Chl. Difficulties in isolation and crystallization

---

Dedicated to Prof. L.N.M. Duysens on the occasion of his retirement

of hydrophobic complexes of green plants with a high molecular weight until now did not allow their organization to be studied by X-ray diffraction. The application of the CD method can greatly contribute to solving the problem. As is known CP1 consists of 65 dKa subunits and four 10–25 kDa polypeptides; the Chl:P700 ratio varies between 120 and 25 [1, 2, 11]. However, the structural organization of CP1 is still unclear. The aim of the present paper is to investigate the molecular arrangement of P700, antenna Chl and the protein part of CP1 using CD measurements.

Materials and methods

CP1 complexes were isolated from chloroplast membranes of pea and maize and from the cyanobacterium *Spirulina platensis* by means of chromatography on hydroxylapatite [16] and gel filtration. CP1-140 was isolated after membrane solubilization with 2% Triton X-100 in 0.05 M Tris HCl buffer, pH 8.2 (detergent: Chl = 8) and sedimentation at 20 000 × g of non-destroyed grana. The resultant supernatant after chromatography on hydroxylapatite (without washing with buffer and detergent) was gel-filtrated through Biogel A-25 m, balanced with 0.02 M Tris HCl buffer + 0.05 M Triton X-100, and again subjected to chromatography on hydroxylapatite. CP1-140 was eluted with 0.02 M phosphate buffer, pH 7.2 with 0.05% Triton X-100 after washing the column with 0.05% Triton X-100 in 0.05 M Tris HCl buffer (4–5 volumes of the column). CP1-100 and CP1-150 were isolated from CP1-140 measuring the amount of free Chl, washed off the column with hydroxylapatite, containing 0.05% or 1% Triton X-100 in Tris HCl buffer. By this method it is possible to obtain complexes with any Chl:P700 ratio (from 140 to 30). Judged by the staining of electrophoretic gels with Coomassie G-250, the 65 kDa polypeptide predominated (80–90%) in all complexes studied.

The absorption spectra and their second derivatives were registered by Beckman-35 and Hitachi-356 spectrophotometers, the CD spectra were measured by a dichrograph JASCO-40AS. The light-induced CD spectra were measured with the dichrograph using a home developed phosphoroscope attachment [18]. The light-induced absorption spectra were measured using a single-beam difference spectrophotometer [8].

P700 content was calculated using as differential extinction coefficients $6.4 \cdot 10^4$ M^{-1} cm^{-1} (pea, maize) and $7.0 \cdot 10^4$ M^{-1} cm^{-1} (*Spirulina*) at 698 nm [6], respectively. Light-induced CD spectra of P700 (dark minus light) were measured point by point (intervals 1–2 nm), using a data processor. These spectra are presented in terms of the differential dichroic coefficient $\Delta\epsilon = \epsilon_1 - \epsilon_r$.

The concentration of the protein in CP1 in the presence of Triton X-100 was derived from the equation proposed by us: $C = 1.98 (A_{210}^{0.1} - A_{214}^{0.1})$, where $A_{210}^{0.1}$ and $A_{214}^{0.1}$ are the absorptions at 210 and 214 nm in a 0.1 cm cell, and C is the protein concentration (mg/ml). The coefficient 1.98 ± 0.06 was calculated

from measurements with bovine albumin, ribonuclease A, glutamate dehydro-
genase and glutamine synthetase. The concentration of the protein was
determined by Scops method [15]. Absorptions of the micellar solution of
Triton X-100 at 210 and 214 nm are identical, whereas the absorption by Chl
in this region is negligible.

The protein secondary structure was determined from the basic spectra
of α-helix, antiparallel and parallel β-structure, β-bands and the irregular
region calculated in [3, 4]. Experimental CD spectra were approximated
by the sum of the basic spectra using a non-linear least squares method
under the condition that $\Sigma f_i = 1$, where f_i is the molar fraction of the amino
acid residues in i-th structure. A BESM-6 computer was used. A mean weight
of the amino acid residues of 111 kDa was determined from the amino acid
composition of the 65 kDa protein [19].

Results and discussion

The composition of the studied complexes is summarized in Table 1. In
each case there are 700 000–800 000 g of protein per mole of P700. Similar
values were obtained for CP1 isolated by SDS-electrophoresis [20]. Since
isolated CP1 complexes contain two main 65 kDa polypeptides, it is obvious
that 80–85% of the CP1 complexes contain no P700. The methods used did

Table 1. Ratio of protein, Chl and P700 in different CP1 complexes isolated from maize

Sample	Chl/P700 mole/mole	Protein/Chl g/g	Protein/P700 g/mole	Chl/65 kDa protein mole/mole
CP1-140	140	6.1	760 000	12
CP1-100	100	8.8	790 000	8.3
CP1-50	50	16	720 000	4.5
CP1, isolated by electrophoresis [20]	65	13–17	780 000– 1 000 000	4.1–5.4
PS1 particles [7]	13	16	160 000	4.5

not allow us to separate CP1-RC from CP1-LH. However, centrifugation in
the sucrose density gradient of the PS1 particles, extracted with ether, resulted in a preparation with protein:P700 ratio of about 160 000 g/mole [7].

The number of Chl molecules per one 65 kDa polypeptide in CP1-140 is
about 12, i.e. intact CP1 complexes contain about 25 Chl molecules. This
explains the instability of P700 in a complex with a ratio Chl:P700 less than
25 [2], as well as the fact that inhibition of one P700 at high detergent
concentrations leads to the destruction of the spatial organization of 20–25
Chl molecules [9, 18]. Apparently, the enrichment of CP1 with RC by strong
detergent treatment is achieved by a more rapid Chl solubilization from complexes without P700.

Spectral properties of CP1 are presented in Figure 1. The absorption spectra and their second derivatives are almost equal for CP1-140, CP1-100 and CP1-50 and are characterized by 6 main narrow (7–9 nm) bands at 663, 670, 677, 684 and 698 nm, four of them predominating in CD spectra of all used CP1 complexes. However, the shape of the CD spectra noticeably deformed in proportion to the removal of antenna Chl because of a drop of

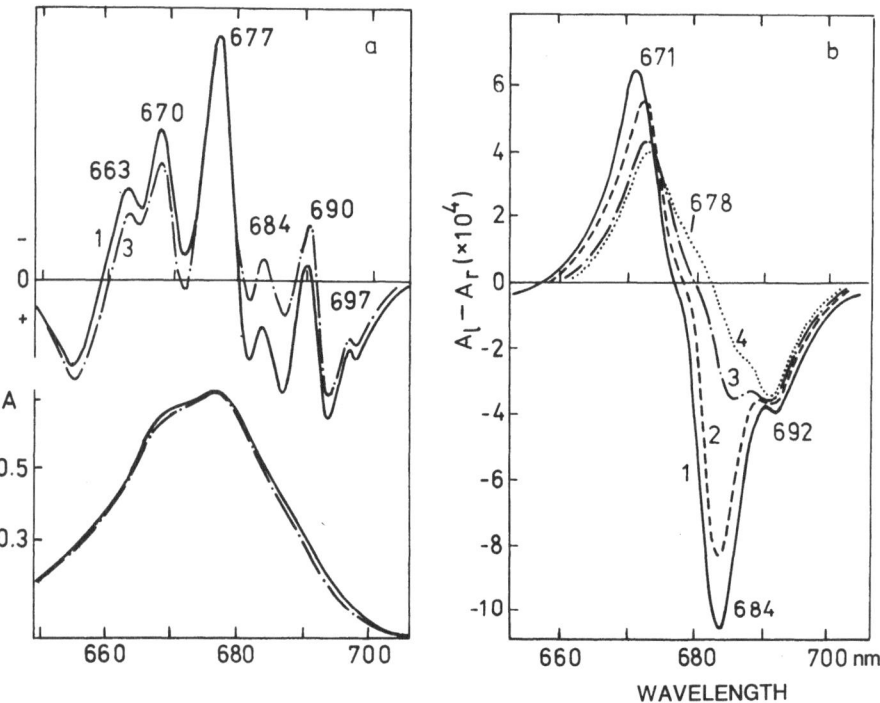

Figure 1. Absorption spectra and their second derivatives (a) and CD spectra (b) of CP1-140 (1), CP1-100 (2), CP-50 (3) and CP1-140 (4) at 77 K; CP1 isolated from maize.

rotational strength at 684 nm (without significant changes of the intensity of the band in the absorption spectrum). The observed deformation can be explained by a change of the ratio of CP1-RC and CP1-LH in the sample.

Thus, the removal of 60% of Chl from CP1-140 does not result in the disappearance of any bands in the electronic spectra of the Q transition of Chl and does not change the ratio between the intensities of the absorption bands. Moreover, the same set of spectral bands remains in the absorption spectra of PS1 particles with a Chl:P700 ratio of about 8–10 [18]. This fact is difficult to explain if one assumes that layer-to-layer removal of part of the antenna Chl from every complex leads to enrichment of CP1 by P700 [11]. Besides, this indicates a significant similarity of the spatial organization of Chl molecules in CP1-RC and CP1-LH. The destruction of the spatial organization of interacting pigment molecules under the effect of Triton X-100 occurring in each type of CP1 complex thus represents a cooperative transition. We have shown the 'cooperativeness' of the same process for CP1-RC, measuring the time course of disappearance of the CD signals of Chl in CP1-40 treated with Triton X-100 [18], the $t_{1/2}$ of the process being about 3–4 h, and for CP1-LH under the same conditions (1–2 min). We can conclude that the two processes are independent. Due to the difference in the stability of the complexes against detergents it is possible to enrich them in P700 without separation of the protein characteristics of CP1-LH.

Since the amount of low molecular proteins (10–25 kDa) in the studied preparations is essentially less than that in highly purified CP1-RC [2], it is obvious that these proteins are absent in CP1-LH (Figure 2). These proteins seem to bring about a higher stability of Chl in CP1-RC against detergent as well as of the activity of P700.

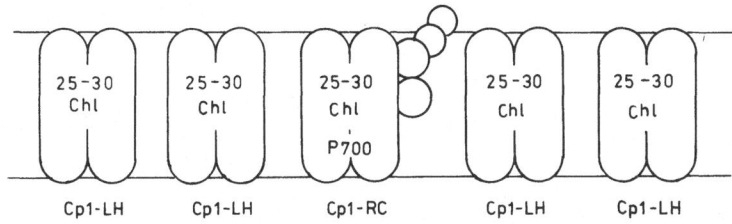

Figure 2. A scheme for the organization of CP1-RC and CP1-LH complexes in the thylakoid membrane.

The secondary structure of the 65 kDa protein was calculated from the CD spectra in the region 190–250 nm, presented in Figure 3. The shape of the spectra is independent of the Chl content and practically identical for CP1-140, CP1-100 and CP1-40. Accordingly, their secondary structures completely coincide (Table 2). Thus, Chl has no significant effect on the stability of the secondary structure of the CP1 protein. Pigments from the complexes could be removed by substituting detergent for them without protein conformational changes. The secondary structure of the 65 kDa protein is rather stable against SDS. A 3-hour incubation of CP1-50 with 0.1% SDS at 22–22 °C (detergent: protein: Chl = 150:10:1) resulted in destroying a considerable amount of pigments, but caused no significant changes in the secondary structure of the 65 kDa protein. Only a slight decrease of the proportion of the parallel β-structure from 0.24 to 0.18 was observed. The secondary structure of the CP1 complexes studied becomes similar to that of the CP1 isolated by SDS electrophoresis [9]. Even the disturbance of the

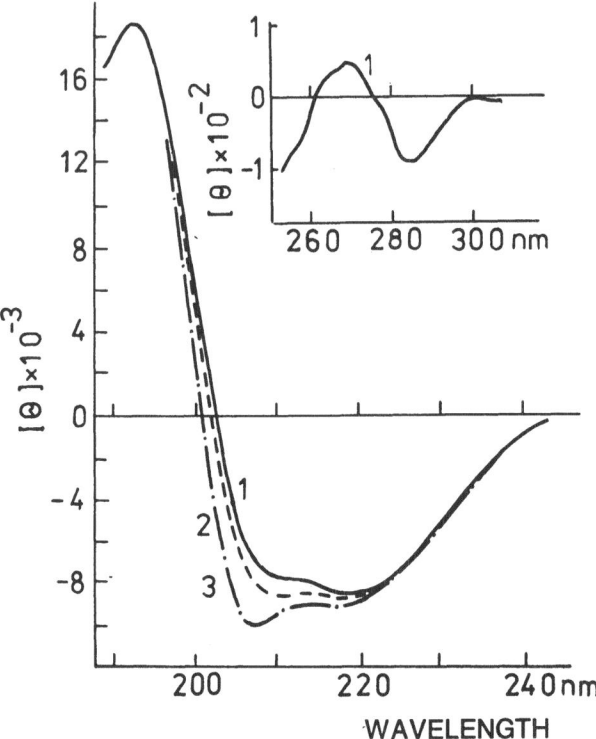

Figure 3. CD spectra of CP1-50 without additions (1) and after treatment with 0.1% SDS (2) or 1% SDS (3); 0.1 m Na-phosphate buffer, pH 7.0; CP1 isolated from maize.

quarternary structure and Chl solubilization with 1% SDS (5 h, 20–22 °C) failed to cause complete denaturation of the protein (Figure 2, Table 2).

The presence in CP1 of a significant amount of parallel β-structure indicates that the 65 kDa protein consists of parallel α/β domains. Thus, the structure of CP1 protein differs to a great extent from the bacteriorhodopsin structure, where α-helices predominate, as well as from water-soluble bacteriochlorophyll protein of green sulfur bacteria, where antiparallel β-layers predominate [14]. It is noteworthy that, when the secondary structure of the protein was calculated only on the basis of the shape of the CD spectra (without determining the molar ellipticity), the absence of β-structure in CP1 was suggested [12]. The observed qualitative correlation between the destruction of the parallel β-structure and the disturbance of the Chl organization by SDS allows us to suppose Chl molecules to be located in parallel β-bands.

Since the parallel α/β domains consist of 150–250 amino acid residues (from which 15–20% of the amino acids form β-layers [14]), we may expect that the 65 kDa protein consists of 2–3 such domains. In this case 12–13

Table 2. Secondary structure of CP1 protein with various content of antenna Chl (f, fraction of component); CP1 isolated from maize

Sample	$f_{\alpha\text{-helices}}$	$f_{\beta\text{-antiparallel}}$	$f_{\beta\text{-parallel}}$	$f_{\beta\text{-bends}}$	$f_{\text{irregular}}$	Root-mean-square error $(\text{degree} \cdot \text{cm}^2 \cdot \text{dmol}^{-1})$
CP1-140	0.34	0	0.25	0.23	0.18	820
CP1-100	0.34	0.02	0.24	0.20	0.20	730
CP1-50	0.32	0.01	0.24	0.22	0.21	650
CP1-50 + 0.01% SDS	0.30	0.04	0.18	0.21	0.25	624
CP1-50 + 1% SDS	0.32	0.06	0.09	0.26	0.32	320
CP1-70 [9]	0.34	0.04	0.14	0.19	0.29	640

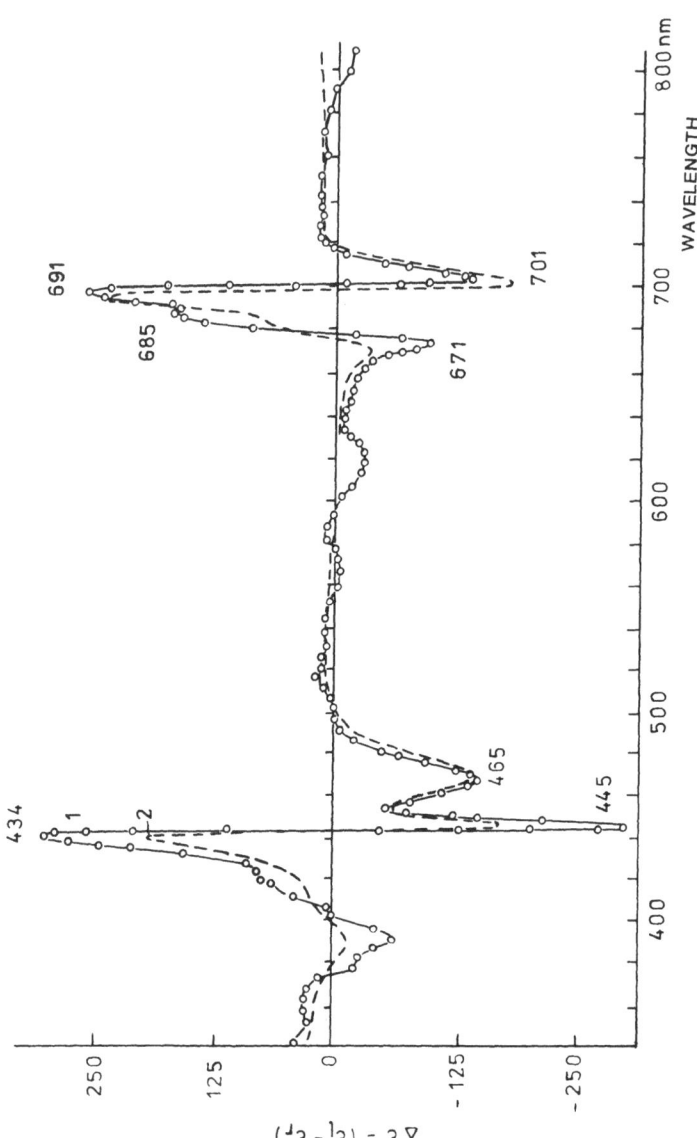

Figure 4. Light-induced (dark minus light) CD spectra of P700 in CP1-40, isolated from pea (1), and CP1-70, isolated from *Spirulina* (2); Tris HCl buffer, pH 8.0, 0.01 M Na-ascorbate, 10 μM methylviologen.

Chl molecules bound to the 65 kDa protein probably form clusters of 4—6 molecules in each domain. This estimate of the number of Chl molecules in each domain coincides with the number of the observed spectral bands. The similarity in the domain organization explains the degeneration of the structure in CP1-25.

In the light-induced CD spectra of P700 (Figure 4) besides intense bands at 690 and 700 nm, components of the dimer splitting of the Q_y transition [13, 17], and bands at 434 and 445 nm, components of the splitting of the Soret band, also broad bands of the cation radical of P700 appeared at 465 and 720—800 nm [17], as well as minor components at 663, 672, 677 and 685 nm [9, 18]. The coincidence of the positions of these minor bands with those of the main forms of antenna Chl indicates that the former are due to a change of the rotational strength of the antenna pigments that are nearest to P700. On oxidation of P700 the rotational strength of the (+) 672 nm band increases and that of the (+) 667 and (−) 684 nm bands decrease. It is possible that the 691 nm band in the light-induced CD spectra of P700 is due to a change of the rotational strength of antenna Chl absorbing at 692 nm. The same changes are also possible, if P700 forms a complex (aggregate) with 4—6 antenna Chl molecules. Thus, P700 may be considered as neither a dimer, nor a monomer, but it is part of a spatially organized group composed of 5—7 Chl molecules, which are located on one of the domains of the 65 kDa protein.

References

1. Anderson JM (1980) Biochim Biophys Acta 591:113—126
2. Bengis C and Nelson N (1977) J Biol Chem 252:4564—4569
3. Bolotina IA, Chekhov VO, Lugauskas VYu and Ptitsyn OB (1980) Mol Biol (USSR) 14:902—909
4. Bolotina IA, Chekhov VO, Lugauskas VYu and Ptitsyn OB (1981) Mol Biol (USSR) 15:167—175
5. Duysens LNM (1952) Doctoral thesis, University of Utrecht
6. Hiyama T and Ke B (1972) Biochim Biophys Acta 267:160—172
7. Ikegami I and Ke B (1984) Biochim Biophys Acta 764:70—79
8. Karapetyan NV, Litvin FF and Krasnovsky AA (1963) Biofizika (USSR) 8:191—200
9. Karapetyan NV, Shubin VV, Rakhimberdieva MG, Vashchenko RG and Bolychevtseva YuV (1984) FEBS Lett 173:209—212
10. Krasnovksy AA, Kosobutskaya LM and Voinovskaya KK (1953) DAN USSR 92:1201—1204
11. Mullet JR, Burke JJ and Arntzen CJ (1980) Plant Physiol 65:814—822·
12. Nabedryk E, Biaudet P, Darr S, Arntzen CJ and Breton J (1984) Biochim Biophys Acta 767:640—647
13. Philipson KD, Sato VL and Sauer K (1972) Biochemistry 11:4591—4595
14. Richardson JS (1981) In: Anfinsen CB, Edsall JT and Richards PM (eds) Advances in protein chemistry, pp. 167—337. New York: Academic Press
15. Scops RK (1974) Anal Biochem 59:277—282
16. Shiozawa JA, Alberte RS and Thornber JF (1974) Arch Biochem Biophys 165:388—397

17. Shubin VV, Efimovskaya TV and Karapetyan NV (1981) J Phys Chem (USSR) 55: 2916–2921
18. Shubin VV, Vashchenko RG and Karapetyan NV (1985) Mol Biol (USSR) 19: 841–850
19. Thornber JP (1975) Annu Rev Plant Physiol 26:127–158
20. Vierling E and Alberte RS (1983) Plant Physiol 72:625–633

Photosynthesis Research 9, 13–20 (1986)
© *1986 Martinus Nijhoff/Dr. W. Junk Publishers, Dordrecht.*

Optical effects of sodium dodecyl sulfate treatment of the isolated light harvesting complex of higher plants

DEMET GÜLEN[1], ROBERT KNOX[1] and JACQUES BRETON[2]

[1] University of Rochester, Department of Physics and Astronomy, Rochester, New York 14627, USA and [2] Service de Biophysique, Département de Biologie, Centre d'Etudes Nucléaires de Saclay, 91191, Gif-sur-Yvette Cedex, France

(*Received 29 July 1985*)

Key words: light-harvesting complex (peak); cPII, optical absorption, linear dichroism, circular dichroism, excitons

Abstract. The light-harvesting complex (LHC) of higher plants isolated using Triton X-100 has been studied during its transformation into a monomeric form known as CPII. The change was accomplished by gradually increasing the concentration of the detergent, sodium dodecyl sulfate (SDS). Changes in the red spectral region of the absorption, circular dichroism (CD), and linear dichroism spectra occurring during this treatment have been observed at room temperature. According to a current hypothesis the main features of the visible region absorption and CD spectra of CPII can be explained reasonably successfully in terms of an exciton coupling among its chlorophyll (Chl) *b* molecules. We suggest that the spectral differences between the isolated LHC and the CPII may be understood basically in terms of an exciton coupling between the Chl *b* core of a given CPII unit and at least one of the Chl *a*'s of either the same or the adjacent CPII. We propose that this Chl *a*–Chl *b* coupling existing in LHC disappears upon segregation into CPII, probably as a result of a detergent-related overall rotation of the strongly coupled Chl *b* core which changes the relative orientations of the two types of pigments and thus the nature of their coupling.

Abbreviations

Chl, Chlorophyll; CD, Circular dichroism; LD, Linear dichroism, LHC, Light-harvesting complex; SDS, Sodium dodecyl sulfate; CPII, A solubilized form of LHC obtained with SDS polyacrylamide gel electrophoresis

Introduction

Green plant photosynthesis is carried out by a highly organized collection of protein building blocks embedded in the photosynthetic membrane. Each of these has specialized functions leading to the conversion of solar energy to chemical energy. The efficiency of this conversion depends on the arrangement of these building blocks with respect to each other, as well as the precise molecular interactions resulting from the specific arrangement of the pigments within each of them.

In this study one such complex, the light-harvesting complex (LHC), will

Dedicated to Prof. L.N.M. Duysens on the occasion of his retirement

be considered. LHC is a Chl a/b protein complex, containing about half of the total Chl a of the thylakoid, known to serve as antenna mostly for photosystem II. A solubilized form of LHC obtained with SDS polyacrylamide gel electrophoresis called CPII, contains 3 Chl b and 3–4 Chl a molecules. CPII has received attention theoretically because of its small number of chromophores. A reasonably successful CPII model based on absorption, CD, fluorescence emission, and fluorescence polarization excitation spectra assumes that 3 Chl b's arranged in C_3 symmetry are exciton coupled and Chl a–Chl b and Chl a–Chl a interactions are much weaker [6, 13, 15]. The nature of the interactions between the pigments of LHC is not known yet, either in vivo or in vitro. However, from a comparison of several spectral properties of chloroplasts and of isolated LHC, it recently became clear that the state of the pigments of LHC in vivo and in vitro are identical [8, 14]. A study of the state of the pigments in isolated LHC will therefore be relevant for LHC in vivo.

Here, in order to gather information about the features underlying the disaggregation of LHC into CPII, the detergent SDS has been used. Explicitly, isolated LHC is segregated into CPII by gradually increasing the concentration of SDS, and the effects of this treatment on absorption, CD, and LD characteristics are determined. Possible origins of the observed spectral changes upon detergent treatment are discussed. The predictions reflect the nature of the interactions that enable the pigments to transfer energy within LHC.

Experimental methods

LHC was isolated from pea as described in [7]. LHC at the concentration of 0.1 mg Chl per ml in Tris-HCl buffer (20 mM, pH 8) was treated for 30 minutes at room temperature with various concentrations of SDS. Absorption and CD spectra were recorded in 1-mm cuvettes using a Cary 17 (Varian) spectrophotometer and a Mark V (Jobin-Yvon) dichrograph, respectively. LD spectra were recorded on samples oriented in polyacrylamide gels as described in [7].

Results

In Figure 1, room temperature absorption spectra of LHC treated with different SDS concentrations ranging from 0.02% to 0.5% are given. The amplitude of the main Chl a absorption band (~ 675 nm) decreases considerably as the amount of detergent increases. There were no discernable absorbance changes in the Chl b absorption region (~ 650 nm) for any of the aggregation states studied.

The CD spectra corresponding to the absorption spectra of Figure 1 are displayed in Figure 2. The CD signal has changed more dramatically with the SDS treatment. At the lowest detergent concentration (0.02%), the behavior is typical of LHC as previously reported [5]. In the spectral range 630–690 nm

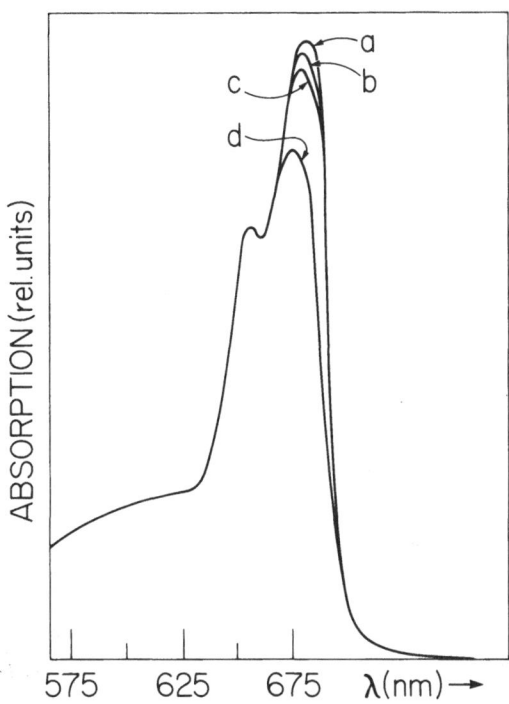

Figure 1. Absorption spectra in the red spectral region at 300 K of LHC treated with different SDS concentrations. The SDS concentrations are: a (0.022%); b (0.05%); c (0.2%); and d (0.05%).

this spectrum is also remarkably similar to the CD spectrum of isolated thylakoids [3, 4]. A transition to the well-known CPII structure, i.e., a nearly conservative CD doublet at 652 nm(−) and 665 nm(+), has taken place at some SDS concentration between 0.2% and 0.5%. A split CD feature which is the characteristic of the highest energy CD band (~ 650 nm) common to all low SDS-treated samples (0.02%–0.2%) has disappeared in CPII (0.5% SDS). The rotational strength of the 685 nm CPII CD band has been attributed earlier to the intrinsic CD of a long wavelength absorbing Chl *a* [13]. Similar systematic decreases with increasing detergent happen to a lesser extent in the rotational strength associated with the 665 nm band.

The 300 K LD spectrum of LHC (Figure 3a) exhibits a small characteristic LD band around 650 nm which is resolved into two positive components (~ 645 nm and 653 nm) and one extremely sharp negative (~ 648 nm) component at 100 K (Figure 3b). Upon addition of 1% SDS (Figure 3c), the linear dichroic signal at this wavelength region is strongly affected, resulting in a single positive band around 650 nm. It has been reported earlier that the LD spectrum of 1% SDS-treated LHC is identical to that of CPII oriented in the polyacrylamide gel [2]. Detergent treatment also causes a small 665 nm

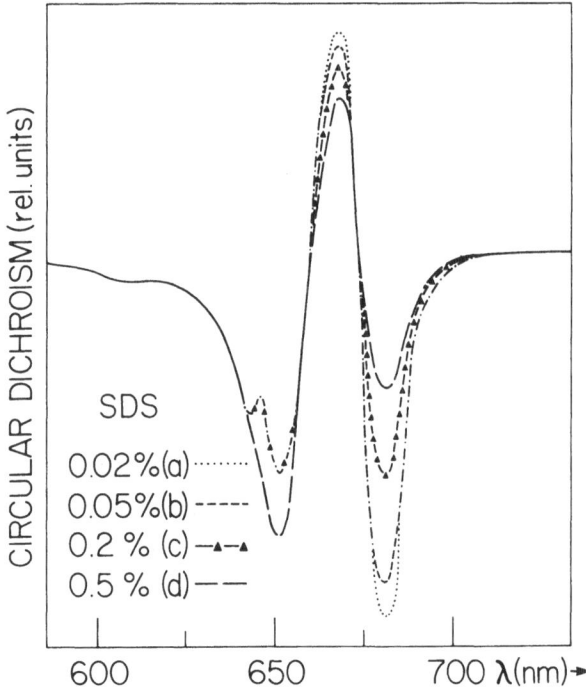

Figure 2. 300 K CD spectra of LHC at different SDS concentrations in the red spectra region.

shoulder observed in the 100 K LD spectrum to decrease strongly (data not shown). The LD signal peaking at 675 nm at 100 K (which is shifted to 680 nm at 300 K) is almost the same in LHC and CPII (1% SDS-treated LHC), Figure 3a, c.

In Figure 4, room temperature CD spectra of 0.5% SDS treated LHC (CPII) are shown at two different times. If the measurements are taken after keeping the sample in the dark at room temperature for 3 hours (Figure 4b), the 685 nm(−) CD signal almost disappears.

Discussion

According to the CPII model mentioned earlier [6, 13, 15], there exist excitonic couplings among the transition dipole moments corresponding to the Q_y, B_y and B_x bands of adjacent Chl b's arranged as a trimer in C_3 symmetry. This model predicts that the Chl b absorption in the red is due to the transitions to three delocalized states, two of them being degenerate at 652 nm and carrying a total rotational strength equal but opposite to the positive-strength non-degenerate one at 665 nm. Excitation is assumed to diffuse

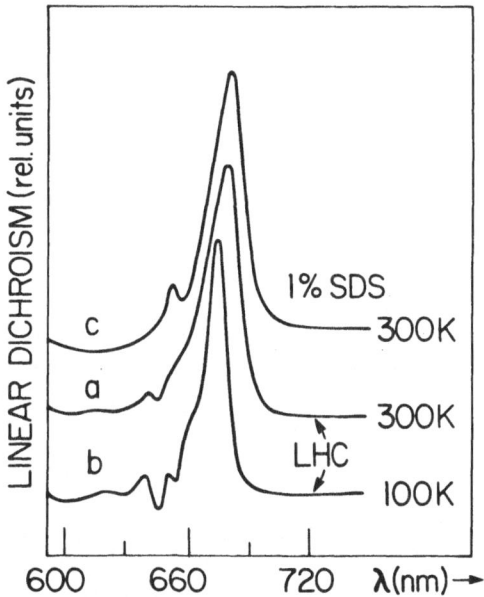

Figure 3. 300 K (a) and 100 K (b) LD spectra of LHC and 300 K LD spectrum of 1% SDS-treated LHC (c) in the red spectral region.

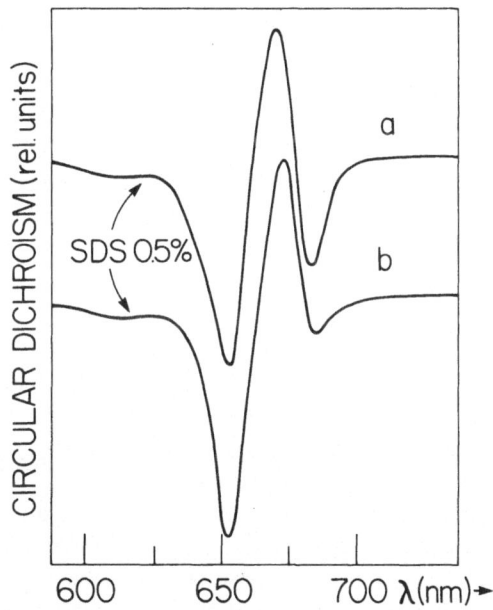

Figure 4. Room temperature CD spectra of LHC treated with 0.5% SDS taken immediately (a) and after keeping sample in the dark at room temperature for 3 hours (b).

among the Förster-coupled Chl a's and similarly between the Chl a's and the Chl b trimer.

We have some evidence that the 685 nm CD signal of CPII may not be attributed to an intrinsic CD of Chl a, but to some Chl a–Chl b or Chl a–Chl a exciton couplings. The evidence is that this signal is observed to decrease strongly in time as shown in Figure 4. However, since there exist no other models that are specific in regard to the predicted optical properties of CPII, in the following we will refer to the Van Metter-Shepanski-Knox model as necessary.

In discussing the changes occurring in the structure during the LHC to CPII transition, we first focus on the CD data of Figure 2. We will base our arguments on the suggestion that the general red region CD features of the LHC can be viewed essentially as a superposition of two different sets of CD structures: 1) A CPII-like CD doublet with 652 nm(−) and 665 nm(+) components around 645–650 nm(+) and 685 nm(−). This superposition can be best appreciated by comparing the CD spectra of (c) and (d) with nearly identical 665 nm(+) CD bands. It is interesting to note the general correspondence of the signs, positions and to a certain extent the magnitudes of the 652–665 nm CD doublets of LHC (c) and CPII (d). It is almost as if the 650 nm CD band of (c) with a split feature has been the result of subtracting a positive rotational strength centered around 645–650 nm from the corresponding CPII band. Furthermore, if a rotational strength of about the same magnitude but opposite handedness has been added to the 685 nm(−) CD signal of CPII, the CD band of (c) in this wavelength region will almost have been accounted for.

We suggest that, on the basis of energy considerations, the second CD doublet with the components around 650 nm and 685 nm can be only a result of some exciton interactions between the Chl b's and Chl a's. This suggestion can easily be understood by considering two molecules with zero-order energies E_1 and $E_2 = E_1 + \delta$. If an exciton coupling V is introduced, the transition energies are

$$E_\pm = E_1 \frac{\delta}{2} \pm \frac{1}{2}(\delta^2 + 4V^2)^{1/2} \tag{1}$$

Therefore, if $\delta \gg 2\,V$, the energies $E_\pm$ approach their unperturbed values $E_{1,2}$. Although the details will not be elaborated upon here, we have examined the feasibility of having a set of transition energies implied by the CD spectrum of LHC. For example, assuming Chl b–Chl b interactions following the above-mentioned CPII model, our calculations show that it is indeed the zero-order energy gap between the Chl a's and Chl b's which determines the position of the transitions, not the extent of the Chl a–Chl b coupling (within reason, i.e., if the a–b coupling is comparable with or smaller than the b–b coupling).

The CD signal due to the exciton-coupled molecules is directly related to and very much sensitive to the relative orientations of the constituents. If viewed in terms of the two CD doublets suggested above, the relative orientations of the 3 Chl b's should not be affected by SDS-treatment. However, it may be argued that changes in the relative orientations of the Chl b core and at least one of the Chl a's belonging either to the same CPII subunit or/and to the different CPII subunits can occur. Either the magnitude of the Chl b–Chl a coupling is weak enough or a special geometrical arrangement may exist which does not disturb the CPII-like nature of the first CD doublet, or both.

Lutz et al. [11] have reported resonance Raman data concerning the different states of Chl b's in thylakoids and CPII, suggesting that some changes in the bonding pattern of the carbonyls of the Chl b's occur upon isolation of CPII. This might seem at first to disagree with our hypothesis on the invariance of the relative orientations of Chl b's in LHC and CPII. However, it does not necessarily imply different electronic couplings within the Chl b core, rather, it can be due to differently bound Chl b cores in LHC and CPII. On the other hand, both LD (Figure 3) and the resonance Raman spectra [11] infer no large changes in the state of Chl a's upon SDS treatment. Based on these observations, it is reasonable to speculate that an overall rotation of the strongly coupled Chl b core drives the change in relative orientations of Chl b's and Chl a's, leading to the modification of the observable CD doublet because of enhanced Chl a–Chl b coupling.

The splitting of a doubly degenerate Chl b trimer level around 650 nm (which follows from a Chl a–Chl b coupling) is also a good candidate to interpret the complex structure of the LD spectrum in the 640–650 nm region. As pointed out earlier [2], the extreme sharpness of the 648 nm(−) LD signal strongly suggests that the complexity of the observed set of signals might arise from a superposition of two close LD signals of opposite dichroism.

A peculiar distribution of dipole strengths (i.e., enhancements only near the Chl a absorption maximum, but no decrease near the Chl b peak) is observed at lower SDS concentrations. Very small changes of the relative orientations of b's and a's are not necessarily to alter the structure of 650 nm band, but may still affect only the 665 and 685 nm bands, because it is the overall excitonic CD which always obeys the CD sum rule. For example, it has been shown that the intermolecular admixture of Soret (B_x, B_y) and red (Q_y) bands leads to the intensity borrowing between the red and Soret dipole and rotational strengths of CPII [6]. This kind of borrowing also explains the non-conservative CD characters of some bacteriochlorophyll dimers in the red and Soret spectral regions [12].

Through structural studies of two-dimensional crystals of LHC, it is now known that in these regular arrays two 'LHC molecules' (one LHC molecule probably consists of 3 CPII-like units) are related by a two-fold symmetry axis [9, 10]. In the native membranes all the 'LHC molecules' are reported to face the same way up [1]. The close resemblance of the CD of in vivo LHC [3, 4] to that of in vitro [5, and Figure 2a) then can be taken for some special geometries to imply no excitonic couplings between neighboring LHC molecules. However we must again emphasize that it cannot be said whether $a-b$ coupling is inter- or intra-CPII within each 'LHC molecule'.

We conclude that the modifications observed in the absorption, CD, and LD spectra upon transformation of isolated LHC into CPII by SDS treatment may be taken to imply the breakdown of an existing exciton coupling between the CPII-like Chl b core (causing the $652 \, \text{nm} [-]$ and $665 \, \text{nm} [+]$ CD doublet which may or may not be a trimer in C_3 symmetry as suggested by the Van Metter-Shepanski-Knox model [6, 13, 15]) and at least one of the Chl a's belonging either to the same and/or to the different CPII-like units. The breakdown of the Chl a–Chl b exciton coupling may be attributed to a change in the relative orientations of the Chl b's and the Chl a's while the relative orientations of the Chl b's are unaffected by the detergent treatment.

Acknowledgements

One of us (JB) wishes to thank A.M. Bardin for her help in recording the CD spectra, and one of us (DG) appreciates the hospitality of the CEN-Saclay group in February 1985. Research supported in part by grant 82-CRCR-1-1128 from the US Department of Agriculture and in part by AFME grant 3-20-1402.

References

1. Andersson B, Anderson JM and Ryrie IJ (1982) Eur J Biochem 123:465–472
2. Breton J (1983) Vol 3, pp 11–17 In Sybesma C, editor. Proceedings of 6th International Conference of Photosynthesis. The Hague: Martinus Nijhoff/Junk
3. Breton J and Hilaire M (1972) CR Acad Sci Paris, D 274:678–681
4. Canaani OD and Sauer K (1978) Biochim Biophys Acta 501:545–551
5. Gregory RPF, Demeter S and Faludi-Daniel A (1980) Biochim Biophys Acta 591: 356–360
6. Gülen D and Knox RS (1984) Photobiochem Photobiophys 7:277–286
7. Haworth P, Arntzen CJ, Tapie P and Breton J (1982) Biochim Biophys Acta 679: 428–435
8. Haworth P, Tapie P, Arntzen CJ and Breton J (1982) Biochim Biophys Acta 682: 152–159
9. Kühlbrandt W (1984) Nature 307:478–480
10. Kühlbrandt W, Thaler Th and Wehrlie W (1983) J Cell Biol 96:1414–1424
11. Lutz M, Brown JS and Remy R (1979) Ciba Foundation Symposium 61, pp 105–125. Elsevier/North Holland, Amsterdam
12. Scherz A and Parson WW (1984) Biochim Biophys Acta 766:668–678
13. Shepanski JF and Knox RS (1981) Isr J Chem 21:325–331
14. Tapie P, Haworth P, Hervo G and Breton J (1982) Biochim Biophys Acta 682: 339–344
15. Van Metter RL (1977) Biochim Biophys Acta 462:642–658

Photosynthesis Research 9, 21–32 (1986)
© 1986 Martinus Nijhoff/Dr. W. Junk Publishers, Dordrecht.

Interactions of the bacteriochlorophylls in antenna bacteriochlorophyll-protein complexes of photosynthetic bacteria

AVIGDOR SCHERZ[1] and WILLIAM W. PARSON[2]

[1] Biochemistry Department, The Weizmann Institute of Science, Rehovot 76100, Israel and
[2] Department of Biochemistry, University of Washington, Seattle, WA 98195, USA

(*Received 2 August 1985*)

Abstract. Several models have been proposed for the arrangements of the bacterio-chlorophylls in the antenna complexes of purple photosynthetic bacteria, but none of the models has accounted fully for the spectroscopic properties of the bacteriochlorophyll-protein complexes. We suggest a model involving strong exciton interactions within a bacteriochlorophyll dimer, and weaker interactions of each dimer with other, relatively distant dimers. The model is shown to account for the spectroscopic properties of the complexes, and to be consistent with other available information.

Introduction

In 1951, Duysens [19] showed that most of the bacteriochlorophyll (BChl) in photosynthetic bacteria acts as an antenna that absorbs light and transfers energy to special sites of photochemical activity. Recent work has led to the characterization of the antenna BChl-protein complexes from several species of purple nonsulfur bacteria [8, 9, 10, 11, 14, 15, 16, 17, 21, 26, 27, 31, 33, 38, 42, 43, 44, 45, 46, 47]. The proteins have been found to contain two small polypeptides, α and β, in equivalent amounts. Both polypeptides have polar amino acids in their N- and C-terminal regions, flanking central stretches of approximately 21 hydrophobic residues. The hydrophobic segments probably form α-helices that span the phospholipid bilayer of the chromatophore membrane [29, 39, 45]. A histidine residue occurs in a conserved location in the hydrophobic region of each polypeptide, and is a likely binding site for a BChl [29, 42, 45, 46, 47].

The BChl-protein complexes have characteristic optical absorption spectra that differ significantly from those of monomeric BChl in vitro. Their long-wavelength absorption bands are shifted farther into the near infrared and sometimes have a substantially larger dipole strength (hyperchormism) and a nonconservative circular dichroism (the rotational strengths of the positive and negative CD bands do not sum to zero) [24, 34]. The B800-850 complex of *Rhodopsuedomonas sphaeroides* strain 2.4.1 has major absorption bands at approximately 800 and 855 nm, compared to 770 nm for BChl in acetone.

Dedicated to Prof. L.N.M. Duysens on the occasion of his retirement

The B850 complex of *Rps. sphaeroides* strain R-26.1 absorbs near 855 nm; the B880 complex of *Rhodospirillum rubrum* near 875 nm. It is of considerable interest to understand how these spectroscopic properties result from the interactions of the BChl molecules in the complex with each other and with the protein.

Several models have been proposed for the arrangement of the BChls in the antenna complexes. Bolt and Sauer [1, 2, 3, 4] suggested that the B850 complex contains two neighboring BChls that are about 16 Å apart, with their Q_y transition dipoles forming an angle of 78 °, and the Q_x dipoles nearly parallel. (The Q_y transition dipole, which accounts for the long-wavelength absorption band of monomeric BChl in vitro, lies approximately on the axis from ring I to ring III of the BChl macrocyles [23]. The Q_x dipole is associated with a transition in the 600-nm region, and lies approximately on the axis from ring II to ring IV.) The estimate of 16 Å for the distance between the molecules was based on the CD spectrum in the Q_y-region. The complex has a positive CD band centered near 850 nm and a negative band near 875 nm, and the splitting between the two bands was attributed to exciton interactions between the two BChls. However, the corresponding splitting between the CD bands in the Q_x-region is larger, and suggests a shorter distance between the two molecules [1]. Bolt et al. [1, 2, 3] also considered the linear dichroism (LD) of complexes that were oriented in stretched films. The LD appeared to change slightly across the 855 nm absorption band, suggesting that the band includes two, closely spaced optical transitions. Measurements of the *4th* derivative of the absorption spectrum at low temperature also have suggested that the band contains several transitions [15], although no change in LD across the 855-nm band was seen in studies by Kramer et al. [26].

In a model suggested by Theiler and Zuber [39, 46, 47], two BChls are centered 10 to 15 Å apart, with the Q_y transition dipoles forming an angle of about 45 °, and the Q_x dipoles an angle of about 20 °. This model is based on an attempt to pack the α and β polypeptides closely together with their helical regions approximately parallel, and to optimize the stacking of aromatic side chains and coulombic interactions between polar amino acids. The conserved histidine residues are assumed to be axial ligands of the BChls, so that each $\alpha\beta$-dimer in the B850 complex binds two BChl molecules. The BChls are placed on the outside of the polypeptide dimer, rather than in between the two polypeptides. There is, however, some uncertainty as to whether interacting pairs of BChl are bound to $\alpha\beta$-heterodimers, as distinguished from $\alpha\alpha$- or $\beta\beta$-dimers. In an early paper on the effects of partial proteolysis on the absorption spectrum, Feick and Drews [22] concluded that the BChls that are responsible for the 855-nm absorption band in *Rps. capsulata* are bound to the α polypeptides, and not to β. A model based on $\alpha\alpha$-dimers has been proposed by Loach et al. [27] and is discussed below.

Kramer et al. [26, 43] have advanced a model for the B800-B850 complex of *Rps. sphaeroides*. They place a pair of BChls on the order of 15 Å apart on an $\alpha\beta$-polypeptide dimer, as in Theiler and Zuber's model. However, they propose that the two Q_y dipoles are perpendicular to the Q_y dipoles. Kramer et al. suggest further that pairs of $\alpha\beta$-BChl$_2$ units are assembled into [$\alpha\beta$-BChl$_2$]$_2$ complexes, with the Q_y dipoles of four BChls arranged roughly in a square. This model is intended to explain the fluorescence polarization of the complexes, which suggests that the Q_y transition dipoles all lie approximately parallel to the plane of the chromatophore membrane but have a circular degeneracy in that plane [4, 5, 6, 7, 26]. The positions and orientations of the BChls on the individual $\alpha\beta$-units are chosen to account for the CD and LD spectra of the complex. Kramer et al. [26] do not discuss the exciton interactions between the BChls on different $\alpha\beta$-units, whose Q_y dipoles form opposite sides of the square, although in their model the distance between these molecules is similar to that between the BChls on the same $\alpha\beta$-units.

The 800-nm absorption band of the B800-B850 complex has been attributed to an additional BChl molecule that interacts only weakly with the pair of neighboring BChls [1, 2, 3, 4, 15, 21, 26, 33, 42, 43]. The Q_y transition dipole of the third BChl is approximately perpendicular to those of the other BChls. It has been suggested that the third BChl is bound to a second conserved histidine residue in the β polypeptide [26, 39, 45], but this would seem surprising because the second histidine is in a region of polar and charged amino acids.

None of the foregoing models accounts explicitly for the bathochromic wavelength shift or the hyperchromism of the Q_y absorption bands. Hyperchromism can be explained by exciton interactions if one considers the mixing of the Q_y transitions with higher-energy transitions [34, 35], but the exciton interactions that have been postulated are too weak for this effect to be significant. From the small separation between the CD bands, and between the two transitions suggested by the LD spectrum, Bolt and Sauer [1, 3, 4] estimated the dipolar interaction energy between the Q_y transitions of the two BChls to be in the range of 100 to 125 cm^{-1}; Kramer et al. [26] considered 100 cm^{-1} to be a maximal value. Since an interaction energy of this magnitude would cause the symmetric and antisymmetric exciton bands to separate by only about 15 nm, the shift of the band from 770 nm to longer wavelengths was attributed to interactions of the BChls with the protein. Molecular orbital calculations [20, 44] have shown that charged amino acids close to the BChls could, in principle, cause such a shift. In most cases, however, the amino acid sequences do not provide support for this possibility. In the B850 and B800–850 complexes, the closest charged amino acids to the histidines that serve as the putative binding sites for the neighboring BChls are 11 residues away in the β polypeptide, and 13 in the α [38, 45].

Loach et al. [27] have proposed a model for the *Rds. rubrum* B880 complex that emphasizes interactions of charged and polar amino acids with the substituents on the periphery of the BChl macrocycles. They suggest that each α polypeptide binds two BChls, one to the conserved histidine residue that features in the models described above, and the second to a threonine near the opposite end of the hydrophobic region. In the case of the B880 complex, the amino acid sequences include charged amino acids close enough to both of these sites so that they might shift the BChl's absorption band to longer wavelengths. BChls bound to the two sites would be far enough apart so that their exciton interactions with each other are weak. Loach et al. [27] propose, however, that the α-BChl$_2$ complexes dimerize to form $[\alpha$-BChl$_2]_2$ structures, in which the BChls in corresponding positions on the two α polypeptides interact closely. The β polypeptides are assumed to play a role in the aggregation of these structures into larger arrays. Loach et al. speculate that interactions between the neighboring BChls account for the CD spectrum of the complexes, but they do not explore this point in detail. (Their Figure 3 suggests that the BChls form a sandwich-like dimer. This structure would result in a hypsochromic shift and hypochromism of the main Q_y exciton band, which would seem inconsistent with the absorption spectrum.)

The idea that the shift of the Q_y absorption band from 770 nm to longer wavelengths is due primarily to something other than exciton interactions was bolstered by experiments by Rafferty et al. [32]. These workers studied the photooxidation of the B850 complex that occurred when chromatophores of *Rps. sphaeroides* strain R-26.1 were illuminated with strong light in the presence of oxygen. The double CD bands in the 855-nm region disappeared as the BChl was destroyed, as one would expect if exciton interactions between neighboring molecules were lost, but the absorption band of the surviving BChl moved only about 8 nm to shorter wavelengths. Rafferty et al. [32] assumed that the spectrum approached that of isolated, monomeric BChl bound to the protein. There is, however, another possible interpretation of their observations, which seems to us to merit consideration.

We have shown recently that oligomers of BChl or bacteriopheophytin in certain mixed solvents or in detergent micelles have absorption and CD spectra similar to those of the BChl-protein complexes found in the photosynthetic bacteria [34; A. Scherz, V. Rosenbach, and S. Malkin, submitted]. The main features of the spectra, including the large bathochromic shift of the Q_y band, can be explained on the basis of exciton interactions between neighboring BChl molecules, provided that one considers the mixing of the Q_y transitions with the higher-energy transitions of the molecules. In addition to the Q_y and Q_x transitions, BChl has strong absorption bands in the Soret region. Depending on the orientations of the BChls, mixing of the different excited states in a dimer can lead to hyper- or hypochromism, and can make the CD spectrum markedly nonconservative [35, 40]. In some of the geometries that are consistent with a large bathochromic shift of the dominant Q_y exciton band, the dipole and rotational strengths of the weaker Q_y

exciton band are calculated to be much smaller than those of the dominant band, so that the absorption and CD spectra both will have effectively only a single band in this region. This suggests that the pair of bands seen in the CD spectrum of the antenna complexes might not be due simply to interactions between closely neighboring BChls. Instead, they could reflect weaker exciton interactions of the BChl dimers with other, relatively distant dimers. Long-range interactions of this sort could be important even in pigment-protein complexes that are isolated by dissociating chromatophore membranes with detergents, because such preparations probably are still aggregates containing 8 or more BChl molecules [8, 26, 36, 41]. We shall show that models based on strong exciton interactions between neighboring BChls and weaker inter-actions between different dimers appear to be consistent with the experiments of Rafferty et al. [32], as well as with the other information summarized above.

Methods

Spectroscopic properties of BChl dimers were calculated as described pre-viously [35]. BChl 1 is placed at the origin of a Cartesian coordinates system, with its Q_x and Q_y transition dipoles on the x and y axes (Figure 1A). BChl 2 is centered at $\vec{R} = (r_x, r_y, r_z)$, with the orientation of its Q_x dipole specified by direction angles α_x, β_x, and γ_x, and that of its Q_y dipole by α_y, β_y, and γ_y. The calculations consider the mixing of the 4 main excited states

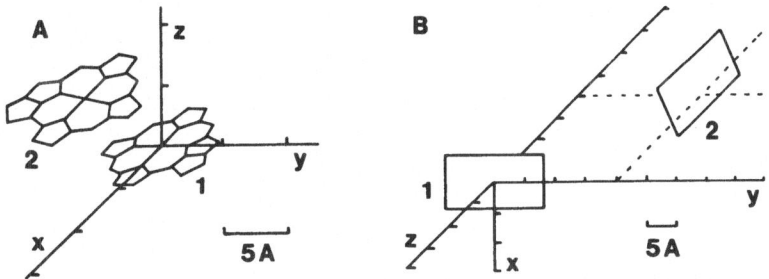

Figure 1.A. A possible structure of the BChl dimer in the B850 complex. BChl 1 is centered at the origin; BChl 2 is at $\vec{R} = (-0.5, -6.5, 3.6)$ Å and has direction angles $\alpha_x = 6°$, $\beta_x = 93.3°$, $\gamma_x = 95°$, $\alpha_y = 92.4°$, $\beta_y = 170°$, and $\gamma_y = 80.3°$.

B. Structure of an oligomer of two BChl dimers. Each rectangle represents a dimer with the structure shown in Fig. 1A; its long and short sides indicate the dimensions of the dimer in the local y and x directions, which are defined by the dimer's dominant Q_y and Q_x exciton transition dipoles. The coordinate system is rotated 90° about the y-axis relative to its orientation in A. Dimer 1 is centered at the origin; dimer 2 is at $(-2, 20, -20)$ Å with direction angles $\alpha_x = 27.1°$, $\beta_x = 63.1°$, $\gamma_x = 87°$, $\alpha_y = 93.9°$; $\beta_y = 89°$, and $\gamma_y = 4°$. In the chromatophore membrane, the B850 oligomer would be oriented with the x-axis approximately normal to the plane of the phospholipid bilayer. However, the structure shown here is for discussion only; the experimental data are not sufficient to determine the actual structure of the oligomer.

of each BChl in the excited states of the dimer, and the mixing of doubly-excited states in the dimer's ground state. The 8 absorption bands of the dimer are identified by index J, odd values of which refer to transitions that are predominantly symmetric combinations of corresponding $\pi\pi^*$ transitions in the two BChls, and even values to those that are predominantly anti-symmetric [35]. The wavelengths and transition dipole moments used for the monomeric BChls were as follows: Q_y, 800 nm and 6.3 debye; Q_x, 600 and 3.5; B_x, 390 and 7; and B_y, 360 and 9. (B_x and B_y are the Soret transitions.) For simplicity, charge-transfer states were not included. Doing so would move the main Q_y exciton band of the dimer to longer wavelengths and decrease its dipole strength, but these effects are expected to be relatively minor over the range of geometrical parameters considered here [30].

We used essentially the same procedure to consider the interactions between a pair of widely separated dimers. Each dimer was assumed to have two optical transitions polarized along the dimer's local x-axis, and two polarized along the local y-axis. The energies and dipole moments of the transitions were taken to be those of the four dominant exciton transitions found for the dimer shown in Figure 1A: Q_y, 855 nm and 10.34 debye; Q_x, 595 and 4.53; B_x, 382 and 10.63; and B_y, 378 and 11.82. The use of only four transitions for this dimer is a reasonable approximation, because the other four exciton transitions are very weak (see below). The dimer's local axes were defined by the dominant Q_x and Q_y exciton transition dipole vectors. This is less satisfactory for the Soret transitions than for the Q_x and Q_y transitions, because the exciton transitions in the Soret region have mixed polarizations with respect to the x- and y-axes. Although a more rigorous treatment of a four-molecule system is straightforward [30], the simplified approach used here allows a useful distinction between long-range interactions of separate dimers and the stronger interactions within each dimer. To calculate the CD spectrum of the pair of dimers, the rotational strengths of all eight transitions of the individual dimers are added to the rotational strengths that result from interactions between the dimers.

Results and discussion

Figure 2 shows some of the spectroscopic properties calculated for a BChl dimer, as functions of the orientation of the molecules. The BChls are centered about 7.5 Å apart with their Q_y dipoles making a fixed angle (β_y) of 170°, as shown in Figure 1A. The abscissa gives the angle (γ_y) between the Q_y dipole of BChl 2 and the z-axis. Over the range of γ_y that is considered, the main absorption band in the near IR (band 2) is calculated to be near 855 nm. This band is due largely to an antisymmetric combination of the Q_y transitions of the two BChls, but also includes significant contributions from the higher-energy transitions. It is polarized essentially along the y-axis in Figure 1A. The dipole strength of band 2 is calculated to be about 107 debye^2, which

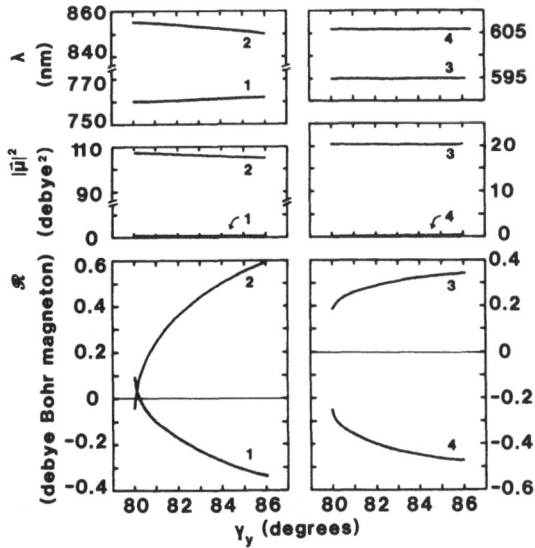

Figure 2. Calculated wavelengths (λ), dipole strengths ($|\vec{\mu}|^2$), and rotational strengths ($\mathscr{R}$) for the first four absorption bands of a BChl dimer, as functions of γ_y. R, β_y, and γ_x are held constant as specified in Figure 1 A; the other angles change with γ_y.

is 1.34-times the sum of the dipole strengths of the two monomeric BChls that make up the dimer. The rotational strength of the band is positive for values of $\gamma_y \geqslant 80.1°$. The weaker, predominantly symmetric Q_y exciton band (band 1) is calculated to be near 760 nm and to have very little dipole strength. The rotational strength of band 1 goes through zero when $\gamma_y \approx 80.3°$.

The stronger of the Q_x exciton bands is band 3, which is due mainly to a symmetric combination of the Q_x bands of the monomers and is polarized essentially along the x-axis. Band 3 is calculated to occur near 595 nm (Figure 2). The Q_x exciton bands exhibit hypochromism; the sum of their dipole strengths is about 20 debye^2, compared to 25 debye^2 for two monomeric BChls.

We have assumed that the Q_y bands of the monomeric BChls would be at 800 nm. 800 nm was chosen instead of 770 nm as the starting point because the third BChl in the B800—850 complex of *Rps. sphaeroides* appears to absorb at this wavelength. As mentioned above, this BChl probably is far enough away from the other two BChls so that exciton interactions do not perturb its absorption spectrum greatly. The shift from 770 to 800 nm presumably is due mainly to interactions of the BChl with the protein. The choice of 800 nm for calculations on other complexes is arbitrary, however, because the binding sites of the BChls probably differ [12, 13]. In principle, it should be possible to calculate the transition energies for the monomeric BChls when sufficient information becomes available on the distribution of polar groups nearby [43, 20].

Results similar to those in Figure 2 can be obtained with BChl 2 located at other positions in the same region. The essential features of the geometry are that BChl 2 is displaced from 1 mainly in the y-direction, and to a smaller extent the z-direction, and that the Q_y transition dipole of BChl 2 is aligned nearly along R. Similar results are obtained if either or both of the BChls are rotated by 180° about their Q_x or Q_y axes. The absorption and CD spectra of the BChl-protein complexes do not allow one to distinguish between dimers of the type shown in Figure 1A, in which ring I of the molecule overlaps with ring I of the other, and those in which the overlap involves ring III of one or both molecules. The structure shown here is similar to the 'special pair' of BChls found in the reaction center of *Rps. viridis* [18], but differs in that it does not have C_2 symmetry; the Q_y dipole of BChl 2 is aligned more closely to R.

Figure 3A shows absorption and CD spectra calculated with the geometrical parameters used in Figure 1A. The calculated stick spectra have been dressed with gaussians with widths similar to those of the absorption bands of the B850 complex. The choices of the bandwidths are somewhat arbitrary and can affect the appearance of the spectra, particularly in regions where the dimer has overlapping bands. The calculated absorption spectrum is similar to that of the isolated B850 complex (Figure 3C); the calculated CD spectrum

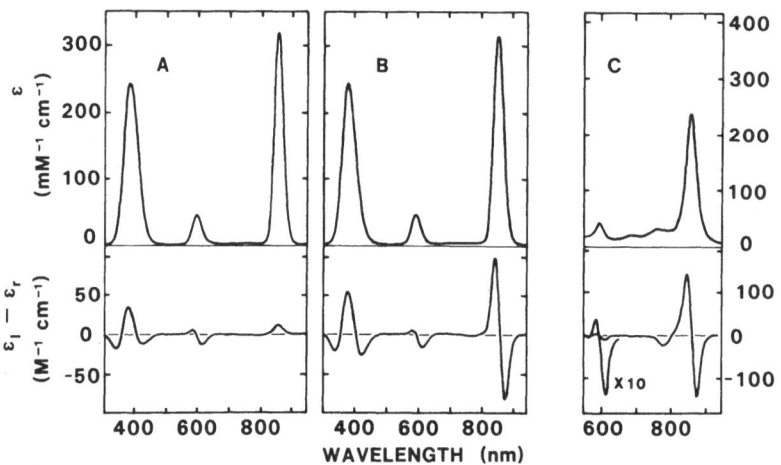

Figure 3.A. Absorption and CD spectra calculated for the BChl dimer shown in Figure 1A.
B. Spectra calculated for the oligomer shown in Figure 1B.
C. Absorption and CD spectra of the B850 complex isolated from *Rps. sphaeroides* R-26.1, redrawn from [1]; only the Q_x and Q_y regions of the spectrum are shown because the CD in the Soret region has not been described. The weak, negative CD band near 760 nm is not seen in all preparations [33]; it can be reproduced in the calculations by increasing γ_y (Figure 2). In all three panels, the vertical scales are based on the molarity of BChl dimers.

differs from the experimental spectrum in the Q_y region in having a single, positive band near 855 nm, instead of a pair of positive and negative bands. In the Q_x region, the calculated CD spectrum is similar to the experimental. Note that the calculations reproduce the nonconservative nature of the CD in this region: the rotational strength of the negative band near 605 nm is greater than that of the positive band near 595 nm. In Bolt and Sauer's [1, 3] analysis, this asymmetry is attributed to an underlying CD of the monomeric BChls; in our treatment it arises from the mixing of other transitions with the Q_x transitions.

We suggest that the double CD bands seen at 850 and 875 nm in the B850 spectrum are due to exciton interactions between the BChls of different dimers. Because the dipole strength of the 855-nm band of a dimer is about 2.7-times greater than the dipole strength of a single BChl molecule (Figure 2), exciton interactions between dimers can persist to relatively large distances. For discussion, Figure 1B shows two dimers that are centered about 28 Å apart with their local y axes approximately perpendicular. To calculate the spectroscopic properties of such an oligomer rigorously requires diagonalizing an interaction matrix of the excited states of all four BChls. It is possible to obtain an approximate solution, however, by treating the individual dimers first and then considering interactions between the dimers, provided that the distance between the dimers is sufficiently great. To consider the interactions between a pair of dimers, one can view each dimer as behaving like an individual BChl molecule, with strong transitions at 855 nm and in the Soret region and a weaker transition at 595 nm (see Methods). Figure 3B shows absorption and CD spectra calculated for the pair of dimers illustrated in Figure 1B. The CD spectrum now has bands near 840 and 870 nm, much like those found experimentally (Figure 3C). The calculated absorption spectrum is almost indistinguishable from that of the individual dimers (Figure 3A). The long-wavelength absorption band of the oligomer consists of two perpendicular transitions with nearly identical dipole strengths, consistent with the circular degeneracy of the fluorescence polarization [1, 2, 4, 5, 6, 7]. The two transitions are too close together to be resolved in Figure 3B (853 and 858 nm), but might be seen in higher derivatives of the spectrum if the band is sharpened by lowering the temperature, in agreement with the observations on the B800–850 complex [15].

In the oligomer shown in Figure 1B, the dipolar interaction energy between the dominant Q_y exciton transitions of the two dimers is only 35 cm^{-1}. This is much smaller than the interaction energy of 728 cm^{-1} between the individual Q_y transitions of the BChls in each dimer, as calculated for the dimer in Figure 1A. The sensitivity of the CD spectrum to weak interactions between dimers stems partly from the fact that the rotational strengths of the oligomer's bands is proportional to the distance between the dimers, even though the splitting of the bands decreases [35]. The calculated CD spectrum of course depends strongly on the geometry of the oligomer. Unfortunately, the experimental spectroscopic information is not sufficient to determine the actual arrangement of the dimers in the B850 or B800–850 complex. The flourescence polarization can be explained by energy transfer between two dimers with approximately perpendicular y axes, as in Figure 1B, or among a larger number of dimers with orientations that are closer to parallel. Models similar to that of Loach et al. [12], in which two BChl dimers are displaced by about 20 Å in the x direction on a $\alpha\alpha$ polypeptide dimer, can be reconciled with the CD data by rotating one of the dimers about the x-axis so that the local y-vectors make an angle of about 80°. Other possibilities would be arrays of three or six dimers with C_3 or C_6 symmetry [47]. Such arrays would lend themselves to the hexagonal arrangements of antenna complexes that have been seen by electron microscopy in some species, such as *Rps. viridis* [25, 28, 37].

The results described by Rafferty et al. [32] can be explained on the assumption that photooxidation destroys both BChl molecules of a dimer together as a unit. It is important to note that the photooxidation occurs with a quantum yield of only about 1.7×10^{-5} [32], and that its mechanism and products are not known. For example, the oxidative process that breaks down one of the BChls in a dimer could change the folding of the polypeptides in such a way that the second BChl becomes labile. We suggest that the disappearance of the double CD band and the 8-nm shift of the absorption band to shorter wavelengths reflect the loss of long-range interactions between different dimers as the dimer population is decreased. The spectra that remain then would be due to isolated dimers that survive, and not to monomeric BChl as Rafferty et al. assumed.

Observations on oligomers of BChl in Triton-X100 micelles lends support to this interpretation of the B850 complex. At relatively high concentrations, the oligomers absorb near 860 nm. Their CD spectrum has a positive band at 851 nm and negative bands at 830 and 870 nm. If the concentration of oligomers in the micelles is decreased, or if the oligomers are oxidized partially with mild $K_3Fe(CN)_6$, the negative CD band at 870 nm disappears, while the absorption band shifts only slightly to shorter wavelengths [V. Rosenbach and A. Scherz, unpublished results].

Acknowledgements

This work was supported by a grant from the National Science Foundation (PCM-8016593). We are indebted to Drs. J. Breton, P. Loach, M. Luzt, D. Middendorf, R. Pearlstein, A. Warshel, N. Woodbury and H. Zuber for helpful discussions and for sharing the results of unpublished work.

References

1. Bolt J (1980) Thesis, Univ. California, Berkeley
2. Bolt J, Hunter CN, Niederman RA and Sauer K (1981) Photochem Photobiol 34: 653–661

3. Bolt J and Sauer K (1979) Biochim Biophys Acta 545:54–63
4. Bolt J and Sauer K (1981) Biochim Biophys Acta 637:342–347
5. Breton J, Farkas DL and Parson WW (1985) Biochim Biophys Acta, in press
6. Breton J and Vermeglio A (1982) in Photosynthesis: Energy Conversion by Plants and Bacteria (Govindjee, ed.), Vol 1, pp 153–194, Academic Press, New York
7. Breton J, Vermeglio A, Garrigo M and Paillotin G (1981) Proc 5th International Congr Photosynthesis (Akoyunoglou G, ed.) Vol 3, pp 445–459, Balaban International Sciences Services, Philadelphia, PA
8. Broglie RM, Hunter CN, Delepelaire P, Niederman RA, Chua N-H and Clayton RK (1980) Proc Natl Acad Sci USA 77:87–91
9. Brunisholz RA, Cuendet PA, Theiler R and Zuber H (1981) FEBS Lett 129:150–154
10. Brunisholz RA, Suter F and Zuber H (1984) Hoppe-Seyler's Z Physiol Chem 365:675–688
11. Brunisholz RA, Wiemken V, Suter R, Bachofen R and Zuber H (1984) Hoppe-Seyler's Z Physiol Chem 365:689–701
12. Bruno R and Lutz M (1985) Biochim Biophys Acta 807:10–23
13. Bruno R, Vermeglio A and Lutz M (1984) Biochim Biophys Acta 766:259–262
14. Clayton RK and Clayton BJ (1981) Proc Natl Acad Sci USA 78:5583–5587
15. Cogdell RJ and Crofts AR (1978) Biochim Biophys Acta 502:409–416
16. Cogdell RJ, Lindsay JG, Valentine J and Durant I (1982) FEBS Lett 150:151–154
17. Cogdell RJ, Zuber H, Thornber JP, Drews G, Gingras G, Niederman RA, Parson WW and Feher G (1985) Biochim Biophys Acta 806:185–186
18. Deisenhofer J, Epp O, Miki K, Huber R and Michel H (1984) J Mol Biol 180:385–398
19. Duysens LNM (1951) Nature 168:548–550
20. Eccles J and Honig B (1983) Proc Natl Acad Sci USA 80:4959–4962
21. Feick R and Drews G (1978) Biochim Biophys Acta 501:499–513
22. Feick R and Drews G (1979) Z Naturforsch 34c:196–199
23. Gouterman MP (1961) J Molec Spectros 6:138–163
24. Hayashi H, Nazawa T, Hatano M and Morita S (1982) J Biochem 91:1029–1038
25. Jay F, Lambillotte M, Stark W and Mühlethaler K (1984) EMBO J 3:773–776
26. Kramer JM, Van Grondelle R, Hunter CN, Westerhuis WHJ and Amesz J (1984) Biochim Biophys Acta 765:156–165
27. Loach PA, Parkes PS, Miller JF, Hinchigeri S and Callahan PM (1985) Cold Spring Harbor Symp, in press
28. Miller KR (1982) Nature 300:53–55
29. Nabedryk E and Breton J (1981) Biochim Biophys Acta 635:515–524
30. Parson WW, Scherz A and Warshel A (1985) in Antennas and Reaction Centers of Photosynthetic Bacteria – Structure, Interactions and Dynamics (Michel-Beyerle ME, ed.) Springer-Verlag, in press
31. Picorel R, Belanger G and Gingras G (1983) Biochem 22:2491–2497
32. Rafferty CN, Bolt J, Sauer K and Clayton RK (1979) Proc Natl Acad Sci USA 76:4429–4432
33. Sauer K and Austin LA (1978) Biochemistry 17:2011–2019
34. Scherz A and Parson WW (1984) Biochim Biophys Acta 766:653–665
35. Scherz A and Parson WW (1984) Biochim Biophys Acta 766:666–678
36. Shiozawa JA, Welte W, Hodapp N and Drews G (1982) Arch Biochem Biophys 213:473–485
37. Stark W, Kuhlbrandt W, Wildhaber I, Wehrli E and Mühlethaler K, EMBO J 3:777–783
38. Theiler R, Suter F, Wiemken V and Zuber H (1984) Hoppe-Seyler's Z Physiol Chem 365:703–719
39. Theiler R and Zuber H (1984) Hoppe-Seyler's Z Physiol Chem 365:721–729
40. Tinoco I (1962) Adv Chem Phys 4:113–157
41. Van Grondelle R, Hunter CN, Bakker JCG and Kramer HJM (1983) Biochim Biophys Acta 723:30–36
42. Van Grondelle R, Kramer HJM and Rijgersberg CP (1982) Biochim Biophys Acta 682:208–215

43. Van Grondelle R (1985) Biochim Biophys Acta 811:147–195
44. Warshel A (1979) J Am Chem Soc 101:744–746
45. Youvan DC and Ismail S (1985) Proc Natl Acad Sci USA 82:58–62
46. Zuber H (1985) in Photosynthesis III: Photosynthetic Membranes. Encyclopedia of Plant Physiology, Vol 19 (Arntzen CJ and Stahaelin LA, eds.) in press
47. Zuber H (1985) in Optical Properties and Structure of Tetrapyrroles (Blauer G and Sund H, eds.), pp 425–441, W. de Gruyter, Berlin.

Photosynthesis Research 9, 33–45 (1986)
© 1986 Martinus Nijhoff/Dr. W. Junk Publishers, Dordrecht.

Pigment organization and energy transfer in the green photosynthetic bacterium *Chloroflexus aurantiacus*

II. The chlorosome

R.J. van DORSSEN, H. VASMEL and J. AMESZ

Department of Biophysics, Huygens Laboratory of the State University,
P.O. Box 9504, 2300 RA Leiden, The Netherlands

(*Received 30 August 1985*)

Key words: bacteriochlorophyll, chlorosome, energy transfer, fluorescence polarization, green photosynthetic bacteria, linear dichroism, (*Chloroflexus aurnatiacus*)

Abstract. The transfer of excitation energy and the pigment arrangement in isolated chlorosomes of the thermophilic green bacterium *Chloroflexus aurantiacus* were studied by means of absorption, fluorescence and linear dichroism spectroscopy, both at room temperature and at 4 K. The low temperature absorption spectrum shows bands of the main antenna pigments BChl *c* and carotenoid, in addition to which bands of BChl *a* are present at 798 and 613 nm. Fluorescence measurements showed that excitation energy from BChl *c* and carotenoid is transferred to BChl *a*, which presumably functions as an intermediate in energy transfer from the chlorosome to the cytoplasmic membrane. Measurements of fluorescence polarization and the use of two different orientation techniques for linear dichroism experiments enabled us to determine the orientation of several transition dipole moments with respect to each other and to the three principal axes of the chlorosome. The Q_y transition of BChl *a* is oriented almost perfectly perpendicular to the long axis of the chlorosome. The Q_y transition of BChl *c* and the γ-carotene transition dipole are almost parallel to each other. They make an angle of about 40° with the long axis and of about 70° with the short axis of the chlorosome; the angle between these transitions and the BChl *a* Q_y transition is close to the magic angle (55°).

Abbreviations

BChl, bacteriochlorophyll; CD, circular dichroism; LD, linear dichroism

Introduction

The study of Duysens [5] on excitation energy transfer in algae and purple bacteria represents a major step in the elucidation of the function of photosynthetic pigments. Since then, numerous investigations have detailed the function and structural organization of pigments in these organisms and in higher plants. However, the antenna systems of green photosynthetic bacteria have been less extensively studied [2].

The present communication deals with the antenna system of *Chloroflexus aurantiacus*. *C. aurantiacus* has been classified as a green photosynthetic

---

Dedicated to Prof. L.N.M. Duysens on the occasion of his retirement.

bacterium on basis of its morphological characteristics and pigment composition [16]. Like the 'classical' green bacteria, the green sulfur bacteria, it contains chlorosomes (when grown photoautrophically), and its major bacteriochlorophyll is BChl c.

Methods have been developed to prepare isolated chlorosomes and membrane fragments [6] and studies with these preparations have shown that the cytoplasmic membrane contains only BChl a, which is organized in the B808-866 light-harvesting complex and in the reaction center, whereas BChl c is exclusively situated in the chlorosomes [6, 13]. The chlorosomes contain approximately 5000–20 000 BChl c molecules, as can be calculated from the data given in refs. 6 and 14, and are located alongside the membrane in the cytoplasma. Freeze fracture studies [14] have shown that the internal structure of the chlorosome consists of long polypeptide rods (approximately 5.2 nm in diameter). The BChl c binding polypeptide (5.6 kDa) has been sequenced recently [19] and presumably binds 7 BChl c molecules per copy, all located on one side of an α-helical segment.

In the preceding communication of this series [18] we reported the results of a study of energy transfer and pigment organization in the photosynthetic membrane of *C. aurantiacus*. The present publication deals with the optical and structural properties of the isolated chlorosome. In addition to the main light-harvesting pigment, BChl c, chlorosomes of *C. aurantiacus* contain small amounts of BChl a [6, 13], the location of which is unknown, and carotenoids (mainly β- and γ-carotene [14]). The chlorosomal BChl a absorbs near 792 nm, i.e. at about 50 nm longer wavelength than BChl c, and it has been suggested that BChl a would function as an intermediate in the transfer of energy from BChl c to the cytoplasmic membrane [2, 3]. In this communication it will be shown that in isolated chlorosomes light energy absorbed by BChl c is transferred to BChl a 792. A tentative model for the three-dimensional organization of both pigments of the chlorosome will be presented, based on fluorescence polarization and linear dichroism spectra. Also the organization and light-harvesting properties of the carotenoids will be considered.

Materials and methods

Cells of *Chloroflexus aurantiacus* strain J-10-fl were grown in medium D as described by Pierson and Castenholz [12]. For the preparation of chlorosomes [6] cells grown under low light conditions (4 W/m^2 of white light) were used within an A_{740}/A_{866} ratio of about 20. The chlorosomes were suspended in 10 mM Tris-HCl buffer, pH 8.0, containing 0.5 M sucrose. Glycerol (50% v/v) was added to prevent crystallization at low temperature.

Low temperature absorption, fluorescence and fluorescence polarization were measured using a single beam spectrophotometer [10]. The optical path length of the vessel was 2.5 mm. Linear and circular dichroism were recorded on an apparatus described by Meiburg [11]. For measurements of linear dichroism the chlorosomes were suspended in a polyacrylamide gel (32% (w/v) acrylamide)[1].

Results

Absorption spectra

The room temperature absorption spectrum of chlorosomes of *Chloroflexus aurantiacus* is shown in Figure 1. It is similar to that published earlier by Feick et al. [6] and shows bands at 742 and 460 nm of BChl *c*, a shoulder near 510 nm, which is presumably due to carotenoid, and a small band on the long-wavelength side of the Q_y bands of BChl *c* that has been attributed to

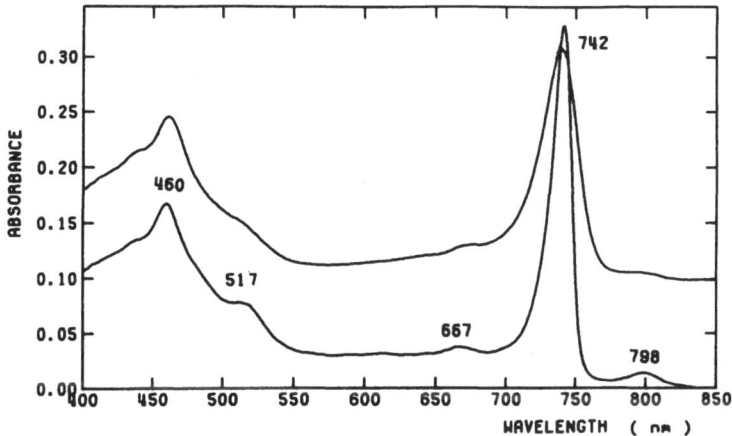

Figure 1. Absorption spectra of isolated chlorosomes of *Chloroflexus aurantiacus* measured at 4 K (———) and at 293 K (– – – –). The room temperature spectrum is shifted by 0.05 absorbance units to enhance clarity.

BChl *a* [6, 13]. Upon cooling to 4 K the bands sharpened, but did not shift appreciably (Figure 1). In the long wave region the BChl *a* absorbance was now observed as a well separated band located at 798 nm. This bacteriochlorophyll will be denoted as BChl *a* 798. By expansion of the spectrum (not shown) the position of the BChl *a* Q_x band was determined to be at 613 nm. The location of the absorption maximum of BChl *c* varied between 739 and 742 nm for different preparations. The second derivative spectrum showed the presence of an additional BChl *c* band near 725 nm which is only barely visible in the normal absorption spectrum. By the same method the maxima of two carotenoid bands were determined to be 487 and 517 nm. The band at 667 nm may be ascribed to a small amount of degradation products of BChl *c*, perhaps bacteriopheophytin *c*. A similar, but much stronger band in this region was observed in isolated chlorosomes of *Chlorobium limicola* [13]. The relative contents of BChl *c* and BChl *a* were determined after extraction by addition of 10 volumes of acetone-methanol (7:2 v/v). On basis of specific extinction coefficients of 74 [15] and 65 mM cm^{-1} [4] for BChl *c* and BChl *a*, respectively, at the Q_y band maxima in this mixture, a molar ratio of 29 ± 2 was obtained, in reasonable agreement with ref. 6. The extinction coefficient of BChl *c* at 742 nm in vivo was determined to be 102 ± 2 mM cm^{-1} at room temperature.

The circular dichroism spectrum recorded at 77 K is shown in Figure 2. In the near-infrared region it shows an enhanced resolution as compared to the spectrum measured at room temperature [3]. The main signal is due to BChl *c* with a nearly conservative signal centered at 746 nm. In addition weak negative bands are visible at 468, 721 and 807 nm. Thus, there is no evidence for a strong exciton interaction between BChl 798 molecules and no carotenoid CD signal was observed. A positive signal at 430 nm as reported in ref. [3] was not observed.

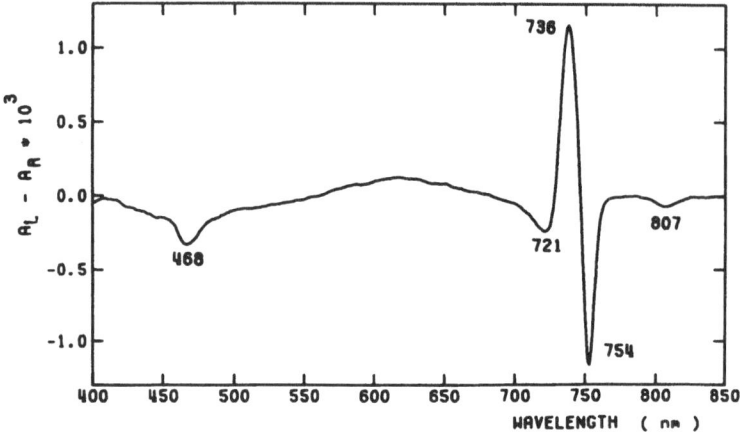

Figure 2. Circular dichroism spectrum of isolated chlorosomes at 77 K. The absorbance at 742 nm was 0.35 (293 K).

Fluoresence emission spectra

Upon excitation at 460 nm in the Soret band of BChl *c* the room temperature emission spectrum shows peaks at 751 and 805 nm (Figure 3). These bands probably correspond to those at 748 and 802 nm in the uncorrected emission spectrum published earlier [3]. The relatively strong fluorescence from BChl *a* upon excitation of BChl *c* shows that there must be significant energy transfer from the latter pigment to BChl *a* 798. Some fluorescence was emitted around 675 nm, presumably arising from the pigment absorbing at 667 nm. Lowering the temperature resulted in a fairly large red shift of the

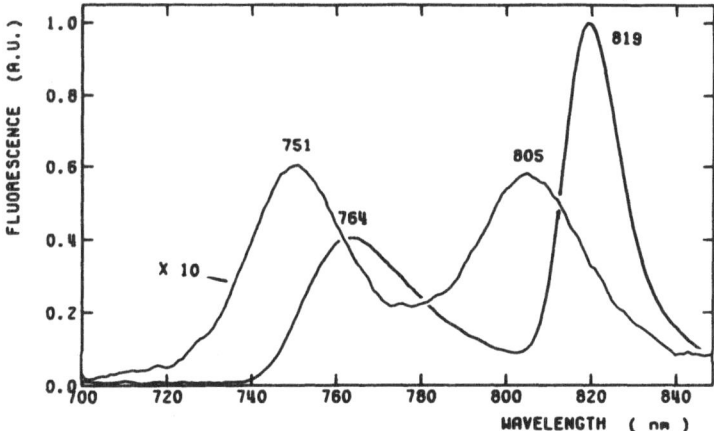

Figure 3. Fluorescence emission spectra of isolated chlorosomes measured at 4 K (——)
and at 293 K (– – – –). The latter spectrum is shown on a 10-fold expanded scale
($A_{742} = 0.2$).

two long wave emission bands. At 4 K BChl c had its peak emission at 764 nm
and BChl a at 819 nm. Cooling also resulted in an increase of the BChl a
emission relative to that of BChl c and a strong increase in the overall fluorescence yield. Changing the excitation wavelenth to 600 nm to obtain more
direct excitation of BChl a resulted in a slight decrease (about 10%) of the
short wave emission relative to that at 819 nm.

Emission spectra were measured at different temperatures between 300
and 4 K. No evidence was found that would indicate a thermal equilibrium
between the populations of excited BChl c and BChl a, not even at room
temperature. In fact, a plot of the logarithm of the ratio of the corresponding
emission intensities against the reciprocal of the temperature showed a
maximum near 200 K, rather than a straight line as predicted in the case of
thermal equilibrium [20], and the BChl c emission at 4 K was five times
stronger than at room temperature.

Excitation spectra

Excitation spectra measured at 4 K are shown in Figure 4. Comparison of
the excitation spectrum of the BChl c emission with the absorption ($1 - T$,
where T is the transmittancy) spectrum showed that energy transfer from
excited carotenoid to BChl c occurred with an average efficiency of 65%.
The bands in the carotenoid region were not only lower than those in the
absorption spectrum but were also shifted to longer wavelengths. The long-
wave band was now situated at 521 rather than at 517 nm, indicating the
presence of at least two different pools of carotenoid with different absorption spectra. The main carotenoids in *C. aurantiacus* chlorosomes are
β- and γ-carotene, which are present in an approximately 1:1 ratio [13].

If both carotenoids display about equal red shifts in vivo as compared to organic solution, then the excitation spectrum indicates that γ-carotene is more efficient than β-carotene in transferring excitation energy to BChl *c*.

The main feature of the excitation spectrum for the BChl *a* emission (Figure 4) is again the strong Q_y band of BChl *c*. Comparison with the height of the band of BChl *a* near 800 nm indicated that the efficiency of energy transfer from BChl *c* to BChl *a* was about 55% at 4 K. The transfer efficiency for carotenoid was lower by approximately the same factor of 0.55 with respect to the excitation spectrum for BChl *c* fluorescence. Thus probably most of the energy transfer from carotenoid to BChl *a* proceeds

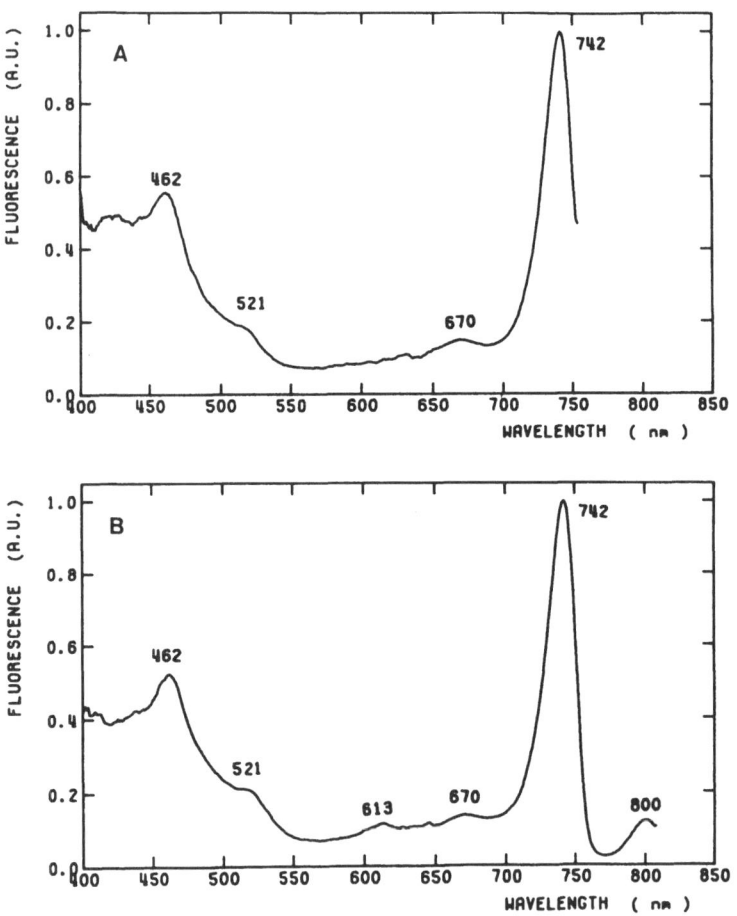

Figure 4. Fluorescence excitation spectrum of isolated chlorosomes at 4 K. A: Fluorescence detected at 770 nm. B: Fluorescence detected at 835 nm. The absorbance at 742 nm was 0.33 (4 K).

via BChl *c*. Due to the rather low rate of energy transfer from BChl *c* to BChl *a* the BChl *a* bands are better resolved in the excitation spectrum than in the absorbance spectrum and the Q_x band of BChl *a* at 613 nm is now easily seen.

Fluorescence polarization

Polarized fluorescence excitation spectra were measured at 4 K for both the BChl *c* and BChl *a* fluorescence. The polarization is defined as $p = (I_{\parallel} - I_{\perp})/(I_{\parallel} + I_{\perp})$, where $I_{\parallel}$ is the fluorescence intensity emitted under 90°, polarized parallel to the excitation beam and $I_{\perp}$ refers to the fluorescence polarized perpendicular to this beam. The polarization values thus obtained are given in Table 1. The fairly high polarization value at 517 nm indicates a more or less parallel alignment for the carotenoids transitions and the BChl *c* Q_y

Table 1. Polarization at various wavelengths of excitation for fluorescence emitted by BChl *c* and BChl *a*

Species	Detection wavelength (nm)	Polarization value p at (nm):				
		434	460	517	742	798
BChl *c*	770	+ 0.16	+ 0.25	+ 0.20	+ 0.35	
BChl *a*	820	− 0.03	+ 0.03	− 0.01	− 0.03	+ 0.22

The standard deviation is ± 0.02 for all detection wavelengths.

transitions, unless it is assumed that little energy transfer between different molecules occurs, which seems unlikely in view of the light-harvesting function of the chlorosome. In contrast to these results stand the low *p*-values observed for BChl *a* 798 emission. In this case significant polarization was only observed upon direct excitation of BChl *a*. Since a more or less random orientation for BChl *c* and carotenoid and depolarization by energy transfer are unlikely in view of the high polarization values for BChl *c* emission, we conclude that the average angle between the optical transitions of these pigments and BChl *a* 798 is close to the magic angle.

Linear dichroism

The chlorosomes were oriented by means of two different pressing methods as described in ref. [17]. Uniaxial orientation was achieved by pressing the gel between two parallel prisms from an initial thickness of 4 mm to a final one of 1 mm. In the second method (biaxial pressing) the gel was pressed in two mutually perpendicular directions with a maximum reduction in one direction of 50%.

The magnitude of the linear dichroic ratio $(A_{\parallel} - A_{\perp})/A$ depends both on the form of the oriented particles and on the orientation method employed. From the electron microscopy studies of Staehelin et al. [14] the chlorosomes appear as oblong ellipsoids with a, b and c axes of approximately

100, 30 and 10 nm, respectively. Ideally, uniaxial pressing should result in an alignment of the short axes (the c-axes), while biaxial pressing aligns the chlorosomes with their long axes (a-axes) parallel. Thus, in the first case the orientation axis is the c-axis, while in the latter it is the a-axis. The relationship between the dichroic ratio $(A_\parallel - A_\perp)/A$ and the angles of an optical transition dipole moment with the axes of orientation can then be described by

$$(A_\parallel - A_\perp)/A = +3/4(3\cos^2\alpha - 1)(3S - 1) \quad \text{(biaxial)} \quad (1)$$

$$(A_\parallel - A_\perp)/A = -3/8(3\cos^2\gamma - 1)(3S - 1) \quad \text{(uniaxial)} \quad (2)$$

where α and γ are the angles with respect to the a- and c-axes, respectively. The degree of orientation is described by the parameter S which contains the orientation distribution function and varies between 1/3 (random sample) and 1.0 (perfectly oriented sample). The reason for the difference in the expression for the uniaxially pressing method by a factor of $-1/2$ with respect to biaxial pressing is twofold. In the first place the sign of the dichroic ratio is reversed, because the axis of orientation is rotated over $90°$. Secondly, due to the fact that the measuring beam makes an angle of $45°$ with the gel in the uniaxial arrangement, the signal is reduced by a factor of 2.

The linear dichroic spectrum at 77 K of chlorosomes oriented by the biaxial pressing method is shown in Figure 5. Clearly the Q_y transitions of BChl c (742 nm) are oriented more or less parallel to the long axis of the chlorosomes and those of BChl a (798 nm) are more perpendicularly oriented. The dichroic ratio of the Q_y transition of BChl c measured at room temperature reached a constant level at about 35% reduction in one direction (see Figure 6). Assuming a perfect orientation at this compression of the gel,

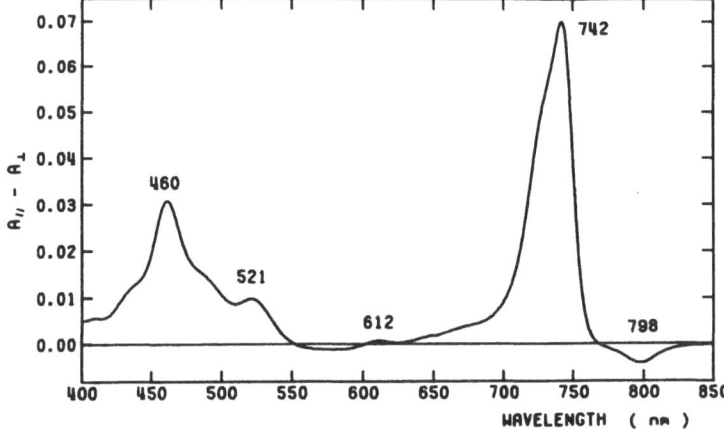

Figure 5. Linear dichroism spectrum of isolated chlorosomes oriented by biaxial pressing, measured at 77 K. Absorbance at 742 nm was 0.14 (77 K).

the angles α between the long axis (a) of the chlorosome and the optical transitions can be calculated by means of equation (1). The results of such calculations are given in Table 2. The fact that this method yields an angle of almost exactly 90° for the orientation of BChl a 798 shows that indeed all chlorosomes are almost perfectly aligned at this compression of the gel. For the BChl c transition at 742 nm this angle was found to be 37°. The shoulder on the short-wave side of the Q_y band at 725 nm is clearly more prominent in the LD than in the absorbance spectrum (Figure 1) corresponding to a significantly smaller angle α (about 16°). The long-wave carotenoid maximum in the LD spectrum of Figure 5, like that in the fluorescence excitation spectrum, is situated at 521 instead of 517 nm. This

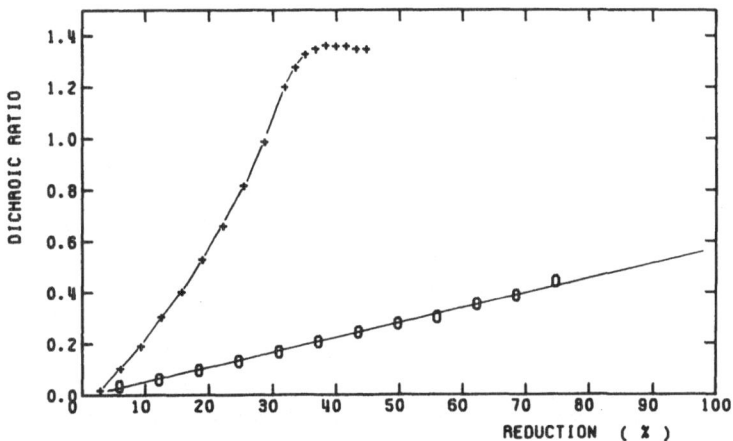

Figure 6. Dichroic ratio over the 742 nm band as a function of the reduction of the thickness of the gel for both orientation methods: biaxial pressing (+ + +), uniaxial pressing (○ ○ ○ ○). Measured at 293 K.

Table 2. Pigment orientation in chlorosomes of *C. aurantiacus*

α and γ are the angles between the optical transition dipole moments and the a- and c-axis of the chlorosome, respectively. Except for the band at 798 nm, no attempt was made to deconvolute the spectra in terms of individual absorption bands. The dichroic ratios $\Delta A/A$ were calculated for perfect orientation, extrapolated by means of the data from Figure 6 from the corresponding LD spectra, measured at 77 K (biaxial pressing) or at room temperature (uniaxial pressing) (see text for details)

| Wavelength (nm) | Biaxial | | Uniaxial | |
	$\Delta A/A$	α	$\Delta A/A$	γ
462	1.22	39 ± 2°	0.41	67 ± 2°
487	0.99	42 ± 2°	0.41	67 ± 2°
521	0.91	43 ± 2°	0.49	70 ± 2°
742	1.37	37 ± 2°	0.51	71 ± 2°
798	− 1.5	86 ± 4°	− 0.9	32 ± 3°

indicates that the angle α of Table 2 applies mainly to γ-carotene, while β-carotene is either randomly oriented or else oriented close to the magic angle with respect to the long axes of the chlorosome.

As shown above the uniaxial pressing method gives information about the angle between the transition dipole and the c-axis of the chlorosome. Linear extrapolation of the dichroic ratio of the 742 nm transition measured at room temperature to zero gel thickness (perfect orientation) yielded a value of 0.50 ± 0.05. Equation (2) then yields an angle $\gamma = 71 \pm 2°$ for this transition with respect to the c-axis. Corresponding angles for the other optical transitions obtained from the room temperature spectrum are given in Table 2. The slower saturation behavior in the case of uniaxial pressing may be due to the small dimensions of the b- and c-axes.

Discussion

We have investigated the energy transfer properties and the spatial arrangement of the pigments present in isolated purified chlorosomes of *Chloroflexus aurantiacus*. Apart from the bulk pigments, BChl *c* and carotenoid, the presence of a small amount of BChl *a* is well established for the green bacteria *Chlorobium limicola* [7] and *Chloroflexus aurantiacus* [3, 6, 13]. The results reported here give further information on the transfer of excitation energy that proceeds via BChl *a* and on the structural arrangements of the pigments within the chlorosome.

Energy transfer

The fluorescence emission and excitation spectra clearly demonstrate the transfer of excitation energy from BChl *c* to BChl *a*. In contrast with a nearly 100% efficiency in chlorosomes of *C. limicola* [17] a value of only 55% was found in *C. aurantiacus*. The same efficiency was observed at room temperature. Addition of dithionite to reduce possible quenchers [9] that could trap the excitations and thus reduce the efficiency of energy transfer had no effect. The lower transfer efficiency of *C. aurantiacus* may be partly due to a smaller overlap of BChl *c* emission and BChl *a* absorption bands, although the efficiency was independent of temperature. The absence of a thermal equilibrium for the BChl *c* and *a* emissions may also indicate that some BChl *c* is dislocated from its original binding place in the chlorosome. It should be noted here, that preliminary experiments with intact cells show a quite high efficiency for excitation energy transfer from BChl *c* to membrane-bound BChl *a*.

Orientation of pigments

The application of two different methods to orient the chlorosomes for LD measurements allows a detailed insight in the spatial arrangement of the pigments of the chlorosome. An elegant and simple way to display these results is by stereographic projection [8] by means of which the direction of a vector in an orthogonal coordinate system can be represented by a single point on the graph. Figure 7 shows such a projection on the plane of the b- and c-axes of the chlorosome. Since the Q_y transition of BChl a 798 lies, within the limit of error, within this plane (Table 2) the point of its vector must be located on the great circle that corresponds to an angle α of 90° with the a-axis, i.e. on the rim of the projection graph. The angle $\gamma = 32°$ derived from the LD measurements (Table 2) defines the position of the transition vector with respect to the c-axis. Only one of the possible symmetrical positions is shown in Figure 7 (open square). The possible orientations of the BChl c Q_y transition vector (742 nm) are defined by the intersections of the small circles of 37° with respect to the a-axis (drawn circle) and of 71° with respect to the c-axis (broken lines). This yields four possible orientations for BChl c with respect to the BChl a transition vector. However, the fluorescence polarization measurements show that the angle between the BChl c and a transition dipoles must be close to the magic angle of 54.7°, which limits the possible location of the BChl c transition vector

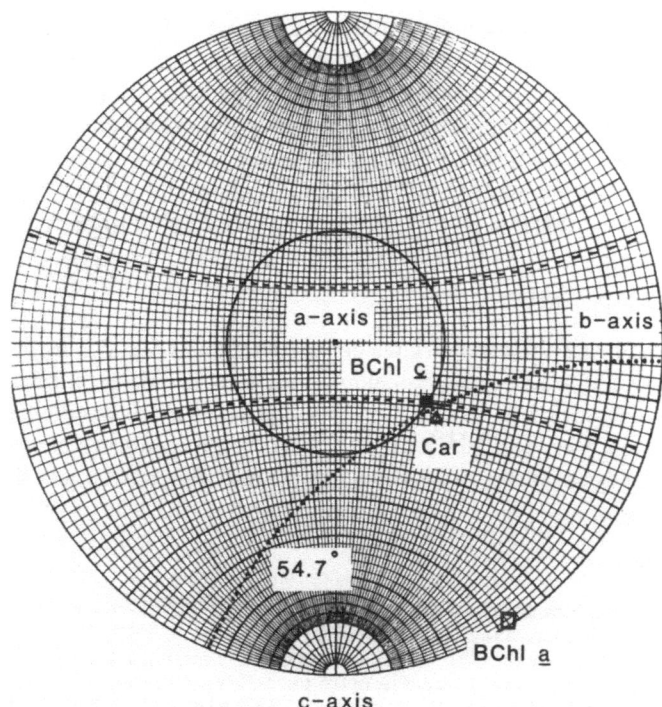

Figure 7. A Wulff stereographic plot of the transition vectors of BChl a (798 nm; □), BChl c (742 nm; ■) and γ-carotene (523 nm, △). Drawn circle, small-circle for $\alpha = 37°$; dashed lines, small-circles for $\gamma = 71°$; dotted line, small-circle for the magic angle (54.7°) with respect to the BChl a transition dipole. See text for further explanation.

to the point indicated by the solid square. In the same way the transition vector for γ-carotene is obtained (triangle).

Several features of this model are of importance for the efficiency of energy transfer in the antenna of *C. aurantiacus*. Energy transfer from carotenoid to BChl *c* is probably optimized by the parallel alignment of their two transition dipole moments. On the other hand, the angle between the Q_y transitions of BChl *c* and BChl *a* 798 seems rather large for efficient energy transfer, but it should be kept in mind that the given value reflects the *average* orientation of the BChl *c* dipole moments, leaving open the possibility that some of the Q_y transitions possess a more advantageous orientation.

Finally we remark that the orientations of the BChl *c* and BChl *a* transitions are very similar for chlorosomes of both *C. aurantiacus* and *C. limicola* [17]. In both types of chlorosomes the Q_y transitions of BChl *c* make a fairly small angle (30–40°) with the long axis of the chlorosome, whereas the Q_y transitions of BChl *a* lie almost perpendicular to this axis; in the case of *C. aurantiacus* investigated here, the angle of the BChl *a* Q_y transition with respect to the shortest axis (c-axis, which is normal to the photosynthetic membrane in vivo [14]) of the chlorosome could also be determined (32°).

We have earlier shown [18] that the Q_y transition of BChl *c* 808, located in the B808-866 light-harvesting complex makes an angle to the normal of the membrane of approximately 44°. Thus the orientations of BChl *a* 798 and BChl *a* 808 as well as their overlap integral (BChl *a* 798 fluoresces at 805 nm at room temperature) are quite favorable for energy transfer. A similar conclusion was recently obtained for green sulfur bacteria [17].

Acknowledgements

We would like to thank F.T.M. Zonneveld and G.J. de Vos for expert help with the preparations of the chlorosomes and A.H.M. de Wit for culturing of the bacteria. The investigation was supported by the Netherlands Foundations for Chemical Research (SON) and for Biophysics, financed by the Netherlands Organization for the Advancement of Pure Research (ZWO).

References

1. Abdourakhmanov IA, Ganago AO, Erokhin YuE, Solov'ev AA and Chunugov VA (1979). Biochim Biophys Acta 546:183–186
2. Amesz J and Vasmel H (1986). In Govindjee, Amesz J and Fork DC, eds. Light Emission by Plants and Bacteria, in the press. New York: Academic Press
3. Betti JA, Blankenship RE, Natarajan LV, Dickinson LC and Fuller RC (1982). Biochim Biophys Acta 680:194–201
4. Connolly JS, Samuel EB and Janzen AF (1982). Photochem Photobiol 36:565–574
5. Duysens LNM (1952). Doctoral thesis, University of Utrecht
6. Feick RG, Fitzpatrick M and Fuller RC (1982). J Bact 150:905–915
7. Gerola PD and Olson JM (1986). Biochim Biophys Acta, in the press

8. Hoff AJ (1985). In Michel-Beyerle ME, ed. Proc Conf Antennas and Reaction Centers of Photosynthetic Bacteria — Structure, Interactions and Dynamics, München, 1985, in the press. Berlin: Springer-Verlag
9. Karapetyan NV, Swarthoff T, Rijgersberg CP and Amesz J (1980). Biochim Biophys Acta 593:254–260
10. Kramer HJM and Amesz J (1982). Biochim Biophys Acta 682:201–207
11. Meiburg RF (1985). Doctoral thesis, University of Leiden
12. Pierson BK and Castenholz RW (1974). Arch Microbiol 100:5–24
13. Schmidt K (1980). Arch Microbiol 136:11–16
14. Staehelin LA, Golecki JR, Fuller RC and Drews G (1978). Arch Microbiol 119: 269–277
14. Stanier RY and Smith JHC (1960). Biochim Biophys Acta 41:478–484
16. Trüper HG and Pfennig N (1978). In Clayton RK and Sistrom WR, eds. The Photosynthetic Bacteria, pp. 19–30. New York–London: Plenum Press
17. van Dorssen RJ, Gerola PD, Olson JM and Amesz J (1986). Biochim. Biophys. Acta, in the press
18. Vasmel H, van Dorssen RJ, de Vos GJ and Amesz J (1985). Photosynth Res, in the press
19. Wechsler T, Suter F, Fuller RC and Zuber H (1985). FEBS Lett 181:173–178
20. Zankel KL (1978). In Clayton RK and Sistrom WR, eds. The Photosynthetic Bacteria, pp. 341–347. New York–London: Plenum Press

Photosynthesis Research 9, 47–54 (1986)
© *1986 Martinus Nijhoff/Dr. W. Junk Publishers, Dordrecht.*

Photosystem I photochemistry: A new kinetic phase at low temperature

PIERRE SÉTIF and PAUL MATHIS

Service de Biophysique, Département de Biologie, Centre d'études Nucléaires de Saclay, 91191 Gif-sur-Yvette, Cedex, France

(*Received 1 August 1985*)

Key words: charge recombination, electron transfer, photosynthesis, photosystem I, reducing conditions

Abstract. A new phase of charge recombination between the oxidized primary electron donor of photosystem I (P700$^+$) and a reduced acceptor has been detected by flash absorption spectroscopy in PS I particles at low temperature. It occurs under highly reducing conditions (the secondary electron acceptors F_A and F_B and one or possibly two 'more primary' acceptors being prereduced) with a $t_{1/2}$ of about 20 μs between 10 and 80 K.

Introduction

In O_2-evolving organisms, photosystem I (PS I) carries out the photodriven reduction of NADP$^+$ and hence comprises low potential membrane-associated electron carriers. Recent EPR studies indicate the existence of five different electron acceptors in PS I: A_0, A_1, F_X, F_B and F_A [1, 5, 8]. Whereas F_X, F_B and F_A are probably three different iron-sulfur centers, the reduced forms of A_0 and A_1 exhibit EPR radical signals in the g = 2.0 region [1, 5] but their chemical nature is unknown at the moment. The forward electron transfer steps in PS I are not yet clear, particularly concerning the three iron-sulfur centers F_X, F_B and F_A. A linear sequence $F_X \rightarrow F_B \rightarrow F_A$ has been proposed, but there is now more evidence for a parallel functioning of F_A and F_B (for recent reviews, see refs. [4, 16]). This parallel model has been recently extended to F_X at low temperature [3, 17], i.e. the electron transfer has been proposed to be linear from the primary acceptor A_0 to A_1 and then in parallel from A_1 to F_X, F_B and F_A. Whether this model can be applied to room temperature electron transfer has yet to be resolved.

Under highly reducing conditions, F_A and F_B can be easily reduced at room temperature and trapped in the reduced state for low temperature experiments, whereas A_0, A_1 and F_X remain in the oxidized state. When PS I particles, in the initial redox state (F_X, F_B^-, F_A^-) are excited by a laser

---

Dedicated to Prof. L.N.M. Duysens on the occasion of his retirement.

flash, the change separation is followed by a recombination reaction at any temperature between 4.2 K and 294 K. At room temperature, this recombination reaction is practically monophasic with a $t_{1/2}$ of about 250 μs [12]. It has been proposed that the back-reaction occurs between the oxidized primary donor P700$^+$ and a reduced species called A_2 which has been tentatively identified with F_X, since its reduction leads to negative absorption changes around 420–450 nm, as expected for an iron-sulfur center [6, 7, 12]. Under the same initial redox conditions, the situation is more complex at low temperature (10–20 K) [17]: whereas 10 to 15% of the PS I reaction centers relax slowly from the (P700$^+$ ... F$_X^-$) state, and some P700 triplet state is photoinduced in a minority (5 to 20%) of reaction centers which are probably damaged, the main proportion of reaction centers (70–80%) relax to the ground state with a $t_{1/2}$ of about 120 μs. This 120 μs decay phase has been proposed to be due to the recombination reaction between P700$^+$ and the acceptor A_1^- [17].

With PS I particles under still more reducing conditions, the electron pathways and their kinetics are not well known. When A_1 is prereduced, these pathways are presumably the same as in PS I particles prepared with SDS, which contain only the primary acceptor A_0. In these CP1 particles, flash-absorption studies have shown that flash excitation produces the state (P700$^+$–A_0^-) which decays in about 100 ns, mostly to the triplet state of P700 [14]. This triplet state has a $t_{1/2}$ of about 800 μs at 10 K [5, 15]. Decay kinetics of spin-polarized transients have been also reported at low temperature (phases with $t_{1/2}$ of 30 and 3 μs [8] and 5 μs [9]) but they reflect intricated relaxation, microwave power induced and chemical decays and are difficult to interpret. In this paper, a new recombination phase is reported from flash-absorption experiments. It can be observed at a low redox potential in PS I particles containing all the membrane-associated electron acceptors. It exhibits a decay halftime of 20 μs at 10 K.

Materials and methods

Biological material

PS I particles, containing about 110 chlorophylls per P700, were prepared according to [11]. This procedure involves the solubilization of thylakoid membranes by digitonin and a one-step fractionation by polyacrylamide gel electrophoresis in the presence of deoxycholate. These particles contain all the membrane-associated PS I acceptors, including the iron-sulfur centers F_X, F_B and F_A, as has been checked by flash-induced absorption changes at room temperature in the presence of sodium ascorbate (recombination reaction with $t_{1/2}$ of about 30 ms) as well as by EPR at 10 K under highly reducing conditions (EPR signals of F$_X^-$, F$_B^-$ and F$_A^-$).

EPR measurements

EPR spectra were recorded with a Bruker ER 200TT spectrometer operating at 100 kHz modulation frequency and equipped with an Oxford Instruments ESR 900 liquid helium cryostat.

Absorption measurements

The general setup for flash-induced absorption changes was the same as described previously [12], using a silicon photodiode and a laboratory-made amplifier capable of resolving components to about $1\,\mu s$ duration. For measurements at room temperature, the material was contained in a square cuvette $(10 \times 10\,mm)$. For low-temperature measurements, the PS I particles were placed in a flat plexiglass cuvette which was inserted in a cryostat cooled with helium gas [13] and positioned at $45°$ of the mutually perpendicular exciting and measuring beams. Excitation was provided by a dye laser which was pumped by a frequency-doubled YAG-laser (duration 20 ns, $\lambda = 594\,nm$, bandwidth at half-intensity 12 nm). The measuring light was filtered before entering the cuvette either by an interference filter or by a red filter (RG 630, Schott). To get a spectrum in the red region, the red filter was used with a constant intensity of the measuring light to ensure an identical, if any, actinic effect of the measuring beam at all wavelengths. Before detection, the light transmitted by the sample was filtered either by an interference filter or by a Bausch and Lomb monochromator with 3 nm (below 720 nm) or 10 nm (above 720 nm) bandwidth at half-height. Signals from the detector were recorded using a transient digitizer (Tektronix R7912) coupled to a multichannel analyzer (Didac 4000, Intertechnique).

Preillumination conditions

PS I particles, after being poised with an excess (about 20 mM) of sodium dithionite at pH 10 were cooled in the dark or under dim light down to 220 K. Some samples were illuminated at 220 K inside the cryostat. The light source was the same one that was used for measuring the absorption changes (halogen quartz lamp, 200–800 W). The light was filtered with red filters (RG 630). The sample was illuminated for times between 10 and 60 s. It was checked that, in our experimental conditions (concentration $30\,\mu g/ml$, focusing of the measuring light onto the sample), the maximum effect of illumination was attained within a period of 60 s with 800 W.

Results

Illumination under highly reducing conditions at 220 K

In PS I particles, in the presence of sodium dithionite at pH 10, the iron-sulfur centers F_A and F_B are fully reduced and the other acceptors (A_0, A_1

and F_X) oxidized, as can be controlled by EPR [17]. In this redox state, which can be easily obtained in the presence of 65% glycerol as well as in its absence, the flash-induced charge separation at low temperature (10–20 K) is followed by a major recombination reaction measured at 820 nm with a $t_{1/2}$ of 120 μs (70–80% of the reaction centers) which has been ascribed to a back-reaction between P700$^+$ and A_1^- [17]. Illumination of the PS I particles at 220 K leads to a higher degree of reduction of the PS I acceptors, as checked by EPR (after a prolonged time of illumination, F_X appears to be fully reduced and some radicals are also formed; EPR and absorption experiments could not be done with the same samples). Different samples of PS I particles were studied at 10 K under such conditions. After a laser flash excitation, the following decay phases were observed:

– a fast decay phase with a $t_{1/2}$ of about 20 μs (Figure 1) appears, without noticeable change in the initial total signal size at 820 nm as well as at 703 nm compared to samples prepared in the redox state $(A_0-A_1-(F_X, F_B^-, F_A^-))$. With a higher degree of illumination at 220 K, either in duration or in intensity, the size of this phase first increases and then decreases. It probably corresponds to a higher reduction of the acceptors, and it never exceeds 60% of the total signal at 820 nm. A quantitative analysis of the absorption transients shows that the increase in the size of the 20 μs phase, in parallel with the degree of illumination, occurs at the expense of slower recombination phases (120 μs and 50–400 ms phases, ascribed to the recombination reactions between P700$^+$ and A_1^-, adn P700$^+$ and F_X^-, respectively) and its decrease occurs for the benefit of:

– a 800 μs phase. Such a decay phase is already present when the PS I particles are in the redox state $(A_0, A_1, (F_X, F_B^-, F_A^-))$ and is probably due to the triplet state of P700 which is photoinduced in some damaged reaction centers (5 to 20% depending on the preparation) [5, 17]. The size of the 800 μs signal appears fairly constant as long as the slower recombination phases are present ($t_{1/2}$ of 120 μs and (50–400 ms)) but increases under the more highly reducing conditions that were attainable in our experimental conditions (Figure 1), at the expense of the 20 μs phase.

The size of the total signal (with a μs time resolution) at 820 nm as well as at 703 nm decreases somewhat when the proportion of the 800 μs phase increases. These quantitative absorption change measurements could only be done with samples containing glycerol; under highly reducing conditions and after a strong illumination at 220 K (maximum light intensity for more than one minute), the absorption change relaxed at 820 nm with 60% of the 20 μs phase and 40% of the 800 μs phase (Figure 1, upper left). Under these conditions, the total signal magnitude induced by a saturating laser flash was about 20% smaller than the signal elicited in a sample with the same chlorophyll concentration in the redox state $(A_0, A_1, (F_X, F_B^-, F_A^-))$. By thawing the sample for a few seconds and freezing it again down to 10 K in the dark, a complete recovery of the total signal size and of the kinetics

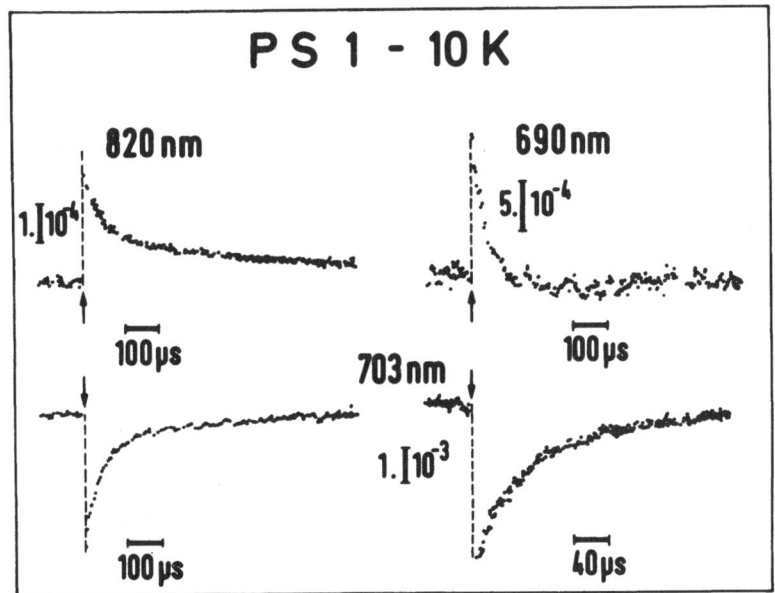

Figure 1. Kinetics of absorption changes induced in PS 1 particles under highly reducing conditions (20 mM sodium dithionite at pH 10) and in the presence of glycerol by a saturating dye laser flash at 10 K. The PS 1 particles were cooled under dim light down to 220 K and illuminated for 1 minute with a strong red light at this temperature before cooling down to 10 K. A 679 nm (45°) = 0.35. For the absorption measurements, interference filters of the appropriate wavelength were inserted between the source of measuring light and the cuvette and between the cuvette and the photodiode.

of decay corresponding to the redox state $(A_0, A_1, (F_X, F_B^-, F_A^-))$ is obtained, thus showing that the sample was not damaged by the light treatment at 220 K.

In PS I samples without glycerol, the same decay phases were observed at 10 K (20 μs and 800 μs) but samples exhibiting only a 800 μs decay phase with no more 20 μs phase could be prepared after a strong illumination at 220 K. However neither quantitative absorption measurements, nor precise spectra could be made because the samples were highly scattering.

The spectrum of the 20 μs decay phase has been measured in the red (Figure 2) and near infra-red regions. In the near infra-red, it shows a broad positive maximum at 815–820 nm, which is characteristic of P700⁺. In the red region, it exhibits three main peaks (two negative maxima at 685 and 703 nm and one positive maximum at 691 nm) and two smaller ones (positive at 680 nm and negative at 667 nm). For the sake of comparison, the spectrum of the 120 μs decay phase, which is due mainly to the (P700⁺–P700) difference [17], is also shown in Figure 2, after normalization at 703 nm. The differences between the two spectra appear to be significant from 650 to 680 nm, a wavelength region wher the absorption changes are relatively

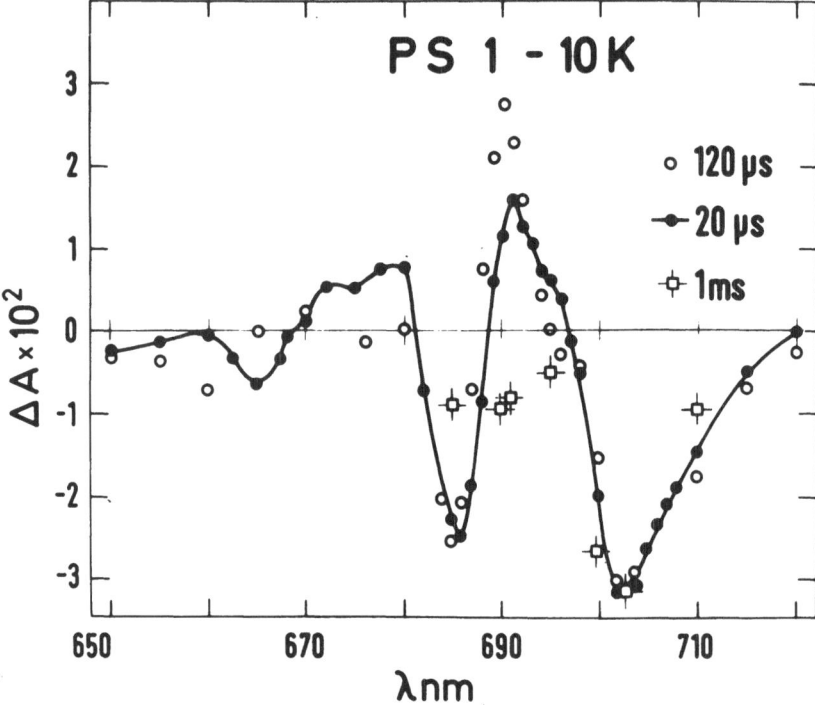

Figure 2. Difference spectra of absorption changes induced at 10 K in PS 1 particles under different redox conditions: – open circles: spectrum of the 120 μs decay phase measured on PS 1 particles that were poised with sodium ascorbate and dichlorophenol indophenol, cooled in the dark down to 10 K and preilluminated at 10 K (17). The vertical scale refers to this experiment (A 679 nm (45°) = 0.5). – full circles: spectrum of the 20 μs decay phase measured on a sample prepared as indicated in Figure 1. – squares: spectrum of the slower (0.8–1 ms) decay phase measured on the same sample. The last two spectra were normalized to the first one at 703 nm.

small, but the two spectra are rather similar above 680 nm: in comparison to the 120 μs spectrum, the 20 μs spectrum exhibits a 1 nm shift of the peaks at 685–686 and 690–691 nm (hardly significant) and a decrease in the size of the 690–691 nm peak. However, the integrated areas of the two spectra from 650 to 720 nm are identical, suggesting that the differences between the two spectra are solely due to electrochromic shifts of surrounding chlorophyll molecules.

The spectrum of the 800 μs phase is considerably different from the 20 μs spectrum: it exhibits, besides a negative maximum at 703 nm, a large flat positive band between 740 and 820 nm (not shown) and no positive maximum around 690 nm (Figure 2). The same spectral features have been already observed in CP1 particles prepared with SDS and have been attributed to the triplet state of P700 [15].

Temperature dependence of the fast phase

After a redox and light pretreatment inducing the appearance of a 20 μs decay phase at 10 K as described in the previous section, the kinetics of decay were studied at increasing temperatures, starting from 10 K: a decay phase with a $t_{1/2}$ of 20 μs and with a constant magnitude could be safely discriminated from slower components up to 80 K. A faster decay phase appears on the same sample above 80 K. However, the temperature dependence of this decay has not been studied in detail owing to the growing complexity of the decay kinetics in this temperature region. This is due to the fact that there is always a significant proportion of P700 triplet, the decay rate of which becomes temperature dependent above 80 K [10].

Discussion

The present data show that, starting from the redox state $(A_0, A_1, (F_X, F_B^-, F_A^-))$, for which the major recombination reaction following charge separation has been proposed to involve P700$^+$ and A_1^- with a $t_{1/2}$ of 120 μs, two decay phases with $t_{1/2}$ of 20 and 800 μs appear in the relaxation of flash-induced absorption changes at 10 K when the acceptors of PS I are increasingly reduced by illumination at 220 K in the presence of sodium dithionite. The spectrum and kinetics of the 800 μs phase indicate that it reflects the decay of the P700 triplet state. This triplet state probably arises, at least for the main part, from a fast sub-μs recombination reaction, as has been already observed in CP1-SDS complexes devoid of secondary acceptors [14], which involves P700$^+$ and the acceptor A_0^-.

The spectrum of the 20 μs phase at 10 K shows that it is due to a recombination reaction between P700$^+$ and an electron acceptor. This decay phase is present at a redox state of the acceptors which is probably intermediate between the redox states $(A_0, A_1, (F_X, F_B^-, F_A^-))$ and $(A_0, A_1^-, (F_X^-, F_B^-, F_A^-))$. It has been shown earlier that the first one of these redox states gives rise to a major decay phase with a $t_{1/2}$ of 120 μs (ascribed to the recombination reaction between P700$^+$ and A_1^-) with a minor slower component ascribed to the recombination between P700$^+$ and F_X^- [17]. The second one of these redox states corresponds probably to a fast sub-μs recombination between P700$^+$ and A_0^- which has been observed in CP1 particles containing only the acceptor A_0 [12]. This back reaction induces the formation of some P700 triplet state which decays at 10 K with an approximate lifetime of 0.8 ms [5, 14, 15, 18]. These properties raise the possibility that there exists another intermediate electron acceptor between A_0 and A_1. When F_X, or both F_X and A_1 become prereduced, charge separation would be followed by a back reaction between P700$^+$ and this reduced acceptor with a $t_{1/2}$ of 20 μs. Another possibility, perhaps more probable, is that the 20 μs decay phase corresponds to the recombination reaction between P700$^+$ and A_1^-, when the acceptor F_X is prereduced. The lifetime for this back-reaction would be decreased from $120\,\mu s$ to $20\,\mu s$ due to Coulomb repulsion by the negative charge carried by F_X:

$$P700^+ \ldots A_1^- - (F_X, F_B^-, F_A^-) \xrightarrow{t_{1/2} = 120\,\mu s} P700 \ldots A_1$$

$$P700^+ \ldots A_1^- - (F_X^-, F_B^-, F_A^-) \xrightarrow{t_{1/2} = 20\,\mu s} P700 \ldots A_1$$

Such a Coulomb effect has already been invoked for the variation of the rate of electron donation to $P680^+$ according to the S-state [2].

Further experiments will be needed to discriminate between these two possibilities but whatever hypothesis is retained, it appears that the electron acceptor involved in the $20\,\mu s$ recombination is not a chlorophyll or pheophytin molecule since, upon reduction, it exhibits no bleaching in the red region of the spectrum.

References

1. Bonnerjea J and Evans MCW (1982) FEBS Lett 148:313–316
2. Brettel K, Schlodder E and Witt HT (1984) Biochim Biophys Acta 766:403–415
3. Crowder M and Bearden A (1983) Biochim Biophys Acta 722:23–35
4. Evans MCW (1982) In Spiro, ed. Iron-sulfur proteins. Vol. 4, pp 249–284. John Wiley & Sons, New York
5. Gast P, Swarthoff T, Ebskamp FCR and Hoff AJ (1983) Biochim Biophys Acta 722:163–175
6. Golbeck JH, Velthuys BR and Kok B (1978) Biochim Biophys Acta 504:226–230
7. Koike H and Katoh S (1982) Photochem Photobiol 35:527–531
8. McCracken JL and Sauer K (1983) Biochim Biophys Acta 724:83–93
9. McCracken JL and Sauer K (1983) In Sybesma C, ed. Proc. VI Int Congr Photosynthesis Vol I, pp 585–588, Junk, The Hague.
10. Mathis P, Sauer K and Remy R (1978) FEBS Lett 88:275–278
11. Picaud A, Acker S and Duranton J (1982) Photosynth Res 3:203–213
12. Sauer K, Mathis P, Acker S and Van Best JA (1978) Biochim Biophys Acta 503:120–134
13. Sauer K, Mathis P, Acker S and Van Best JA (1979) Biochim Biophys Acta 545:466–472
14. Sétif P, Bottin H and Mathis P (1985) Biochim Biophys Acta 808:112–122
15. Sétif P, Hervo G and Mathis P (1981) Biochim Biophys Acta 638:257–267
16. Sétif P and Mathis P (1985) Encyclopedia of Plant Physiol, in the press, Springer Verlag, Heidelberg
17. Sétif P, Mathis P and Vänngärd T (1985) Biochim Biophys Acta 767:404–414
18. Sétif P, Quaegebeur JP and Mathis P (1982) Biochim Biophys Acta 681:345–353

Photosynthesis Research 9, 55–62 (1986)
© *1986 Martinus Nijhoff/Dr. W. Junk Publishers, Dordrecht.*

Thermodynamics of the charge recombination in photosystem II

H.J. VAN GORKOM, R.F. MEIBURG and L.J. DE VOS

Department of Biophysics, Huygens Laboratory of the State University, P.O. Box 9504, 2300 RA Leiden, The Netherlands

(*Received 30 August 1985*)

Key words: electric field effect, luminescence, photosystem II, thermodynamics

Abstract. The temperature dependence of the electric field-induced chlorophyll luminescence in photosystem II was studied in Tris-washed, osmotically swollen spinach chloroplasts (blebs). The system II reaction centers were brought in the state Z^+P^+-$Q_A^-Q_B^-$ by preillumination and the charge recombination to the state $Z^+PQ_AQ_B^-$ was measured at various temperatures and electrical field strengths. It was found that the activation enthalpy of this back reaction was $0.16\,eV$ in the absence of an electrical field and diminished with increasing field strength. It is argued that this energy is the enthalpy difference between the states IQ_A^- and I^-Q_A and accounts for about half of the free energy difference between these states. The redox state of Q_B does not influence this free energy difference within $150\,\mu s$ after the photoreduction of Q_A. The consequences for the interpretation of thermodynamic properties of Q_A are discussed.

Abbreviations

DCMU, $3(3',4'$-dichlorophenyl)-1,1-dimethylurea; I, intermediary electron acceptor; Mops, 3-(N-morpholino)propanesulphonic acid; P, (P_{680}) primary electron donor; PS II, photosystem II; Q_A and Q_B, first and second quinone electron acceptors; Tricine, N-tris(hydroxymethyl)methylglycine; Tris, tris-(hydroxymethyl)aminomethane; Z, secondary electron donor

Introduction

Although redox changes of the photosystem II (PS II) electron acceptor Q_A were recognized as such already 23 years ago by Duysens and Sweers [5], the energetics of these reactions, undoubtedly of crucial importance to the efficiency of photosynthetic energy conversion, are still not clear. The thermodynamic parameters associated with electron transfer reactions in PS II are not easily accessible experimentally. Standard redox potentials under equilibrium conditions, as obtained by chemical titration, are unknown or controversial for nearly all redox components involved: those of the electron

---

*Dedicated to Professor L.N.M. Duysens on the occasion of his retirement

donors are too high to titrate without serious irreversible damage to the system; those of the electron acceptors are rendered uncertain by various kinds of heterogeneous behaviour [17]. Perhaps the only exception is the intermediary acceptor I, a pheophytin a molecule which titrates at about -0.6 V [8, 13] like it does in vitro [6]. But even if these redox potentials were known their relation to the non-equilibrium states actually present during normal PS II turnover would be rather uncertain. Thus the current thermodynamic picture of PS II electron transport is largely based on kinetic data and indirect arguments [15].

One potentially useful tool in this area of research is the effect of a membrane potential on the photosynthetic charge separation. If two reactants are located on opposite sides of the membrane dielectric, electron transfer between them is expected to equilibrate according to the sum of their midpoint potential difference and the membrane potential. By this approach, we have earlier obtained evidence suggesting that the midpoint potential difference between I and Q_A was 0.33 V if in PS II only between these components electron transfer is sensitive to a membrane potential (and less otherwise) [9]. An additional advantage of the method was that it avoids most of the heterogeneity problem: only the predominant, conventional type of PS II seems to be affected by a membrane potential [10]. The present paper extends and supports these results by showing that the temperature dependence of the equilibrium between I^-Q_A and IQ_A^- indeed decreases at large membrane potentials and that the entropy difference between these states is gratifyingly similar to that reported for the corresponding states in purple bacteria. It is also shown that the redox state of the second quinone acceptor, Q_B, does not change the operating potential of Q_A. Its dramatic effect on the equilibrium midpoint potential of Q_A [14] must then be ascribed to a different state of the PS II acceptor side.

Material and Methods

Spinach leaves obtained from local shops were ground in a cooled blender in 40 mM Tricine buffer, pH 7.8, containing 0.4 M sucrose, 10 mM KCl and 2 mM $MgCl_2$. After filtration through a 25 μm mesh nylon cloth, the chloroplasts were sedimented by 5 min centrifugation at 8000 x g and resuspended in 0.8 M Tris buffer, pH 8.3, containing 0.4 M sucrose. After 30 min incubation at 4°C the chloroplasts were sedimented again and resuspended in the the Tricine buffer. Shortly before measurement the sample was diluted 200-fold in 2.5 mM Mops buffer, pH 6.6. Luminescence induced by electrical field pulses was measured with a phosphoroscope-type apparatus as described earlier [10]. The stopped-flow cuvette, the phosphoroscope and the sample stock were all cooled to the same temperature.

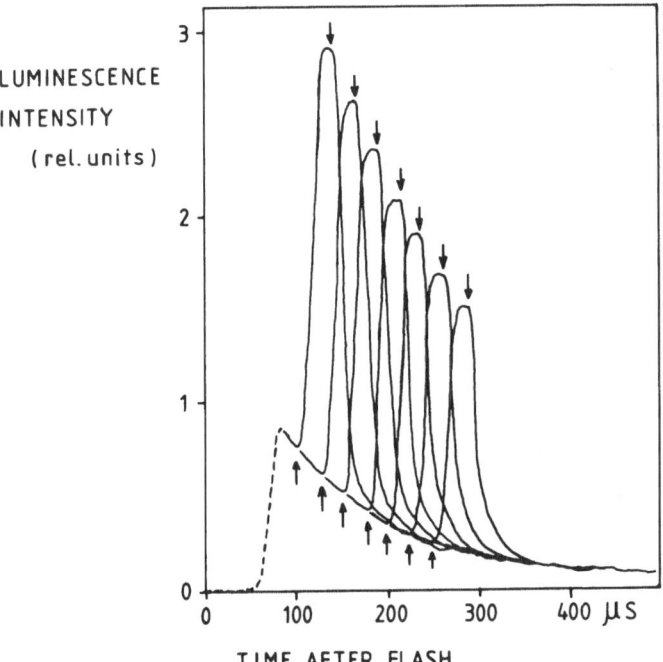

Figure 1. Examples of recorder traces obtained in a typical experiment. The risetime of the luminescence signal reflects the opening of the phosphoroscope shutter. At a variable time after the flash an electrical pulse of 700 V/cm is applied to the suspension, which induces a burst of 'electroluminescence' superimposed on the – undisturbed – normal luminescence decay. Upward and downward arrows indicate field on and off, respectively.

Results

Convenient conditions to study the charge recombination in PS II were obtained by inactivating the oxygen evolving complex by Tris-washing and by oxidizing Z, the only secondary electron donor still active, by preillumination with a saturating flash. At 10 ms after this flash, when reoxidation of Q_A^- by Q_B had proceeded to equilibrium and the rereduction of Z^+ was still negligible, the probing flash was fired, generating the state $Z^+P^+Q_A^-Q_B^-$. The luminescence was recorded from 70 μs to 1 ms later, the limits set by the phosphoroscope chopper window. During this time the charge recombination to the state $Z^+PQ_AQ_B^-$ is expected to occur, with a half time near 0.2 ms [7]. The observed luminescence decayed much faster (half time about 80 μs) and originates from a different precursor, presumably in PS II β centers [10]. By applying a short electrical field pulse, however, the 'normal' charge recombination is greatly accelerated [3], causing a burst of additional,

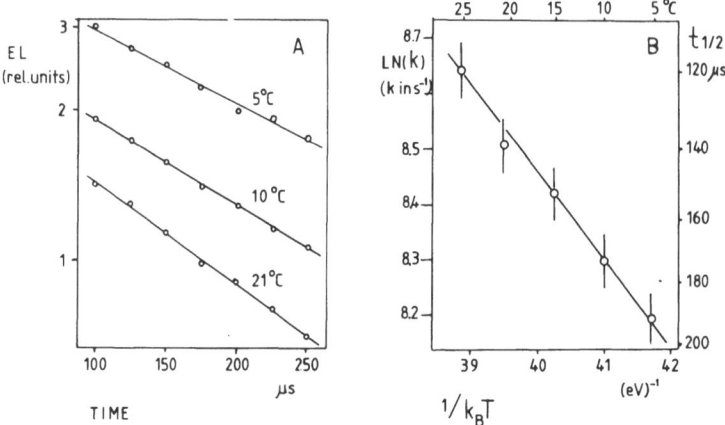

Figure 2. Temperature dependence of the decay of the electroluminescence precursor. A: electroluminescence peak amplitude (EL, normal luminescence subtracted) plotted on a log scale versus the time between flash and pulse, for three temperatures. B: (Arrhenius plot) log of the decay rate constant versus $1/k_B T$ (k_B = Boltzmann's constant; T = abs. temperature). For convenience temperatures and half times are also indicated (upper and right hand scales). The slope of the line yields an activation enthalpy of 0.16 eV for the decay of the electroluminescence precursor.

'electroluminescence' emission. In a series of successive experiments, each time taking a fresh sample to avoid depletion of the precursor, the decay kinetics of the electroluminescence precursor was determined by varying the time between probing flash and electrical pulse, as illustrated in Figure 1. Only the state $Z^+P^+Q_A^-Q_B^-$ and not the state $Z^+PQ_AQ_B^-$ produced a significant electroluminescence signal in the conditions used, as could be verified by omitting the probing flash.

The decay of $P^+Q_A^-$ was measured in this way at various temperatures in the physiological range and the temperature dependence was found to be rather weak (Figure 2A). From an Arrhenius plot (Figure 2B) of the first order rate constants estimated from such data an activation enthalpy of 0.16 eV ± 0.03 eV was calculated.

In order to obtain information on the effect of an electrical field on the activation enthalpy, we determined the decay rate of the electroluminescence during a long pulse at various temperatures. In Figure 3A the kinetics at several field strengths measured at 22 °C and 4 °C are compared; the signal observed in the absence of an electrical field was subtracted and all traces were normalized to the same peak amplitude. In Figure 3B the actual amplitudes of the same traces are shown. It is clear from these data that the temperature dependence of the field-induced luminescence decreases with increasing field strengths. The increase of the emission amplitude with increasing field strengths is in agreement with earlier findings [3]. A corresponding acceleration of the decay was not obvious, suggesting that in

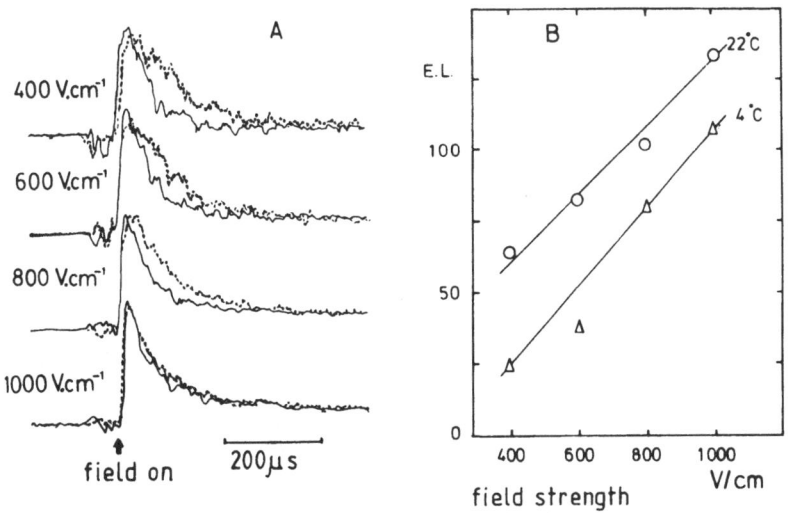

Figure 3. Kinetics (A) and peak amplitudes (B) of the electroluminescence induced by long pulses of different field strengths, at 4 °C (dashed lines in A) and at 22 °C. In A the amplitudes were normalized (after subtraction of the normal luminescence).

in these conditions the emission yield may be field strength dependent, as discussed in ref. [9].

The possibility that the operating potential of Q_A^- is strongly dependent on the redox state of Q_B, as might be inferred from the data in ref. [14], is virtually ruled out by the experiment shown in Figure 4. For this experiment 1 mM ferricyanide was added to the Tris-washed chloroplasts before dilution, in order to decrease the amount of Q_B^- present in the dark. Then $10 \mu M$ tetraphenylboron was added as an electron donor and a sequence of saturating flashes, spaced by 1 s, was fired. During this flash series Q_B was alternatingly reduced and reoxidized, as was verified in separate experiments by measurement of the DCMU-induced increase of the fluorescence yield [16]. The traces in Figure 4 show the electroluminescence induced by a non-saturating (500 V/cm) field pulse applied $150 \mu s$ after the first, second, third, or fourth flash. Presumably the emission is caused by charge recombination of $Z^+PQ_A^-$ Q_B after one or three flashes and of $Z^+PQ_A^-Q_B^-$ after two or four flashes. The amplitude is consistently larger after an even than after an odd number of flashes, but the difference is much smaller than the difference in Q_B^- concentration: at $150 \mu s$ after the first flash almost no Q_B^- is present in these conditions. More likely, the difference can be explained by the somewhat faster reoxidation of Q_A^- by Q_B than by Q_B^- [2]. It is concluded that the redox state of Q_B has no dramatic effect on electroluminescence from the state $Z^+PQ_A^-$, at least at $150 \mu s$ after its formation.

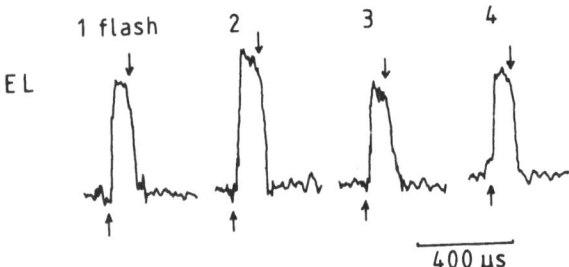

Figure 4. Electroluminescence measured at $150\,\mu s$ after the last 1, 2, 3, or 4 flashes (spaced by 1 s), fired after dark adaptation of Tris-washed chloroplasts in the presence of 1 mM ferricyanide. Blebs were formed just before illumination, by 20-fold dilution in sucrose- and ferricyanide-free buffer containing $10\,\mu M$ tetraphenylboron to reduce endogenous electron donors between flashes. Upward and downward arrows indicate field on and off, respectively. A non-saturating field strength of 500 V/cm was used.

Discussion

From the slope of the Arrhenius plot in Figure 2B it follows that the decay of the state $P^+Q_A^-$ had an activation enthalpy of 0.16 eV. About the same value, 0.15 eV, was found by Reinman and Mathis [11], who studied the temperature dependence of P^+ decay by measuring the decay of its absorbance at 820 nm. Also the decay rates reported in ref. [11] match those shown here, indicating that the same reaction was studied. The chloroplast preparation was also similar, apart from the 200-fold dilution required for our measurements. Approximately, the 0.2 ms recovery phase of the absorbance of P (measured at 685 nm), for which an 0.31 eV activation enthalpy was reported in ref. [4], must also be ascribed mainly to $P^+Q_A^-$ recombination, but in this case no Tris-washing was applied and the identification of the reaction responsible for the signal is less unambiguous. A strong temperature dependent contribution by a different reaction might explain this apparent discrepancy. Thus we conclude that the activation enthalpy of the back reaction of $Z^+P^+Q_A^-Q_B^-$ to $Z^+PQ_AQ_B^-$ is 0.16 eV, in agreement with ref. [11].

Earlier work [3] led to the conclusion that 97% of the reversed electron transport from Q_A^- to P^+ proceeds via an electric field sensitive, energetically uphill electron transfer to 1, followed by an electric field insensitive [9] downhill transfer to P^+, presumably forming the triplet state of P, and subsequent decay to the ground state. Since the forward electron transfer from I^- to Q_A is probably temperature independent, our data then suggest that the enthalpy difference between I^-Q_A and IQ_A^- is 0.16 eV. The 0.33 eV free energy difference between these states [9] then must contain an 0.17 eV entropy contribution, the same value as obtained by Arata and Parson with reaction centers from the purple bacterium *Rhodopseudomonas sphaeroides* strain R-26 [1]. This agreement is reassuring, since the molecules involved in this reaction are very similar in PS II and purple bacteria.

Our value of 0.33 eV for the free energy difference between I^-Q_A and $I\ Q_A^-$ is not in agreement with most titration data on the midpoint potentials of I/I^- and Q_A/Q_A^-, which suggest a value near 0.45 eV or even larger [17]. It was shown in ref. [14], that $Q_AQ_B/Q_A^-Q_B^-$ titrates at a much (0.3 eV) higher midpoint potential than Q_A/Q_A^-. If there is such a strong dependence of the midpoint potential of Q_A on the redox state of Q_B, a huge activation enthalpy of the electroluminescence, if detectable at all, might have been expected under the conditions used here, because Q_B^- was present. The experiment illustrated in Figure 4 clearly shows, however, that the redox state of Q_B has little influence on the electroluminescence induced by a pulse shortly after photoreduction of Q_A. It should be noted, that this observation does not exclude a pronounced influence on the apparent midpoint potential of Q_A in equilibrium titrations. The influence of Q_B^- may depend on a protonation [12] and/or conformational change which does not take place within 150 μs after photoreduction of Q_A. It may be this change which causes the delay in Q_A^- reoxidation when Q_B is in the reduced state [2] and it could very well be a normal and essential event in the process of electron transfer to the plastoquinone pool.

The rationale of such a strongly exothermic change without electron transfer, effectively stabilizing Q_A^- by several orders of magnitude, may perhaps be found in the regulation of electron transport. While permitting a smaller energy dissipation and thereby perhaps higher speed of electron transfer from I^- to Q_A, it nevertheless allows the redox state of the plastoquinone pool to be reflected in that of Q_A^- and hence in excitation trapping and excitation transfer between PS II units — without undue losses by charge recombination of oxidizing equivalents already stored at the donor side.

Acknowledgements

This investigation was supported by the Netherlands Foundation for Chemical Research (SON), financed by the Netherlands Organization for the Advancement of Pure Research (ZWO).

References

1. Arata H and Parson WW (1981) Biochim Biophys Acta 638:201–209
2. Bowes JM and Crofts AR (1980) Biochim Biophys Acta 590:373–384
3. de Grooth BG and van Gorkom HJ (1981) Biochim Biophys Acta 635:445–456
4. Döring G (1975) Biochim Biophys Acta 376:274–284
5. Duysens LNM and Sweers HE (1963) In: Studies on Microalgae and Photosynthetic Bacteria. Special Issue of Plant & Cell Physiol., pp. 353–373. University of Tokyo Press, Tokyo
6. Fujita I, Davis MS and Fajer J (1978) J Am Chem Soc 100:6280–6282
7. Haveman J and Mathis P (1976) Biochim Biophys Acta 440:346–355
8. Klimov VV, Allakhverdiev ST, Demeter S and Krasnovskii AA (1980) Dokl Acad Nauk 249:227–230
9. Meiburg RF, van Gorkom HJ and van Dorssen RJ (1983) Biochim Biophys Acta

724:352–358
10. Meiburg RF, van Grokom HJ and van Dorssen RJ (1984) Biochim Biophys Acta 765:295–300
11. Reinman S and Mathis P (1981) Biochim Biophys Acta 635:249–258
12. Robinson HH and Crofts AR (1984) In: Advances in Photosynthesis Research (Sybesma C, ed) Vol. I, pp. 477–480, Martinus Nijhoff/Dr W Junk Publishers, Den Haag
13. Rutherford AW, Mullet JW and Crofts AR (1981) FEBS Lett 123:235–237
14. Thielen APGM and van Gorkom HJ (1981) FEBS Lett 129:205–209
15. van Gorkom HJ (1985) Photosynth Res 6:97–112
16. Velthuys BR and Amesz J (1974) Biochim Biophys Acta 333:85–94
17. Vermaas WFJ and Govindjee (1981) Photochem Photobiol 34:775–793

Photosynthesis Research 9, 63–70 (1986)
© *1986 Martinus Nijhoff/Dr. W. Junk Publishers, Dordrecht.*

Stoichiometric determination of pheophytin in photosystem II of oxygenic photosynthesis

N. MURATA,[1] S. ARAKI,[2] Y. FUJITA,[1] K. SUZUKI,[3] T. KUWABARA[4] and P. MATHIS[5]

[1] National Institute for Basic Biology, Okazaki 444, Japan, [2] Yamamoto Nori Research Laboratory, 5-4-6 Oomori-Higashi, Oota-ku, Tokyo 143, Japan, [3] Department of Biochemistry, Faculty of Science, Saitama University, Urawa 338, Japan, [4] Department of Chemistry, Faculty of Science, Toho University, Miyama, Funabashi 274, Japan and [5] Service de Biophysique, Département de Biologie, Centre d'Études Nucléaires de Saclay, 91191 Gif-sur-Yvette, France

(Received 24 September 1985)

Key words: chlorophyll, pheophytin, photochemical reaction center II, photosystem II

Abstract. Pheophytin and chlorophyll extracted from oxygen-evolving photosystem II particles, chloroplast thylakoids and cyanobacterial cells were separated by column chromatography with DEAE-Toyopearl, and quantitatively determined by spectrophotometry. The molecular ratio of chlorophyll $a + b$ to pheophytin a was about 100 in spinach photosystem II particles and about 140 in spinach thylakoids. Using flash spectrophotometry of P680 and measurement of flash-induced oxygen yield, the molecular ratio of the chlorophyll to the photochemical reaction center II was determined to be about 200 in the photosystem II particles. These findings suggest that the stoichiometry in photosystem II particles is one reaction center II and two pheophytin a molecules per about 200 chlorophyll molecules. The same stoichiometry for pheophytin to the reaction center II was obtained in the cyanobacteria, *Anacystis nidulans* and *Synechocystis* PCC 6714. A quantitative determination of pheophytin a and the electron donor P700 in stroma thylakoids from pokeweed suggests that photosystem I does not contain pheophytin.

Introduction

Photosystem (PS) II particles prepared from thylakoids with Triton X-100 [9] have enabled us to establish that the stoichiometry of components in the oxygen-evolving complex is one molecule each of the 33-kDa, 24-kDa and 18-kDa extrinsic proteins, four Mn atoms and two cytochrome b-559 per about 220 chlorophyll (Chl) molecules [12, 13, 14]. The molecular ratio of Chl to the reaction center II in PS II particles has been estimated to be about 220:1 by co-electrophoresis of the intrinsic polypeptides from the particles and purified reaction center II complex [14]. To confirm this stoichiometry, a further study is necessary to quantitatively determine the reaction center II by different methods, such as spectrophotometry of P680 and measurement of flash-induced oxygen yield.

Klimov et al. [7] demonstrated that pheophytin (Pheo) a is the intermediate

---

Dedicated to Prof. L.N.M. Duysens on the occasion of his retirement.

electron acceptor between P680 and the primary quinone acceptor of the photochemical reaction in PS II. However, determination of nonphotoreducible Pheo had been difficult until a simple technique to separate Pheo from Chl, such as ion-exchange column chromatography [16], was developed, since the presence of a much greater amount of Chl made it impossible to determine Pheo by simple spectrophotometry. By a combined use of ion-exchange column chromatography and spectrophotometry [17], we determined the Pheo a content in a purified reaction center II preparation and suggested that there are two Pheo a molecules per reaction center II.

In order to establish the stoichiometry of Pheo a, the photochemical reaction center II and Chl, we determined the molecular ratios of Pheo a and the reaction center II to Chl in various samples from organisms that carry out oxygenic photosynthesis.

Materials and methods

Spinach (*Spinacia oleracea*) was purchased from a local market. Chloroplast thylakoids were prepared by grinding the leaves in a medium containing 400 mM sucrose, 10 mM NaCl and 50 mM Na/phosphate buffer (pH 7.8), followed by differential centrifugation. PS II particles were prepared from the thylakoids with Triton X-100 as described previously [9]. Stroma thylakoids and grana thylakoids (corresponding to SD-26 and SD-41, respectively) were prepared from pokeweed (*Phytolacca americana*) growing in the campus of Saitama University according to the method described previously [22].

Anacystis nidulans (TX 20) was obtained from the Algal Collection in the Institute of Applied Microbiology, University of Tokyo. The cells were grown in the medium of Kratz and Myers [8] under aeration with 1% CO_2 in air. *Synechocystis* PCC 6714 (*Aphanocapsa* 6714), kindly provided by Dr. C. Astier (Laboratoire de Photosynthèse, CNRS, Gif-sur-Yvette), was grown in the medium of Herdman et al. [4] as modified by Astier et al. [1] under aeration with 1% CO_2 in air. Both cultures were grown at 28 °C under continuous incandescent illumination at an intensity of 3,000 lux. Cell at the middle-logarithmic phase were harvested for use in the experiments.

Pigments were extracted from the thylakoids and PS II particles of spinach and from the stroma and grana thylakoids of pokeweed with 80% acetone, and from the cyanobacterial cells with 90% methanol. Pigments were also extracted from spinach leaves by disrupting the leaves in nine volumes of methanol with a blender. In all cases, extraction was repeated with the same solvent mixtures, and the extracts were combined. The extracted pigments, corresponding to about one mg Chl, were transferred to diethylether [20], and the solvent was evaporated to complete dryness under reduced pressure.

The resultant crude pigment preparation was dissolved in 10 ml dry acetone, and one ml of the solution was applied to a DEAE-Toyopearl column. Pheo a and carotenoids were eluted with 7 ml dry acetone, after which Chl a (and b) were eluted with 10 ml acetone/H_2O (99:1, v/v). The effluent was diluted with 80% acetone to a suitable concentration for spectrophotometric determination. The DEAE-Toyopearl 650 M obtained from Toyo Soda Industry Co. was suspended in 500 ml of 1.0 M Na/acetate buffer (pH 7.0). It was then washed twice with 500 ml H_2O by filtration, then three times with 200 ml acetone, and finally packed in a glass column having an internal diameter of 10 mm to a height of 20 mm. The column was washed with 100 ml dry acetone.

Amounts of Chl a and b were determined using the absorption coefficients of Mackinney [10]. Amounts of Pheo a were determined using the absorption coefficient of Wilson et al. [29].

P680 in PS II particles was quantitatively determined by flash-induced absorbance change at 820 nm as described previously [19]. The particles were treated with 1.0 M Tris/HCl (pH 8.0) or 1.5 mM NH_2OH [19] before measurement to totally block oxygen evolution; thus maximizing the flash-induced absorbance change. For measurement, they were suspended at a Chl concentration of 21 μM in 300 mM sucrose, 10 mM NaCl, 0.3 mM phenyl-p-benzoquinone (PBQ), 2.0 mM $K_3Fe(CN)_6$, and 25 mM Mes/NaOH (pH 6.5). The difference absorption coefficient at 820 nm used was 7 mM^{-1} cm^{-1} [11].

P700 in stroma thylakoids and grana thylakoids from pokeweed was quantitatively determined by ferricyanide-minus-ascorbate difference absorption at 700 nm using a Hitachi 557 spectrophotometer. For measurement, the thylakoids were suspended at a Chl concentration of 20 μM in 20 mM Hepes/NaOH (pH 6.5), and 20 mM $K_3Fe(CN)_6$ was added until the P700 change reached the maximum level with reference to the thylakoids in the presence of 10 mM Na/ascorbate. The difference absorption coefficient at 700 nm used was 64 mM^{-1} cm^{-1} [5].

The amounts of reaction center II in PS II particles and cyanobacteria were also determined from the flash-induced oxygen yield according to the method of Kawamura et al. [6] with 4 μs xenon repetitive flashes given at 10 Hz. A far-red background light (wavelength longer than 680 nm, 1.6 W m^{-2}) was provided in measurements with the cyanobacteria. For PS II particles the reaction medium was 300 mM sucrose, 10 mM NaCl, 5 mM $CaCl_2$, 25 mM Mes/NaOH (pH 6.5) and 0.3 mM PBQ, while for the cyanobacterial cells, fresh culture media were used.

Results and discussion

Figure 1 shows the elution pattern of DEAE-Toyopearl column chromatography of the crude pigment extract from spinach PS II particles. When the crude pigment extract dissolved in acetone was applied to the column, all carotenoids and Pheo a were eluted with dry acetone giving the first fraction. Chl a and b were then eluted with acetone:H_2O (99:1, v/v) giving

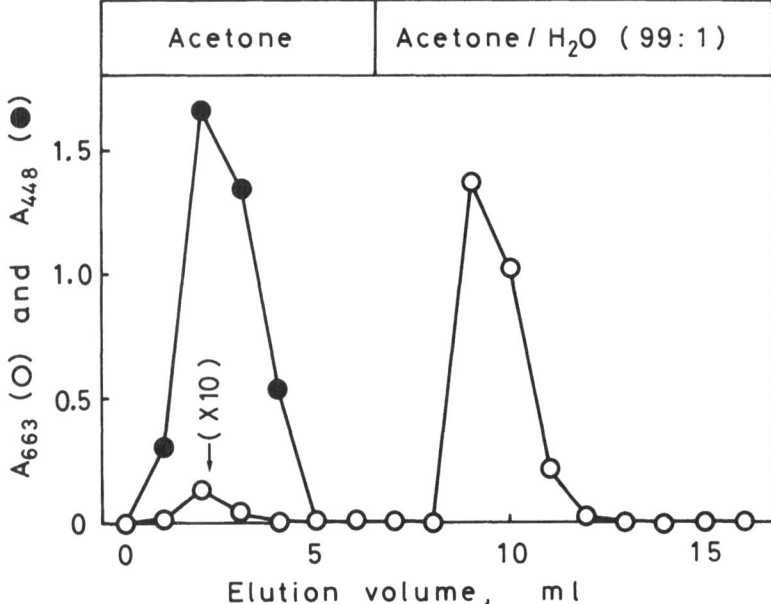

Figure 1. Column chromatogram of pigments extracted from PS II particles. The sample dissolved in acetone was applied to a DEAE-Toyopearl column. Carotenoids and Pheo a were eluted with dry acetone, and Chl $(a + b)$ with acetone/H_2O (99:1, v/v). Flow rate was 2 ml/min.

the second fraction. From the absorbance in the red region, Pheo a was determined in the first fraction, and Chl a and b in the second.

Quantitative determination of these pigments in the other samples was carried out in a similar manner. The ratios of Chl a (and b) to Pheo a are listed in Table 1. The ratio of Chl to Pheo was $107(\pm 8):1$ in spinach PS II particles, and $142(\pm 12):1$ in spinach thylakoids. This difference in the ratios can be explained if the Pheo is associated with PS II but not with PS I. The ratio of Chl to Pheo in spinach leaves was determined to be $152(\pm 10):1$, which was similar to that in isolated spinach thylakoids. In pokeweed, the ratio was $282(\pm 12):1$ in PS I-enriched stroma thylakoids, and $123(\pm 2):1$ in PS II-enriched grana thylakoids. These observations again suggest that Pheo is associated with PS II, but not, or only slightly, with PS I. In the cyanobacteria, the ratio of Chl to Pheo was $110(\pm 9):1$ in $A.$ $nidulans$ and $103(\pm 5):1$ in $Synechocystis$ PCC 6714.

The flash-induced absorbance change of P680 at 820 nm was measured in PS II particles which had been treated with 1.0 M Tris/HCl (pH 8.0) or 1.5 mM NH_2OH. Since the oxygen-evolving complex had been destroyed by these treatments, only the slow reduction of P680$^+$ with a half-time of 11 μs was observed, as previously observed in thylakoid membranes [3, 23]. Under these conditions, the flash produced the maximum absorbance change

Table 1. Molecular ratios of Chl (*a* and *b*) to Pheo *a* and of Chl *a* to Chl *b*. The numbers in parentheses represent the numbers of experiments

Sample	Chl (*a* + *b*)/Pheo	Chl *a*/*b*
Spinach		
PS II particles	107 ± 8 (4)	2.0
Thylakoids	142 ± 12 (4)	2.6
Leaves	152 ± 10 (4)	2.6
Pokeweed		
Stroma thylakoids	282 ± 12 (8)	5.5
Grana thylakoids	123 ± 2 (2)	2.4
Anacystis nidulans	110 ± 9 (5)	–
Synechocystis PCC 6714	103 ± 5 (4)	–

Table 2. Molecular ratios of Chl (*a* and *b*) to reaction center II. The reaction center II was determined by spectrophotometry of P680 and by polarography of flash-induced oxygen yield. The numbers in parentheses are the numbers of experiments

Sample	Chl (*a* + *b*)/reaction center II measured by	
	P680	O$_2$ evolution
Spinach PS II particles	206 ± 16 (4)	199 ± 15 (5)
A. nidulans	–	211 ± 15 (15)
Synechocystis PCC 6714	–	211 ± 7 (13)

of P680. The molecular ratio of Chl to P680 determined by this method was 206 (± 16):1 (Table 2).

Using the flash-induced oxygen yield, the molecular ratio of Chl to the reaction center II in spinach PS II particles was determined to be about 199 (± 15):1 (Table 2). These results obtained by flash spectrophotometry of P680 and flash-induced oxygen yield, which show that the molecular ratio of Chl to the reaction center II is about 200:1, are consistent with our previous estimate by densitometry of sodium dodecylsulfate polyacrylamide gel electrophoresis of the 47 kDa and 43 kDa polypeptides of reaction center II complex [14]. The molecular ratios of Chl to the reaction center II in cyanobacterial cells were determined from the flash-induced oxygen yield to be 211 (± 15):1 in *A. nidulans* and 211 (± 7):1 in *Synechocystis* PCC 6714 (Table 2).

The results presented in Tables 1 and 2 clearly indicate that the molecular ratio of Chl to reaction center II was about twice as large as that of Chl to Pheo. This leads to the conclusion that each reaction center II contains two Pheo *a* molecules. This is consistent with our previous result obtained with the purified reaction center II preparation [17].

However, these stoichiometric determinations of Chl, Pheo and the reaction center II depend on the absorption coefficients of Chl and Pheo. If the absorption coefficients of Vernon [25] for Chl *a* and *b* are used instead of Mackinney's [10], the values for the Chl to Pheo ratio (Table 1) and for the

Chl to reaction center II ratio (Table 2) should be reduced by about 10%. If the absorption coefficient of Watanabe et al. [26] for Pheo a is used instead of Wilson's [29], the values for the Chl to Pheo ratio (Table 1) should be increased by about 10%. However, these variations in the molecular ratio do not significantly alter the stoichiometry of Pheo a and the reaction center II. The quantitiative determination of P680 contains an uncertainty, since the difference absorption coefficient of P680 at 820 nm used in the present study, e.g., 7 mM^{-1} cm^{-1}, was based on the assumption that the absorption coefficient of P680 was the same as that of the cation radical of Chl a in solution [11]. Thus, we determined the reaction center II also by flash-induced oxygen yield. The two independent methods provided very similar numbers for the ratio of Chl to the reaction center II, thus confirming the molecular ratio of Pheo to the reaction center II being 2:1.

It has long been known that Chl looses Mg from the porphyrin ring and changes to Pheo under acidic conditions [25]. To evaluate the present study, it is necessary to estimate the extent of this conversion during the procedures of extraction, concentration and column chromatography. For this purpose we added purified Chl a at the extraction step to give various ratios of added to native Chl, and performed the same procedure as used in the Pheo determination. The result indicated that the amount of Pheo did not increase with the addition of Chl, suggesting that the conversion from Chl to Pheo was negligible small during the determination of Pheo. Therefore, it is concluded that the Pheo detected in the various samples in the present study was natural, and not an artifactual product.

Similarity between the reaction center II of higher plant chloroplasts and the reaction center of purple bacteria has been suggested by the components at the electron acceptor side. In the higher plants, the first quinone acceptor is the one-electron carrier plastoquinone [24], and the secondary acceptor is the two-electron carrier plastoquinone [18], whereas in purple bacteria the first quinone acceptor is the one-electron carrier ubiquinone [2] and the secondary is the two-electron carrier ubiquinone [15]. The intermediate electron acceptor before the first quinone acceptor in higher plants is presumably Pheo a [7], and in the purple bacteria bacteriopheophytin (Bpheo) a [21]. The present study provides further evidence for the similarity, in that the higher plant reaction center II, like the bacterial reaction center, contains two Pheo molecules. It should be noted that one of the two Pheo a molecules in the plant reaction center II [7] is photoreducible while the other is not.

In order to estimate whether PS I contains Pheo, the content of reaction center I was studied by oxidation-reduction difference absorption of P700 in the PS I-enriched stroma thylakoids from pokeweed. The molecular ratio of Chl to P700 in this sample was 148($\pm$ 12):1. Since the ratio of Chl to Pheo was 282($\pm$ 12):1, the Pheo to P700 ratio was calculated to be 0.524($\pm$ 0.020):1. This suggests that PS I does not contain a stoichiometric amount of Pheo. Since the stroma thylakoids prepared from pokeweed contain cytochrome b-559 [22], which is a PS II component, it is reasonable to assume that the small amount of Pheo in this preparation originates from the small proportion of PS II present. The lack of Pheo a in PS I was also recently suggested in a PS I preparation from spinach [27].

The present study shows that Pheo a can be quantitatively determined by a simple technique. The size of the DEAE-Toyopearl column used was only 20 mm in height and 10 mm in internal diameter, and the chromatography took less than 10 min. Since it has been established that the stoichiometry of the molecular ratio of Pheo a to the reaction center II is 2:1, this technique can be used as a convenient and practical method for quantitative estimation of the reaction center II in the preparations from the photosynthetic membranes. However, this technique cannot be applied when the pigments are extracted directly from leaves which contain acidic substances [28], since Chl is rapidly converted to Pheo at low pH [25].

Acknowledgements

The authors are grateful to Dr. M. Miyao for the supply of spinach PS II particles. This work was supported by Grants-in-Aid for Scientific Researches from the Japanese Ministry of Education, Science and Culture.

References

1. Astier C, Joset-Espardellier F and Meyer I (1979) Arch Microbiol 120: 93–96
2. Cogdell RJ, Brune DC and Clayton RC (1974) FEBS Lett 45:344–347
3. Conjeaud H, Mathis P and Paillotin G (1979) Biochim Biophys Acta 546:280–291
4. Herdman M, Delaney SF and Carr NG (1973) J Gen Microbiol 79:233–237
5. Hiyama T and Ke B (1972) Biochim Biophys Acta 267:160–171
6. Kawamura M, Mimuro M and Fujita Y (1979) Plant Cell Physiol 20:697–705
7. Klimov VV, Klevanik AV, Shuvalov VA and Krasnovsky AA (1977) FEBS Lett 82:183–186
8. Kratz WA and Myers J (1955) Amer J Bot 42:282–287
9. Kuwabara T and Murata N (1982) Plant Cell Physiol 23:533–539
10. Mackinney G (1941) J Biol Chem 140:315–322
11. Mathis P and Setif P (1981) Israel J Chem 21:316–320
12. Murata N and Miyao M (1985) Trends Biochem Sci 10:122–124
13. Murata N, Miyao M and Kuwabara T (1983) In Inoue Y et al, eds. The Oxygen Evolving System in Photosynthesis, pp. 223–228. Tokyo: Academic Press
14. Murata N, Miyao M, Omata T, Matsunami H and Kuwabara T (1984) Biochim Biophys Acta 765:363–369
15. Okamura MY, Isaacson RA and Feher G (1975) Proc Natl Acad Sci USA 72: 3491–3495
16. Omata T and Murata N (1980) Photochem Photobiol 31:183–185
17. Omata T, Murata N and Satoh K (1984) Biochim Biophys Acta 765:403–405
18. Pulles MPJ, van Gorkom HJ and Willemsen JG (1976) Biochim Biophys Acta 449:536–540
19. Satoh K and Mathis P (1981). Photobiochem Photobiophys 2:189–198
20. Seståk Z (1971). In Seståk Z et al, eds. Plant Photosynthetic production, pp. 672–701. The Hague: Martinus Nijhoff/Dr. W. Junk Publishers

21. Shuvalov VA and Klimov VV (1976). Biochim Biophys Acta 440:587–599
22. Suzuki K (1977). 'Photosynthetic Organelles' Special Issue of Plant Cell Physiol, pp. 415–425
23. Van Best JA and Mathis P (1978) Biochim Biophys Acta 503:178–188
24. Van Gorkom HJ (1974) Biochim Biophys Acta 347:439–442
25. Vernon LP (1960) Anal Chem 32:1144–1150
26. Watanabe T, Hongu A, Honda K, Nakazato M, Konno M and Saitoh S (1984) Anal Chem 56:251–256
27. Watanabe T, Kobayashi M, Hongu A, Nakazato M, Hiyama T and Murata N (1985) FEBS Lett in press
28. Watanabe T, Nakazato M, Mazaki H, Hongu A, Konno M, Saitoh S and Honda K (1985) Biochim Biophys Acta 807:110–117
29. Wilson JR, Nutting M-D and Bailey GF (1962) Anal Chem 34:1331–1332

Photosynthesis Research 9, 71–78 (1986)
© 1986 Martinus Nijhoff/Dr. W. Junk Publishers, Dordrecht.

Total recovery of O_2 evolution and nanosecond reduction kinetics of chlorophyll-a_{II}^+ (P-680$^+$) after inhibition of water cleavage with acetate

Ö. SAYGIN, S. GERKEN, B. MEYER and H.T. WITT

Max-Volmer-Institut für Biophysikalische und Physikalische Chemie, Technische Universität Berlin, Strasse des 17. Juni 135, 1000 Berlin 12, BRD

(Received 24 September 1985)

Abstract. Oxygen evolution and reduction kinetics of the photooxidized Chl-a_{II}^+ have been measured in oxygen-evolving complexes from the thermophilic cyanobacterium Synechococcus sp.

1. Incubation of PS II particles with acetate resulted in an inhibition of oxygen evolution and a retardation of the Chl-a_{II}^+-reduction kinetics from the nanosecond range to the microsecond range, indicating a modification of the donor side of photosystem II (PS II).

2. After the first two flashes given to a dark-adapted, acetate treated sample, Chl-a_{II}^+ was re-reduced with a half-life time of $160\,\mu s$ by a component of the donor side of PS II. Under repetitive excitation Chl-a_{II}^+ was re-reduced in $500\,\mu s$ by electron back reaction from the primary acceptor Q_A^- (X-320$^-$). Obviously, in the presence of acetate only two electrons are available from the donor side.

3. Both oxygen evolution and nanosecond reduction kinetics of Chl-a_{II}^+ were restored to the control level when acetate was removed.

4. The results indicate a tight coupling between O_2 evolution and nanosecond reduction kinetics of Chl-a_{II}^+.

5. The reversible inhibition is probably due to a replacement of Cl$^-$ by acetate within the water splitting enzyme.

6. Due to its strongly retarded kinetics, the reversibly modified system may facilitate investigations of the mechanism of the donor side.

Abbreviations

Chl, chlorophyll; PpBQ, phenyl-p-benzoquinone; PS, photosystem

Introduction

The fundamental reaction for water cleavage in photosynthesis is the photooxidation of Chl-a_{II} (P-680) [7,8] . Light excitation of Chl-a_{II} leads to the transfer of one electron to the first stable acceptor, a special plastoquinone, Q_A (X-320) [25, 26]. The photooxidized Chl-a_{II} extracts, via electron carriers, an electron from the O_2-evolving complex. In four single turnover flashes the complex runs through four oxidation states, S_0 to S_3, ultimately ending in the cleavage of 2 H_2O into 4 H^+ and one O_2. The kinetics of the

---

This work is dedicated to Prof. Dr. L.N.M. Duysens on the occasion of his retirement.

$Chl\text{-}a_{II}^{+}$ reduction are a function of the S-states: the $Chl\text{-}a_{II}^{+}$ reduction correlated with the S_0 and S_1 states occurs within 23 ns; whereas, in states S_2 and S_3 a biphasic reduction with 50 ns and 260 ns is observed [2]. The retardation of the electron transfer times in states S_2 and S_3 was explained by Coulombic attraction due to an excess positive charge in states S_2 and S_3 [2]. The existence of a positive surplus charge in states S_2 and S_3 has been shown independently by corresponding electrochromic signals [17, 18]. Under repetitive flash excitation a multiphasic $Chl\text{-}a_{II}^{+}$ reduction is observed [3, 9, 20]. This has been explained quantitatively by a superposition of the individual kinetics correlated with the S_0–S_3 states [2, 21].

The electron transport from H_2O to an artificial acceptor can be blocked more or less reversibly through different types of treatments (Cl^- depletion [11] (for review see [5], protein removal [1–10, 12–15], or bicarbonate depletion [23, 28]. In PS II reaction centers in which oxygen evolution was blocked by various treatments, the nanosecond reduction kinetics of $Chl\text{-}a_{II}^{+}$ disappear and $Chl\text{-}a_{II}^{+}$ is reduced slowly in the time range of microseconds [1, 3, 27].

Incubation with formate or acetate under a CO_2-free atmosphere is a currently used tool for bicarbonate depletion, resulting in a reversible retardation of the electron transfer kinetics both from Q_A to Q_B and from Q_B to the PQ-pool by a factor of about 10 (for recent reviews of bicarbonate depletion see [23, 28]). There is disagreement whether or not bicarbonate depletion also acts – to a minor extent – at the donor side of PS II [23, 28]. Acetate has also been used in experiments together with Cl^--depletion [11, 22]. In our experiments the acetate concentration was 660 mM in the presence of 15 mM Cl^- at pH 5.5. Under these conditions a complete inactivation of oxygen evolution and a retardation of the reduction kinetics of $Chl\text{-}a_{II}^{+}$ to the microsecond time range have been observed. After the removal of acetate, O_2 evolution was totally restored together with the reappearance of the nanosecond reduction of $Chl\text{-}a_{II}^{+}$.

Materials and methods

PS II particles from Synechococcus sp. were prepared according to Schatz and Witt [19] and stored at 193 K in the dark. Under repetitive flash illumination oxygen evolution was measured with a Clark-type electrode. Single flash as well as repetitive flash-induced oxygen evolution was measured with a Zirconia oxygen sensor. Flash-induced absorption changes at 824 nm were measured using a spectrophotometer essentially described in [3] with variations described in [2]. The fluorescence artifact was negligible. At 820 nm we observed an additional light-induced absorption change in flash numbers higher than 2 (10–25% of the total signal) which was not related to $Chl\text{-}a_{II}$. Simultaneous measurements at 334 nm and 824 nm were performed with a spectrophotometer similar to the one described in [26].

Control measurements were made using a suspension of $28\,\mu M$ Chl in a buffer at pH 5.5 containing 0.5 M mannitol, 20 mM MES/NaOH, 2 mM KH_2PO_4 and 10 mM $MgCl_2$ (MMPM). 0.2 mM PpBQ was used as electron acceptor. Submission of a control sample to a gel filtration column (Sephadex G25) which was equilibrated with MMPM buffer, did not affect the oxygen evolving activity, but caused an artifact in the absorption change measurements at 820 nm. An additional, flash-induced absorption increase, which decayed in the ms-range, was observed (5–15% of the signal). This artifact was not DCMU-sensitive. For acetate treatment the particles were incubated in the dark for 15 min with 660 mM Na-acetate at pH 5.5 and 20 °C. The chloride concentration was 15 mM. For further measurements 0.2 mM PpBQ was added as external acceptor. Acetate was removed using Sephadex G25 gel filtration columns at 0–4 °C, which were equilibrated with MMPM buffer. 0.2 mM PpBQ was readded to samples having passed the column. Where indicated MMPM buffer at pH 6.5 was used instead of MMPM pH 5.5.

Results and discussion

Figure 1, left, shows the time course of the Chl-a$_{II}^+$ reduction after the first flash given to a dark-adapted sample without acetate. The observed nanosecond reduction kinetics of Chl-a$_{II}^+$ are typical for the first flash. Deviations from previously published signals in PS II particles from Synechococcus [2] are caused by using pH 5.5 instead of pH 6.8. Figure 1, center, shows that with addition of 660 mM acetate the Chl-a$_{II}^+$ reduction kinetics are retarded from the nanosecond range to the microsecond range. The total decay kinetics have a half-life time $\tau_{Chl}^{820} = 120\,\mu s$ (not shown). In the 2nd flash the same kinetics are observed, while in the 3rd and 4th flash the kinetics are multiphasic. At flash numbers higher than 4 or under repetitive excitation the half-life time is $\tau_{Chl}^{820} = 500\,\mu s$. With the addition of 660 mM acetate O_2 evolution was almost completely inhibited (15% of the control, see Table 1). This inhibition was observed already after the first four flashes given to a dark adapted sample. Figure 1, right, shows the results of acetate removal on a gel filtration column (Sephadex G25). The nanosecond kinetics were restored. Also, the oxygen evolution was recovered almost completely (95% of the control, see Table 1). A necessary condition for the complete reconstitution is that the sample in its inhibited state is not illuminated with more than two flashes; i.e., that only two electron transfers are allowed. With more than two flashes a gradual increase of irreversibly inhibited photosystems is observed.

From the retardation of the reduction kinetics of the Chl-a$_{II}^+$ from the ns to the μs range we conclude that the described acetate treatment results in a modification of the donor side of PS II. It is most likely that the acceptor side is also modified (bicarbonate depletion), but this should not affect our results because we measured O_2 evolution and absorption changes in flashes with a repetition rate of 1 Hz.

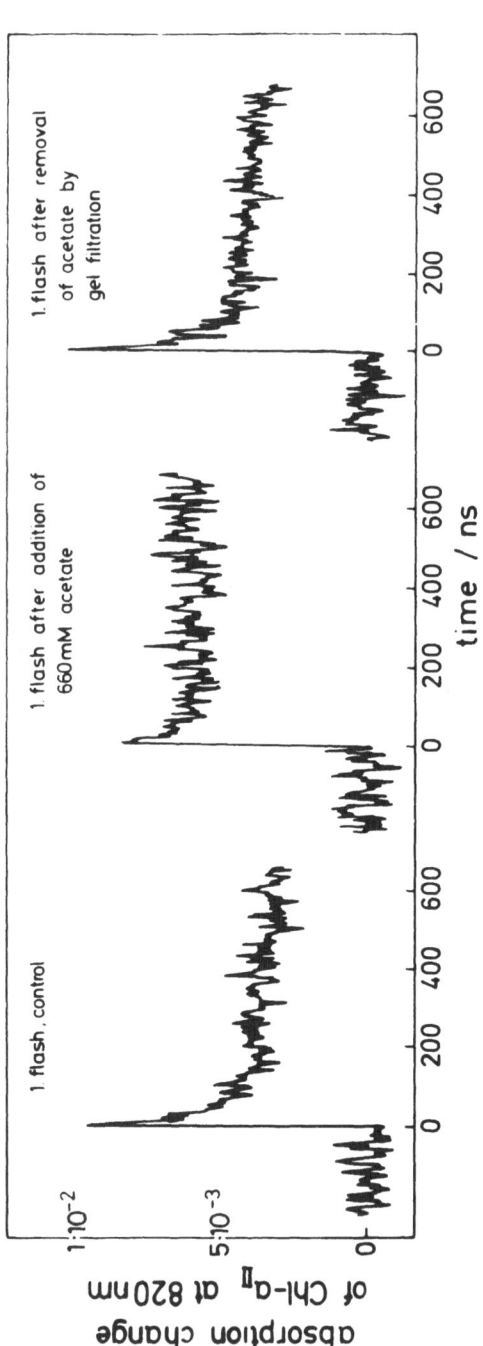

Figure 1. Time course of the redox reaction of Chl-a$_{II}$ after the first flash in dark-adapted PS II complexes from Synechococcus under different conditions. Chl-concentration 28 µM, optical path length 5 cm, coverage of two measurements. *Left:* control; *Center:* after addition of 660 nM Na-acetate; *Right:* after removal of Na-acetate by gel filtration on a column with Sephadex G25.

Table 1. O_2 evolution per flash and Chl-a$_{II}^+$ (824 nm) re-reduction times under different conditions of the PS II complex from Synechococcus

	Control	+ 660 mM acetate	After removal of acetate
O_2 Evolution	$2.4 \cdot 10^{-3}$ O_2/Chl 100%	$3.7 \cdot 10^{-4}$ O_2/Chl 15%	$2.3 \cdot 10^{-3}$ O_2/Chl 95%
Chl-a$_{II}$ kinetics	1st flash $\tau_{Chl} = 25$ ns	1st flash $\tau_{Chl} = 120\,\mu s$	1st flash $\tau_{Chl} = 25$ ns

After the 1st and 2nd flash given to a dark adapted acetate-treated sample, Chl-a$_{II}^+$ decays monophasically with $\tau = 120\,\mu s$. After the 3rd and 4th flash a multiphasic decay is observed. From the 5th flash on or under repetitive excitation a monophasic decay with $\tau = 500\,\mu s$ takes place. Under the latter condition we observed for the Q_A^- reoxidation monitored at 334 nm also 500 μs kinetics (Figure 2). From this we concluded that under repetitive excitation Chl-a$_{II}^+$ is reduced exclusively by the electron from Q_A and this with $\tau = 500\,\mu s$. The 120 μs phase in the first two flashes can be explained as a concurrent reduction of Chl-a$_{II}^+$ by Q_A^- (τ_Q) and a donor (τ_D). In this case and under first order reaction conditions the overall half-life time of Chl-a$_{II}^+$ should be $1/\tau_{Chl} = 1/\tau_Q + 1/\tau_D$. With $\tau_Q = 500\,\mu s$ and $\tau_{Chl} = 120\,\mu s$ it is $\tau_D = 160\,\mu s$. (In the presence of acetate the oxidation of Q_A^- through other acceptors takes place in the order of several ms and can be neglected). A consequence of these concurrent reactions is that the ratio of reoxidized Q_A to reduced Chl-a$_{II}$ should be $Q_A/Chl\text{-}a_{II} = \tau_{Chl}/\tau_D = 1/4$. To check whether this ratio is indeed obtained, we performed double flash experiments. The flashes were spaced 250 μs apart so that Chl-a$_{II}$ was mainly reduced when the 2nd flash was fired. The amplitude at 820 nm induced by the 2nd flash showed that only about 1/4 of the reduced Chl-a$_{II}$ has been photooxidized. This means that only 1/4 of Q_A is present in the reoxidized state; i.e., $Q_A/Chl\text{-}a_{II} = 1/4$. This supports the above-made assumption.

When under repetitive excitation no more electrons can be transferred from the donor side to Chl-a$_{II}^+$, then $\tau_D = \infty$ and Chl-a$_{II}^+$ can be reduced only by Q_A^-; i.e., $\tau_{Chl} = \tau_Q = 500\,\mu s$. This was observed (see above). From the fact that Chl-a$_{II}^+$ reduction occurs monophasically in the 1st and 2nd flash but multiphasically in the 3rd and 4th flash and already in the 5th flash monophasically in a pure back reaction with Q_A^- (500 μs), we conclude that the donor capacity is two under acetate conditions.

The inhibition effect of acetate depends on the pH of the sample. Incubation at pH 6.5 instead of pH 5.5 under otherwise unchanged conditions resulted in an inhibition of the oxygen evolution by only 50%. Stemler and Murphy [24] recently showed that in broken chloroplasts that were treated with 100 mM acetate at pH 6.4 the rate of O_2 evolution could be restored from 61 to 93% of the control by addition of 5 mM HCO_3^-. In our case we were, however, not able to restore oxygen evolution (as measured in flashes with 1 Hz)

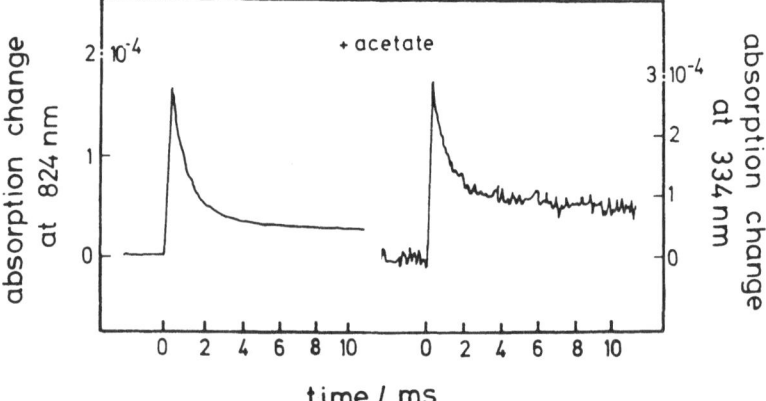

Figure 2. Time course of the absorption changes of Chl-a$_{II}$ at 824 nm and of Q$_A$ at 334 nm, measured simultaneously under repetitive flash excitation. Acetate concentration 660 mM, chlorophyll concentration 2.7 μM, pathlength 1 cm, average of 256 measurements. The slow phase at 334 nm is observed also in the control and depends both on the acceptor and the preparation used.

when we added up to 10 mM HCO$_3^-$ to our PS II particles in the presence of 660 mM acetate at pH 5.5.

By decreasing the chloride concentration from 15 mM to 1 mM we observed that the required acetate concentration for inhibition can be reduced from 660 mM down to 200 mM. On the other hand, addition of 100 mM chloride to a 200 mM treated acetate sample (1 mM Cl$^-$) restores oxygen evolution up to 50% of the control. These results point to the possibility that chloride and acetate compete for the same binding site. Chloride is suggested to act as a bridging ligand in the oxygen evolving complex [5, 11, 16]. Displacement of Cl$^-$ through an ion with a smaller radius than that of Cl$^-$ has been shown for F$^-$ [6]. The fact that the acetate ion is also smaller than the Cl$^-$ ion is in line with our suggestion that acetate binds to the manganese in competition with chloride. Acetate ligand binding with manganese(III) in vitro is well known [4]. A consequence of this interpretation may be that under Cl$^-$-depletion without acetate treatment the Chl-a$_{II}^+$ reduction kinetics should also be retarded to the μs level (corresponding experiments are in progress).

Conclusions

In this work we demonstrate through addition and extraction of acetate an almost totally reversible inhibition of the ns-reduction kinetics of Chl-a$_{II}^+$ as well as of the water oxidation. This result shows the tight coupling between O$_2$ evolution and ns-reduction kinetics of Chl-a$_{II}$. The inhibition is probably the result of a replacement of Cl$^-$ by acetate within the water splitting enzyme. Since the inhibition at the donor side is reversible, the reaction system has been modified by acetate without any destructions. In the modified system simultaneously with the inhibition of water cleavage $Chl\text{-}a_{II}^+$ reduction time is dramatically slowed down from 23 ns to 160 μs.

We are left with the question from which substances these electrons are taken: from the manganese, from the electron carriers between Mn and $Chl\text{-}a_{II}$ or from other intrinsic donors, e.g., from $cyt\text{-}b_{559}$. If in the modified system the unknown immediate donor D_1 of $Chl\text{-}a_{II}^+$ [2] is still the same and if its very fast kinetics with $Chl\text{-}a_{II}^+$ have only been slowed down, e.g., through conformational changes, then the possibility is given to analyze the properties of D_1 under extremely simplified conditions. We hope that future work on the modified system presented here can help to gain further insights into the mechanism of water cleavage.

Acknowledgements

The authors wish to thank D. DiFiore for measuring O_2 evolution (Clark electrode) and I. Geisenheimer for preparing the particles. This work was supported through Sonderforschungsbereich 312 'Gerichtete Membran-prozesse'.

References

1. Akerlund HE, Brettel K and Witt HT (1984) Biochim Biophys Acta 765, 7–11
2. Brettel K, Schlodder E and Witt HT (1984) Biochim Biophys Acta 766, 403–415
3. Brettel K and Witt HT (1983) Photobiochem Photobiophys 6, 253–260
4. Cotton/Wilkinson Anorganische Chemie, 4th ed., Verlag Chemie (1982), p. 757
5. Critchley C (1985) Biochim Biophys Acta 811, 33–46
6. Critchley C, Baianu IC, Govindjee, Gutowski HS (1982) Biochim Biophys Acta 682, 436–445
7. Döring G, Renger G, Vater J and Witt HT (1969) Z Naturforsch 24b, 1139–1143
8. Döring G, Stiehl HH and Witt HT (1967) Z Naturforsch 22b, 639–644
9. Eckert HJ, Renger G and Witt HT (1984) FEBS Lett 167, 316–320
10. Ghanotakis DF, Babcock GT and Yocum CF (1984) FEBS Lett 167, 127–130
11. Kelley PM and Izawa S (1978) Biochim Biophys Acta 502, 198–210
12. Miyao M and Murata N (1983) Biochim Biophys Acta 725, 87–93
13. Miyao M and Murata N (1983) FEBS Lett 164, 375–378
14. Nakatani HY (1984) Biochem Biophys Res Comm 120, 299–304
15. Ono T and Inoue Y (1984) FEBS Lett 166, 381–384
16. Sandusky PO and Yocum CF (1984) Biochim Biophys Acta 766, 603–611
17. Saygin Ö and Witt HT (1984) FEBS Lett 176, 83–87
18. Saygin Ö and Witt HT (1985) FEBS Lett 187, 224–226
19. Schatz G and Witt HT (1984) Photobiochem Photobiophys 7, 1–14
20. Schlodder E, Brettel K, Schatz GH and Witt HT (1984) Biochim Biophys Acta 765, 178–185
21. Schlodder E, Brettel K and Witt HT (1985) Biochim Biophys Acta 808, 123–131
22. Sinclair J (1984) Biochim Biophys Acta 764, 247–252
23. Stemler A (1982) in: Photosynthesis (Govindjee, ed.), Vol II, pp. 513–539, New York: Academic Press
24. Stemler A and Murphy JB (1985) Plant Physiol 77, 974–977
25. Stiehl HH and Witt HT (1968) Z Naturforsch 23b, 220–224

26. Stiehl HH and Witt HT (1969) Z Naturforsch 24b, 1588–1598
27. van Best JA and Mathis P (1978) Biochim Biophys Acta 503, 178–188
28. Vermaas WFJ (1982) in: Photosynthesis (Govindjee, ed.), Vol II, pp. 541–558, New York: Academic Press

Photosynthesis Research 9, 79–88 (1986)
© *1986 Martinus Nijhoff/Dr. W. Junk Publishers, Dordrecht.*

Mechanism of photoinhibition: photochemical reaction center inactivation in system II of chloroplasts

ROBYN E. CLELAND[1], ANASTASIOS MELIS* and PATRICK J. NEALE

Division of Molecular Plant Biology, 313 Hilgard Hall, University of California, Berkley, CA 94720, USA

(*Received 27 September 1985*)

Key words: photosystem II, charge separation, fluorescence, photoinhibition

Abstract. Photoinhibition of photosynthesis is manifested at the level of the leaf as a loss of CO_2 fixation and at the level of the chloroplast thylakoid membrane as a loss of photosystem II electron-transport capacity. At the photosystem II level, photoinhibition is manifested by a lowered chlorophyll *a* variable fluorescence yield, by a lowered amplitude of the light-induced absorbance change at 320 nm (ΔA_{320}) and 540-minus-550 nm ($\Delta A_{540-550}$), attributed to inhibition of the photoreduction of the primary plasto-quinone Q_A molecule. A correlation of the kinetics of variable fluorescence yield loss with the inhibition of Q_A photoreduction suggested that photoinhibited reaction centers are incapable of generating a stable charge separation but are highly efficient in the trapping and non-photochemical dissipation of absorbed light. The direct effect of photoinhibition on primary photochemical parameters of photosystem II suggested a permanent reaction center modification the nature of which remains to be determined.

Abbreviations

Chl, chlorophyll; PS, photosystem; LHC, light-harvesting complex; P_{680}, photochemical reaction center of PSII; P^d, photoinhibited reaction center of PSII; Pheo, pheophytin primary electron acceptor of PSII; Q_A, primary quinone electron acceptor of PSII; DCMU, 3-(3,4-dichlorophenyl)-1,1-dimethylurea.

Introduction

Inactivation of photosynthesis by visible light has been termed photo-inhibition [14, 17]. It occurs when the photochemical dissipation of captured light-energy by the photosynthetic apparatus is impaired and part of the system is damaged [24]. As such it occurs under conditions of excessive light intensities or when energy flow is limited by some other factor, for instance

---

[1] REC was on leave from the Botany Department, Australian National University, GPO Box 4, Canberra City, ACT 2601 Australia
*Address for offprints and all correspondence: A. Melis, Department of Molecular Plant Biology, 313 Hilgard Hall, University of California, Berkeley, CA 94720, USA

Dedicated to Prof. L.N.M. Duysens on the occasion of his retirement

environmental stress. Thus during chilling or water loss, plants suffer a light-dependent damage [2, 4, 30]. Moreover, photoinhibition is enhanced upon removal of CO_2 [28].

The phenomenon has been studied extensively in whole plants and algae (for a review see ref. 25). It is manifested by a lowered rate of light-saturated CO_2 fixation and a lowered quantum yield of photosynthesis [3, 17, 25, 27, 29]. Measuring partial electron transport reactions with electron donors and acceptors in isolated chloroplasts, Critchley [10] demonstrated that the site of photoinhibition was at or near PSII. This contention was supported by room and low temperature fluorescence yield measurements where a specific variable fluorescence quenching was detected upon photoinhibition [11, 18, 26, 32]. Early biochemical studies [14, 31] showed that antenna Chl molecules facilitated the damage although they were not themselves affected. These results indicated that the primary photochemical reactions of photosynthesis were affected.

The PSII complex contains a specialized reaction center Chl molecule(s) (P_{680}), an intermediate primary electron acceptor pheophytin (Pheo), and a bound primary quinone electron acceptor (Q_A) [6]. Associated with the reaction center in higher plants is a Chl ($a + b$) LH antenna complex containing about 230 Chl molecules [20]. Light absorbed by the Chl antenna is transferred to and trapped by P_{680} where the excitation energy brings about the charge separation reaction according to the reaction scheme:

$$P_{680}^* \text{Pheo } Q_A \longrightarrow P_{680}^+ \text{Pheo}^- Q_A \longrightarrow P_{680}^+ \text{Pheo } Q_A^-$$

These reactions are completed in the ps time scale. Under physiological conditions, the positive charge is transferred from P_{680}^+ to the secondary electron donor in the ns time scale. In the presence of the electron transport inhibitor DCMU, the negative charge on Q_A^- is sufficiently stable to allow quantitation of the charge separation process in long ms and near s time scale: photochemical reduction of the primary quinone acceptor results in a semiquinone anion formation accompanied by characteristic changes of absorbance in the 320 nm region of the electromagnetic spectrum [21, 33, 34]. It has been shown that the storage of a negative charge on the primary electron quinone-acceptor Q_A of PSII causes, through the influence of a localized electric field, the electrochromic band shift of the reaction center Pheo molecule, thus giving rise to a typical absorbance difference change in the 530 ato 560 nm region, termed C550 [7, 12, 16, 23, 34].

In the present work, we used direct measurements of variable and nonvariable fluorescence yield, and of absorbance change of Q_A (ΔA_{320}) and C550 ($\Delta A_{540-550}$) to provide insight into the mechanism of photoinhibition. Our results show inhibition of primary photochemistry of system II, possibly originating from photodamage to the reaction center P_{680}. Morever, $PSII_\alpha$ centers are substantially more sensitive to photoinhibition than $PSII_\beta$.

Materials and methods

Spinach plants (*Spinacea oleracea* L.) were grown in the greenhouse on half-strength Hoagland nutrient solution. Chloroplasts were isolated by grinding freshly harvested leaves for 10 s in a Waring blender in a medium containing 0.4 M sucrose, 10 mM NaCl, 5 mM $MgCl_2$ and 50 mM Tricine, pH 8.0. The slurry was filtered through miracloth and chloroplasts were precipitated by centrifugation at 5000 × g for 5 min. The pellet was resuspended in isolation buffer to a Chl concentration of about 1 mM using a Wheaton homogenizer. All operations described were carried out in dim light at 0 °C. Chl concentrations were determined in 80% acetone using the procedure described by Arnon [1].

Photoinhibition treatments were administered at an incident light intensity of 2500 $\mu E\,m^{-2}\,s^{-1}$. White light of uniform field was obtained from a quartz-halogen lamp operated at maximum power. Chloroplast aliquots were suspended at a Chl concentration of 200 μM in a flat Petri dish and illuminated for a period of time ranging from 5 to 60 min. A constant temperature of about 4 °C was maintained throughout the photoinhibition treatment.

Chloroplast fluorescence and absorbance difference measurements in the UV were performed with a laboratory-constructed difference spectrophotometer [21]. The optical pathlength of the cuvette for the measuring beam was 2.06 mm and for the actinic beam it was 1.46 mm. Actinic excitation was provided in the green region by a combination of Corning CS 4-96 and CS 3-69 filters. Absorbance difference measurements in the visible region of the spectrum ($\Delta A_{540-550}$) were obtained with an Aminco DW-2a spectrophotometer operated in the dual beam mode (red actinic light transmitted by Corning CS 2-58 filter). The half-band width of the measuring beam was set at 1 nm and the optical pathlength of the cuvette was 1.0 cm. Quantitation of the stable PSII charge separation was obtained from the amplitude of the absorbance change ΔA_{320} (Q_A) and $\Delta A_{540-550}$ (C550) under continuous illumination. Signal recovery and processing was implemented by a Hewlett-Packard 3437A digital voltmeter interfaced with an on line HP 86B computer. The kinetic analysis of the data, such as integration of the area over the fluorescence induction curves and the semilogarithmic analysis were performed by the computer. The results were plotted on an HP 7475A plotter.

Results and discussion

The effect of photoinhibition on primary system II photochemical parameters was investigated by fluorescence and sensitive absorbance difference spectroscopy. Figure 1 shows a family of fluorescence induction curves of DCMU-poisoned chloroplasts incubated either in the dark (Control) or exposed to high light intensity for various periods of time (5–60 min). Photoinhibition treatment for 5 min (Figure 1, 5 min) lowered the variable yield of Chl

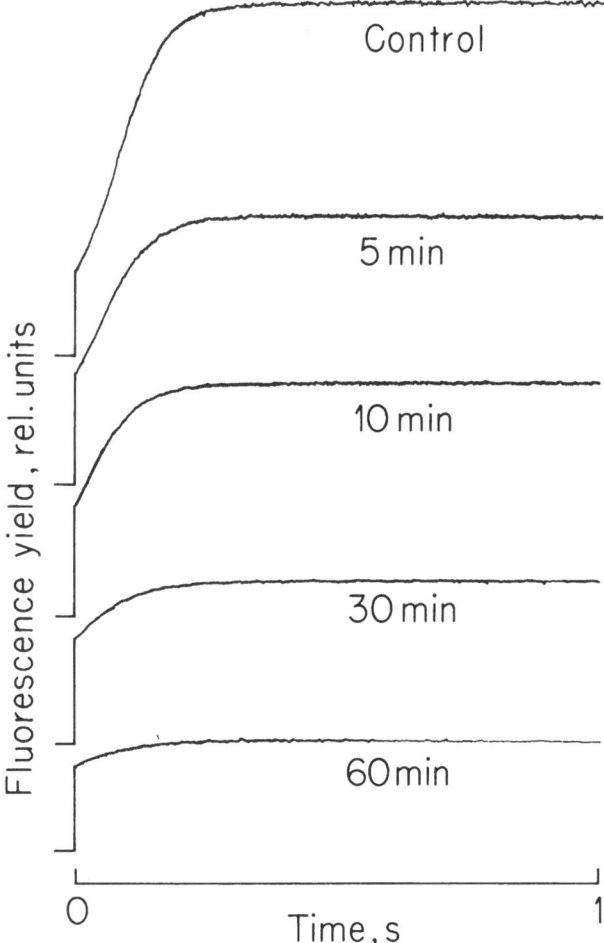

Figure 1. Chlorophyll *a* fluorescence induction kinetics of DCMU-poisoned spinach chloroplasts. The concentration of Chl $(a + b)$ was approximately $200\,\mu M$. Thylakoid membranes were either kept in the dark upon isolation (Control) or were illuminated with strong $(2500\,\mu E.m^{-2}.s^{-1})$ white light for 5, 10, 30 and 60 min prior to the fluorescence induction measurement. Note the progressive loss of variable fluorescence upon photoinhibition.

fluorescence (F_v). At the same time, a slight increase in the non-variable fluorescence yield (F_o) was also observed. The latter probably originates from a light-induced change in the coupling between the Chl antenna and the photochemical reaction center of PSII [8, 15, Björkman, personal communication]. A pronounced lowering of F_v and a less pronounced decreased in F_o occurred with further photoinhibition treatment, as previously reported (see Figure 1, also ref. 11, 18, 19, 26, 32). The time course of variable fluorescence

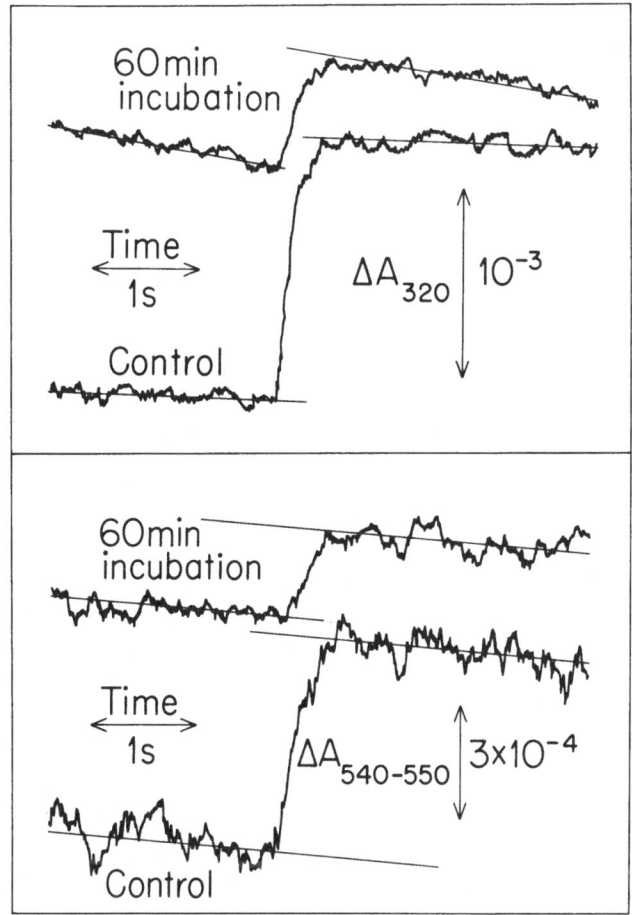

Figure 2. Light-induced absorbance change measurements at 320 nm (ΔA_{320}, upper) and at 540-minus-550 nm ($\Delta A_{540-550}$, lower) of spinach thylakoids suspended in the presence of 20 μM DCMU and 2 mM potassium ferricyanide. The concentration of Chl ($a + b$) was approximately 200 μM. Control thylakoids were kept in the dark and photoinhibited thylakoids were incubated for 60 min at 0 °C under strong white light (2500 μE.m^{-2}.s^{-1}) prior to the measurement. Note the substantially lower amplitude of Q_A photoreduction (ΔA_{320}) and of C550 formation ($\Delta A_{540-550}$) upon photoinhibition.

yield loss is accompanied by a similar lowering of the rate of electron transport through PSII (not shown, see refs. 10, 19).

The substantial lowering of F_v upon photoinhibition suggested a direct effect on the primary photochemistry of system II [5, 15, 26]. To gain insight into the underlying mechanism of this photoinactivation, we measured the primary photochemical activity of PSII directly from the amplitude of the light-induced absorbance change at 320 nm (ΔA_{320}) and at 540-minus-550 nm ($\Delta A_{540-550}$) monitoring the activity of Q_A and C550, respectively

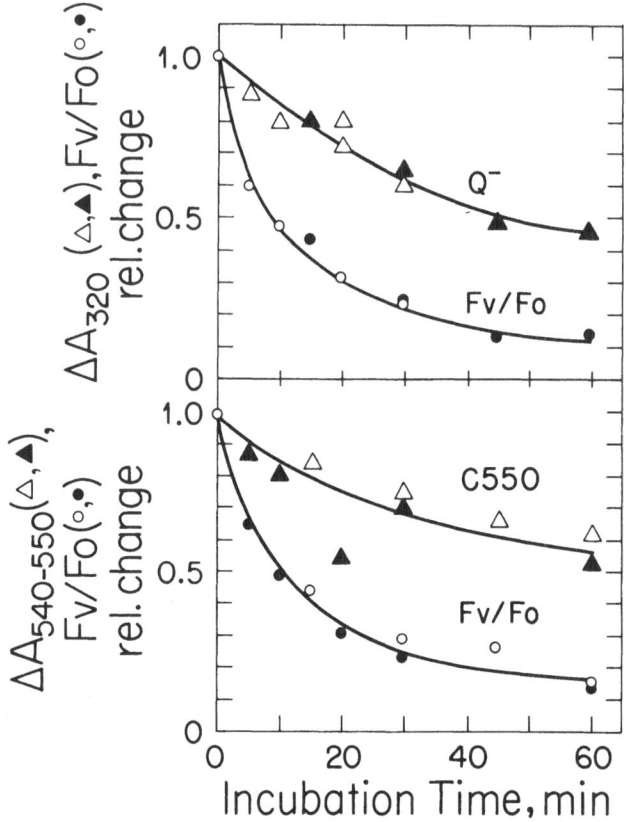

Figure 3. The amplitude of the light-induced absorbance change reflecting Q_A photo-reduction (ΔA_{320}), C550 formation ($\Delta A_{540-550}$) and the variable fluorescence yield (F_v/F_o) as a function of incubation time under strong actinic illumination (photo-inhibition). Open and solid symbols refer to different experiments with separate chloroplast samples. Note the dissimilar dependence of F_v/F_o from that of Q^- and C550.

[21, 23]. Figure 2 (upper) compared the amplitude of the light-induced ΔA_{320} in dark adapted (Control) and thylakoid membranes subjected to 60 min photoinhibition treatment. Similarly, Figure 2 (lower) compared the amplitude of the light-induced $\Delta A_{540-550}$ in dark (Control) and thylakoid membranes illuminated for 60 min under strong white light. It is evident that both the amount of photoreducible Q_A and the amplitude of the C550 signal are lowered in the photoinhibited samples. The time course of the amplitude decrease for the three photochemical parameters (F_v, Q_A and C550) during photoinhibition is presented in more detail in Figure 3. In this presentation, fluorescence data were normalized by plotting the ratio F_v/F_o. The results of Figure 3 show that the activity of Q_A and that of C550 were identical functions of photoinhibitory treatment, whereas a substantially more

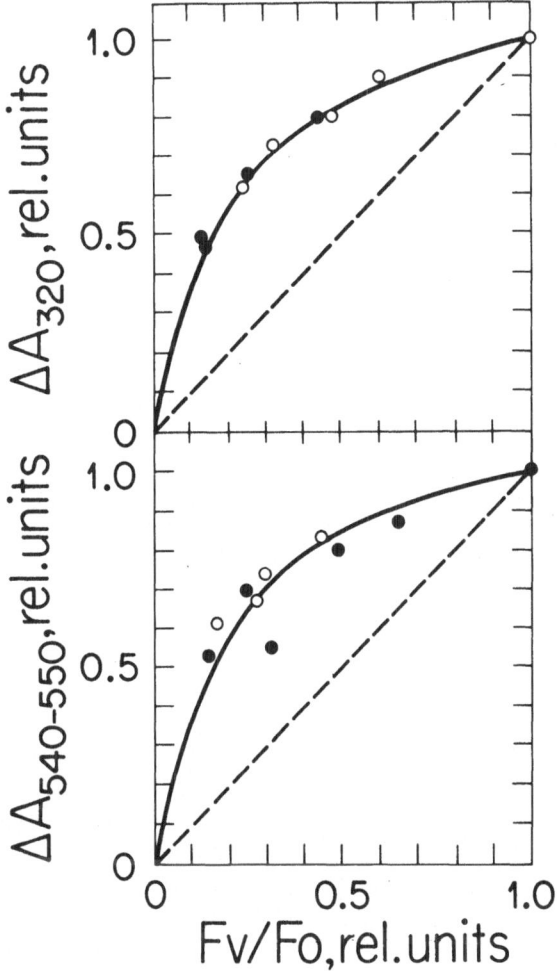

Figure 4. The relationship between the loss of variable fluorescence yield (F_V/F_o) and the loss of primary PSII activity as measured by the activity of Q^- (ΔA_{320}) and C550 ($\Delta A_{540-550}$) formation. Note the non-linear transition during photoinhibition from coordinates (1,1) to coordinates (0,0).

pronounced decrease was detected for the variable fluorescence (F_V/F_o). In Figure 4 the normalized amplitudes of these absorbance changes are plotted versus F_V/F_o. The results in Figure 4 clearly show a large positive deviation from a straight line in the transition from coordinates (1,1) to (0,0) during high-light exposure. The relationship is similar to that occurring during the time course of photoreduction [13, 22, 23] but, in contrast to these previous studies, it is now obtained from the steady-state amplitude of Q_A^-, C550 and F_V/F_o during photoinhibition. Such a non-linear relationship between the fraction of functional PSII reaction centers (Q_A^-, C550) and the yield of the variable fluorescence (F_v/F_o) probably arises from the connected pigment bed organization of the Chl a/b LH antenna associated with PSII$_\alpha$. Approximately 4 PSII$_\alpha$ are interconnected via their peripheral LHC to enable excitation energy sharing. Apparently, PSII reaction centers damaged by photoinhibition (P^dQ) remain functionally coupled to the LH antenna but cannot undergo a stable charge separation. Instead, they must function as permanent excitation energy quenchers.

The relative efficiency of excitation trapping by photodestructed P^dQ and open $P_{680}Q$ centers was studied by comparing the rate of trapping of absorbed light by open $P_{680}Q$ centers in chloroplast control samples (no photoinhibition) and in samples where photoinhibition had inactivated about 50% of all PSII. This was implemented by measuring the values of K_α (rate of photon trapping by PSII$_\alpha$) according to the method of Melis and Anderson [20] in the two chloroplast samples (e.g. Figures 1 and 2, Control and 60 min incubation, respectively). In several such experiments, it appeared that open $P_{680}Q$ centers in photoinhibited samples were trapping absorbed photons at a rate faster by about 1.2 ± 0.1 than open $P_{680}Q$ centers in control samples [19]. Since the overall rate of light absorption remained unchanged, this phenomenon could be realized in a connected pigment bed organization of the LH Chl antenna when open $P_{680}Q$ centers possess a greater effective absorption cross section than P^dQ centers. This suggests a lowered trapping efficiency by P^dQ centers.

The results presented above suggest that photoinhibition is manifested predominantly in PSII$_\alpha$. This contention was tested directly by measurements of the effect of photoinhibition on the variable fluorescence yield Fv_α and Fv_β controlled by PSII$_\alpha$ and PSII$_\beta$, respectively [23]. Figure 5 shows the relative yield of Fv_α and Fv_β as a function of time under strong white illumination. It is evident that photoinhibition is strongly manifested in PSII$_\alpha$ units. PSII$_\beta$ units also show the symptoms of photoinhibition, although at a much slower rate. In our chloroplast thylakoids, electron transport was minimized by omitting artificial electron acceptors during the photoinhibition treatment. Therefore, a differential sensitivity to photoinhibition between PSII$_\alpha$ and PSII$_\beta$ cannot be attributed to different rates of electron transport. The difference in the response to photoinhibition must be attributed to the difference in the LH Chl antenna size associated with PSII$_\alpha$ and PSII$_\beta$. The smaller antenna of the latter results in a much slower rate of light absorption and dissipation at PSII$_\beta$ [22].

Conclusions

The results presented in this work implicate the photochemical reaction center of PSII as the primary site of photoinhibition. Such an interpretation will explain the results presented here and in the work of other investigators.

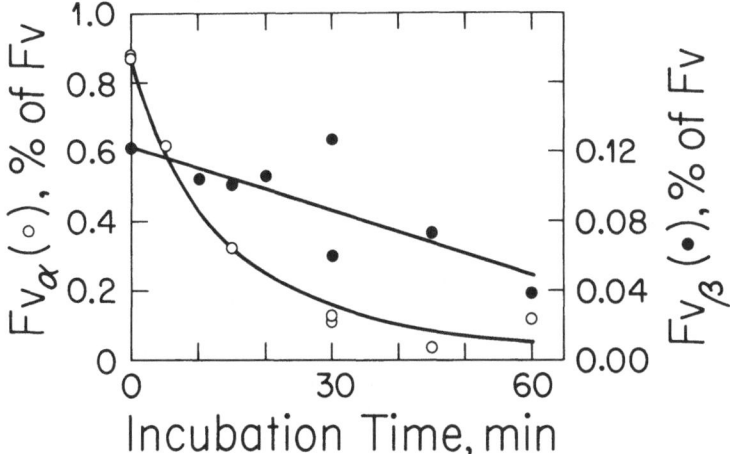

Figure 5. Amplitude of the variable fluorescence yield Fv_α and Fv_β controlled by $PSII_\alpha$ and $PSII_\beta$, respectively, as a function of photoinhibition. Note the fast loss of Fv_α and the much slower loss of Fv_β in the chloroplast samples upon illumination with strong white light.

Foremost, the inability of PSII to generate a stable charge separation and the simultaneous quenching of fluorescence by photoinhibited P^dQ centers suggests a direct effect of photoinhibition on P_{680}. The underlying molecular mechanism could involve a permanent modification of P_{680} the nature of which remains to be elucidated. If the photoinhibited form of the photo-chemical reaction center, P^d, would be an efficient quencher of excitation, like P_{680}^+ [5, 9], this would explain the lower variable fluorescence yield F_v observed. Dissociation of P_{680} from the secondary PSII electron donor molecule Z appears to be ruled out as a possible mechanism since artificial electron donors to P_{680} will not alleviate photoinhibition [10, 19].

Dissociation of P_{680} from the light-harvesting antenna of PSII would also explain the apparent inactivation of PSII reaction centers as manifested by the lower amplitudes of ΔA_{320} and $\Delta A_{540-550}$ (Björkman, personal communication). In this case, however, the lowering of the chlorophyll fluorescence yield must be explained by the additional assumption of light-induced quencher formation in the bulk Chl pigment of PSII. A site of photoinhibition on the reducing site of PSII, as evidenced by the loss of binding sites for (^{14}C) atrazine in partially photoinhibited chloroplasts [19], does not explain the inhibition of the charge separation. The relation of this observation to the effects of photoinhibition shown here remains to be determined.

Acknowledgements

REC gratefully acknowledges the financial support from a Commonwealth

Postgraduate Award, the Australian National University and the Carnegie Institution of Washington. The work was supported by NSF grant DMB-8400169 to AM.

References

1. Arnon DI (1949) Plant Physiol 24:1–15
2. Baker NR, East TM and Long SP (1983) J Exp Bot 34:189–197
3. Björkman O and Holmgren P (1963) Physiol Plant 16:889–914
4. Björkman O and Powles SB (1984) Planta 161:490–504
5. Butler WL (1972) Proc Natl Acad Sci USA 69:3420–3422
6. Butler WL (1977) in Encyclopedia of Plant Physiology, New Series, Photosynthesis I, eds Trebst A and Avron M (Springer-Verlag), Vol 5, pp 149–167
7. Butler WL and Okayama S (1971) Biochim Biophys Acta 245:231–239
8. Butler WL and Strasser RJ (1977) Proc Natl Acad Sci USA 74:3382–3385
9. Butler WL, Visser JMW and Simons HL (1973) Biochim Biophys Acta 292:140–151
10. Critchley C (1981) Plant Physiol 67:1161–1165
11. Critchley C and Smillie RM (1981) Aust J Plant Physiol 8:133–141
12. Erixon K and Butler WL (1971) Biochim Biophys Acta 234:381–389
13. Joliot A and Joliot P (1964) CR Acad Sci Paris 258:4622–4625
14. Jones LW and Kok B (1966) Plant Physiol 41:1037–1043
15. Kitajima M and Butler WL (1975) Biochim Biophys Acta 376:105–115
16. Knaff DB and Arnon DI (1969) Proc Natl Acad Sci USA 63:963–969
17. Kok B (1956) Biochim Biophys Acta 21:234–244
18. Kok B, Gassner EB and Rurainski HJ (1965) Photochem Photobiol 4:215–227
19. Kyle DJ, Ohad I and Arntzen CJ (1984) Proc Natl Acad Sci USA 81:4070–4074
20. Melis A and Anderson JM (1983) Biochim Biophys Acta 724:473–484
21. Melis A and Brown JS (1980) Proc Natl Acad Sci USA 77:4712–4716
22. Melis A and Duysens LNM (1979) Photochem Photobiol 29:373–382
23. Melis A and Schreiber U (1979) Biochim Biophys Acta 547:47–57
24. Osmond CB (1981) Biochim Biophys Acta 639:77–98
25. Powles SB (1984) Ann Rev Plant Physiol 35:15–44
26. Powles SB and Björkman O (1982) Planta 156:97–107
27. Powles SB and Critchley C (1980) Plant Physiol 65:1181–1187
28. Powles SB and Osmond CB (1978) Aust J Plant Physiol 5:619–629
29. Powles SB and Thorne SW (1981) Planta 152:471–477
30. Powles SB, Berry JA and Björkman O (1983) Plant Cell Environ 6:117–123
31. Satoh K (1970) Plant Cell Physiol 11:15–27
32. Satoh K (1971) Plant Cell Physiol 12:13–27
33. Stiehl HH and Witt HT (1968) Z Naturforsch 23:220–224
34. Van Gorkom HJ (1974) Biochim Biophys Acta 347:439–442

Photosynthesis Research 9, 89–101 (1986)
© *1986 Martinus Nijhoff/Dr. W. Junk Publishers, Dordrecht.*

Triplet-minus-singlet absorbance difference spectra of reaction centers of *Rhodopseudomonas sphaeroides* R-26 in the temperature range 24–290 K measured by Magneto-Optical Difference Spectroscopy (MODS)

E.J. LOUS and A.J. HOFF

Department of Biophysics, Huygens Laboratory of the State University, P.O. Box 9504, 2300 RA Leiden, The Netherlands

(*Received 28 October 1985*)

Key words: bacterial photosynthesis, reaction center, triplet state, MOS, (*Rps. sphaeroides* R-26)

Abstract. The recently developed technique of Magneto-Optical Difference Spectroscopy (MODS) [10] has been applied to reaction centers (RC) of the photosynthetic bacterium *Rhodopseudomonas sphaeroides* R-26. Absorbance changes induced by a magnetic field are measured as a function of wavelength yielding the triplet-minus-singlet (T-S) absorbance difference spectrum. (T-S) spectra thus obtained have been measured from 24–290 K. Going from low to high temperature the (T-S) spectra show the following features:
(a) A rapid decrease of positive absorption bands at 809 and 819 nm.
(b) A slow appearance of a band shift at 798 nm.
(c) A shift of the peak wavelength of the Q_y absorbance band of the primary donor P-860 from 992 to 861 nm, and of its Q_x band from 603 to 600 nm.
The spectra at 24, 66, 116, and 290 K have been analyzed by Gaussian deconvolution. The 800 nm region of the spectrum at 24 K can be decomposed in a combination of two band shifts and an appearing band. The temperature dependence of the spectra in this region is well explained by spectral broadening of the two shifting bands combined with a decrease in intensity of the appearing band when the temperature increases.
The two shifting bands in the 800 nm region are identified as the two bands at 803 and 813 nm which together make up the 800 nm band in the absorption spectrum and are assigned to the two accessory RC bacteriochlorophylls (BChls). The band shift of the 813 nm pigment is appreciably larger than that of the 803 nm pigment. The appearing band at 808 nm is attributed to monomeric absorption of ^{3}P-860, the triplet state being localized on one BChl.
We find no evidence for admixture of a charge transfer (CT) state of ^{3}P-860 with one of the accessory BChls at higher temperature.

Abbreviations

ADMR, absorption detected magnetic resonance; BChl, bacteriochlorophyll; BPh, bacteriopheophytin; B-800, accessory BChl molecule; CT, charge transfer; FWHM, full width at half maximum; I, BPh acceptor; MODS, magneto-optical difference spectroscopy; P-860, primary donor; Q_A, primary quinone acceptor; RC, reaction center; *Rps.*, *Rhodopseudomonas*; R-26, *Rps sphaeroides*, mutant R-26; (T-S), triplet-minus-singlet

Dedicated to Prof. L.N.M. Duysens on the occasion of his retirement

Introduction

The recently published X-ray diffraction study of crystallized reaction centers (RC) of *Rhodopseudomonas* (*Rps.*) *viridis* [2] has given a tremendous impetus to efforts to unravel the structure-function relationships that govern the unique electron transport properties of the photosynthetic RC. As a first result the optical absorption spectrum of RC of *Rps. viridis* has been analyzed with some success [17] in terms of singlet-singlet exciton interactions between the pigments, which arise because of electrostatic interactions between the outer electronic orbitals of the pigments. It is important to realize, however, that although the X-ray pictures are marvelously detailed, their resolution is not sufficient to predict with accuracy anything on the electron density in the outer orbitals, as these contribute only little to the X-ray scattering cross section (which is mostly determined by core electrons). Hence, fairly crude approximations have to be made, such as point dipole—dipole coupling, for the calculation of the interactions between the pigments. Yet, knowledge of the detailed electronic structure of the outer, highest energy orbitals in the excited and ground states of the pigments is crucial for the understanding of the very fast (< 3 ps, [12]) and efficient charge separation in (bacterial) photosynthetic RC.

One branch of optical spectroscopy, viz. triplet-minus-singlet absorbance difference spectroscopy, seems to be uniquely suited to gather information on the pigment-pigment interactions that govern electron transport. The reason is, that interactions between the triplet excited state of one pigment and the singlet ground state of another pigment are, like those that determine the matrix element of electron transfer, determined by electronic overlap integrals. One may therefore hope that accurate measurement of triplet-minus-singlet (T-S) spectra, and their interpretation will offer insight in the afore-mentioned electron transfer properties of the reaction center.

Recently, we have presented two new techniques to obtain highly resolved (T-S) spectra of photosynthetic RCs. One makes use of absorbance detected magnetic resonance (ADMR) of the triplet state that is generated on the primary electron donor by charge recombination [3]. It offers the possibility of site-selection [4] and of measuring linear-dichroic (T-S) spectra [5], but the experiments have to be carried out at liquid helium temperatures ($\leqslant 4.2$ K). (T-S) spectra at higher temperatures, up to room temperature, can be accurately and rapidly measured by magneto-optical difference spectroscopy (MODS) [10], which makes use of the effect of a small magnetic field on the yield of triplet formation. Both the ADMR and the MODS technique are several orders of magnitude more sensitive than conventional flash spectroscopy as previously applied to photosyntheitc RCs [15, 16].

Although the ADMR-monitored (T-S) and linear-dichroic (T-S) spectra obtained so far [3, 9] are very detailed, their interpretation is still not fully resolved. One aid is the temperature dependence of the (T-S) spectra [10], and this is one of the reasons why we have undertaken the present study of the temperature dependence of the (T-S) spectra of RC of *Rps. sphaeroides* R-26. Another reason is the previously reported study of this temperature dependence by flash spectroscopy over the temperature range 77–290 K [16], in which it was suggested that at higher temperatures a triplet charge transfer (CT) state, lying 0.15 eV above the triplet state that is localized on the primary donor, contributes considerably to the (T-S) spectrum. In our previous study of the temperature dependence of the (T-S) spectra of RC of *Rps. viridis* [10] no evidence for such a state was found it seemed of interest to see whether the higher accuracy and resolution of MODS as compared to flash spectroscopy would lead to new insight for *Rps. sphaeroides* R-26 also.

We here report (T-S) spectra obtained by MODS over the temperature range 24–290 K. Considerable changes in the bleaching of the long wavelength absorption band of the primary donor were found, including a blue shift and a broadening with increasing temperature. In addition the strong bands observed at 24 K in the 790–825 nm region, where the accessory bacteriochlorophyll *a* molecules absorb, lose almost all their intensity when heating the sample from 24 K to 290 K. We offer a band analysis of the (T-S) spectrum at 24 K (which spectrum is very similar to that previously obtained by ADMR at 1.2 K [3]) on the basis of which a tentative explanation is given of the temperature effect on the bands in the 790–823 nm region. The temperature-induced changes are attributed mostly to band broadening. Specifically, we find no evidence that at the higher temperatures a triplet CT state contributes to any significant extent to the (T-S) spectrum.

Materials and methods

Instrumental

The MODS spectrometer is schematically depicted in Figure 1. Light from a tungsten-iodine lamp (Osram HLX 646555, 24 V, 250 W), functioning both as measuring beam and as excitation light source, is focussed to a 5 × 10 mm spot on the flat sample cell. After passing the sample and a monochromator (Bausch and Lomb, f = 4.2, resolution, 1.6 nm/mm) the transmitted light intensity is detected by a Si photodiode (RCA C-30842). The monochromator can be scanned from 350 to 1050 nm; from 350 to 510 nm it is used in second order, using a Schott BG-38 glass filter to cut off the light beyond 510 nm. The sample cuvette (optical pathlength, 1.5 mm) is placed in a double walled vacuum-isolated helium flow cryostat made of pyrex-borosilicate glass. The top and bottom of the cryostat are cylindrical; the center part is flattened to reduce light scattering at the glass surfaces to a minimum.

The pyrex cryostat is placed on the body of an Oxford helium flow cryostat (ESR-9). The cooling helium stream passes the sample, which is placed in a

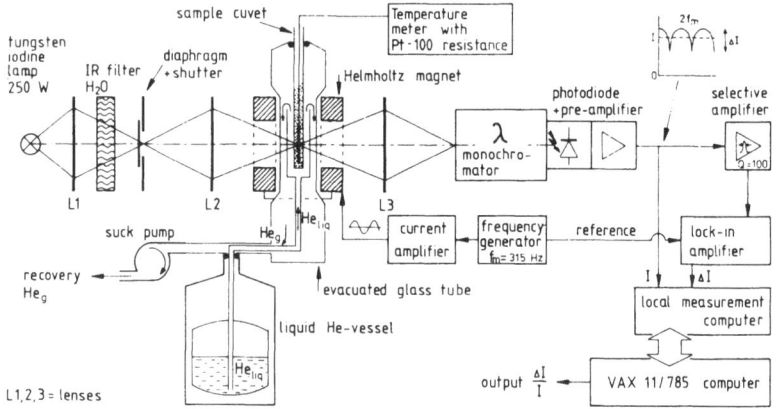

Figure 1. Block diagram of the experimental set-up of the variable temperature MODS spectrophotometer

glass chimney, and is sucked back along the outside of the chimney for pre-cooling and to act as a heat shield. A coated and calibrated platina resistance element (Pt100/15C) is placed inside the sample and directly in the light spot to regulate the average sample temperature down to 24 K, with an accuracy of 0.5 K. Besides the lower attainable temperature, an important advantage of helium cooling above nitrogen cooling is the more effective cooling at higher temperatures, so that a less strong cooling flow is then needed. The reduction in turbulence of the gas then results in a considerably reduced noise level.

A pair of Helmholtz coils flank the glass cryostat producing a sinusoidally varying magnetic field of 10.5 mT at 315 Hz at the location of the sample. The modulated magnetic field creates a modulated change in triplet concentration, which in turn gives rise to modulated changes in transmittance of the sample [10]. These changes have twice the modulation frequency of the current passing the Helmholtz coils, as the magnetic field effect is independent of the sign of the magnetic field. The modulated intensity of the transmitted light is detected by the photodiode. Its signal is divided in two parts after pre-amplification with a bandwidth of 4.2 kHz: One is further amplified by a narrow-banded amplifier (PAR 181, Q = 1000) and lock-in detected (PAR 5101), resulting in a signal ΔI that is stored in a local measurement computer. The other part passes a low-pass filter and gives the total transmitted light intensity I, which is also stored as a function of wavelength. Final data treatment and storage were carried out on a VAX-11/785 computer, yielding the relative change $\Delta I/I$ as a function of wavelength, i.e. the triplet-minus-singlet spectrum [10]. The present sensitivity of our absorption spectrophotometer is $\Delta A/A = 1 \times 10^{-6}$ at 290 K, the signal-to-noise ratio being determined by the turbulence of the cooling gas stream.

The data have been simulated with aid of a least squares fitting procedure

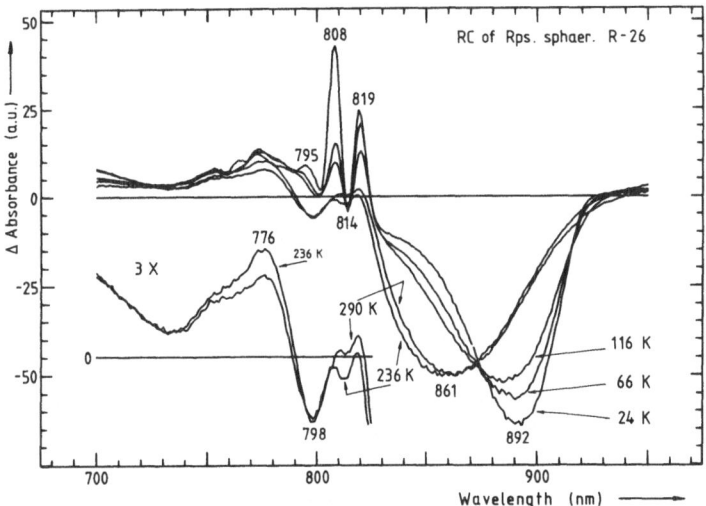

Figure 2. (T-S) spectra of RC of *Rps. sphaeroides* R-26 at several temperatures normalized on the surface of the band between 823 and 950 nm. Normalization factors are 1.00, 1.46, 1.67, 3.80, 3.60 for the temperatures 24, 66, 116, 236 and 290 K, respectively. Lock-in time constant, 1 s. Optical resolution, 1.6 nm except for 236 K, 3.0 nm. Sine-modulated magnetic field, 10.5 mT. Optical density (803 nm, 300 K), 0.3. Light intensity of the focussed spot on the cuvette was 1.5 W/cm² continuously. One arbitrary Δ absorbance unit corresponds approximately with a relative absorbance change ΔA/A of 2.89 10^{-5}

developed by Marquardt [11]. The single absorption bands where approximated by Gaussians.

Preparations

Reaction centers of the photosynthetic purple bacterium *Rps. sphaeroides* R-26 were prepared as described in [7, 3] and suspended in 10 mM 4-morpholine propane sulfonic acid buffer, pH = 8, containing in addition 66% v/v ethyleneglycol. Reduction of the first quinone acceptor Q_A was carried out by the addition of 10 mM sodium ascorbate and illuminating the sample while cooling to 77 K, with the temperature sensor being placed inside the sample cell.

The temperature dependence of the (T-S) spectra was measured by starting at 24 K and stepwise heating up to about 290 K. At the end of such a run the sample was still intact, as checked by its optical absorption spectrum.

Results

Difference spectra were taken from 350 up to 950 nm at several temperatures from 24 to 290 K. In Figures 2 and 3 only the most significant parts of the spectra are shown. At 24 K the shape of the MODS spectrum is very similar to

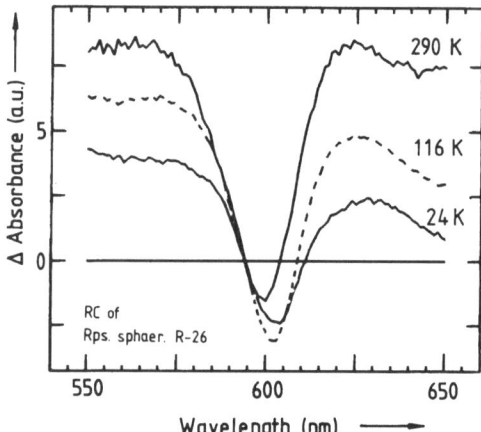

Figure 3. Normalized (T-S) spectra of RC of *Rps. sphaeroides* R-26 between 550 and 650 nm. Temperatures as indicated, conditions as in Figure 2

the (T-S) spectra at 1.2 K measured by Den Blanken et al. [3]. Upon heating from 24 to 290 K the features in the parts of the (T-S) spectra not shown in Figures 2 and 3 were broadened and the relative intensity decreased, but the main features did not change much.

The MODS intensity reflects two temperature-dependent processes:
(a) formation of the triplet state on P-860,
(b) spectral broadening and sharpening, or shifts of the absorption maximum, of the various single absorption bands. Those temperature-induced changes are also seen in the normal absorption spectra.

To correct for the temperature dependence of the triplet concentration, we normalized our spectra to the area under the broad band between 823 and 950 nm that is due to the bleaching of the primary donor (Figures 2 and 3). From Figure 2 it is seen that the bleaching of P-860 shifts from 892 nm at 24 K to 861 nm at 290 K, while broadening considerably and becoming less intense. This agrees with the band shift and narrowing observed for the P-860 band in the absorbance spectrum of R-26 and for the P-860 bleaching in the $P\text{-}860^+ - P\text{-}860$ spectrum when cooling the reaction centers to cryogenic temperatures [1, 6]. The isobestic point at 873 nm in Figure 2 shows that the apparent shift of the 861 nm band to 890 nm upon cooling is actually not a shift but a stabilization at low temperature of a conformation of P-860 that absorbs at the longer wavelength. Of special interest is the short wavelength side of the P-860 bleaching, where at low temperature a shoulder seems to extend well beyond 800 nm.

The most interesting temperature-induced changes are observed in the 760–830 nm region. Going from 24 to 290 K we see a dramatic decrease in the intensity of the 809 and 819 nm peaks, while at 798 nm some bleaching appears at higher temperatures.

Spectral simulations

To explain the temperature dependence of our (T-S) spectra we have endeavoured to decompose the difference spectra at the various temperatures in single features (such as bleachings, appearance of new bands, shifts). Our aim hereby was to select the best possible interpretation with a minimum of (new) bands and a maximum of reasonable restrictions. Thereto we adopted the following assumptions:

(1) The number of the single difference features is independent of temperature.

(2) Single absorption bands can be approximated by Gauss functions.

(3) The long wavelength absorption band of P-860 is heterogeneous. We will see below that we are able to approximate the bandshape at all temperatures with a linear combination of two Gaussians: one that dominates at low temperature, with center wavelength at 892 nm and full width at half maximum (FHWM) of 40 nm, and one that dominates at high temperature with center wavelength at 855 nm and FWHM of 45 nm.

(4) In the low temperature absorption spectrum, the 800 nm band is made up of two bands of unequal intensity and width: the more intense one absorbs close to 803 nm, while the lesser intense band absorbs close to 813 nm. With increasing temperature the two bands broaden, while their surface areas remain constant.

(5) The triplet-induced shift in nm of the 803 and 813 nm bands is independent of temperature.

We start the simulations of our (T-S) spectra by first fitting the absorbance spectra at all temperatures, using assumptions (1)–(4). With the band parameters (wavelength, intensity, width) thus obtained, we then proceed to simulate the MODS-recorded (T-S) spectra at the various temperatures, using assumption (5) and employing the model that was used earlier to explain the ADMR-monitored (T-S) spectra at 1.2 K [3]. Thus, we assume that upon triplet formation on P-860:

(6) The P-860 long wavelength absorption band bleaches.

(7) The 803 nm band of the absorption spectrum shifts to the blue. This shift is analogous to the blue shift of the 834 nm band in the (T-S) spectrum of *Rps. viridis* [3, 10], albeit that in *Rps. sphaeroides* R-26 the blue shift is much smaller.

(8) The 813 nm band of the absorption spectrum shifts to the red.

(9) At 809 nm an absorption band appears.

To improve the fit of the (T-S) spectra we allowed for slight variations in peak wavelength, surface area and width of the absorption bands as determined from the fit of the absorption spectrum, and optimized the shifts of the 803 and 813 nm bands. With the new band parameters obtained from the (T-S) spectrum, a new fit of the absorption spectrum was made, the parameters of which were then used in a new fit of the (T-S) spectrum, etc.

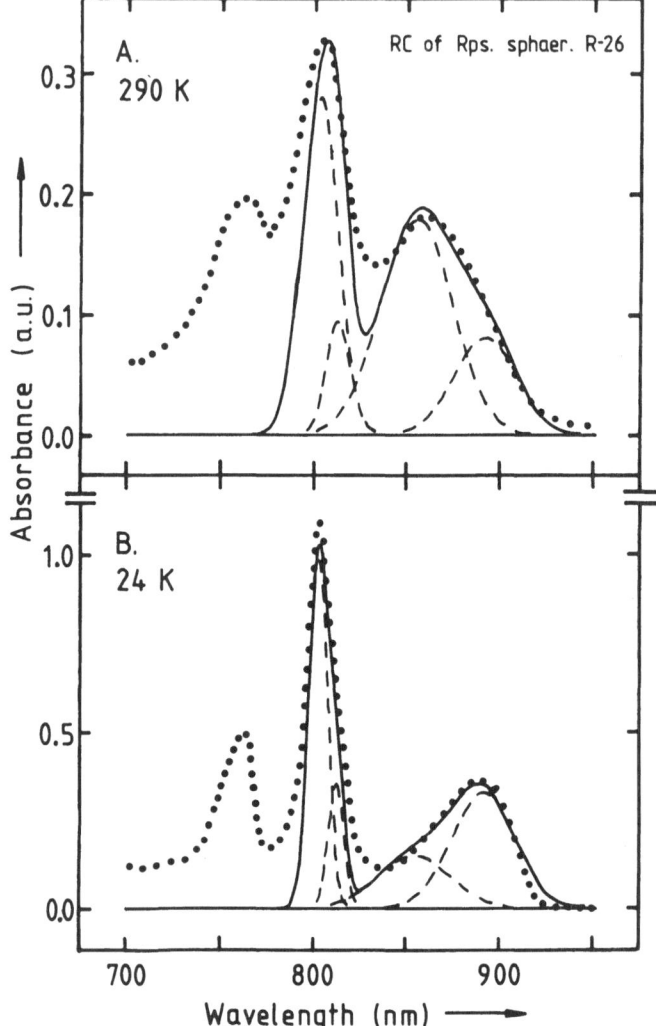

Figure 4. (○-○-○), Absorption spectra (in a.u.) of RC of *Rps. sphaeroides* R-26 at (a) 290 and (b) 24 K. (——), Simulation with 2 Gaussians for the P-band and one Gaussian for each of the B-800 pigments. See Table 1 for parameter values

The above iterative procedure led to an acceptable fit of both the absorbance and the (T-S) spectra at all temperatures (Figures 4 and 5). The parameters of the various bands in Figures 4 and 5 are collected in Table 1.

We emphasize that in the fitting procedure described above we have purposely used Gaussian band shapes to underline the basic assumptions and to keep the results as transparent as possible. Of course, the use of more complicated band shapes, as skewed Gaussians, etc., would have somewhat

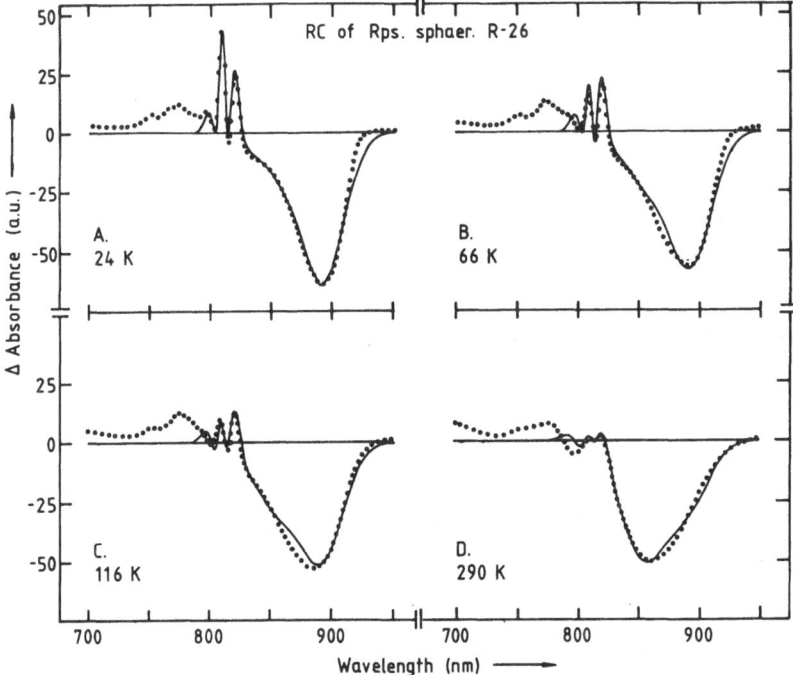

Figure 5. Simulation of the (T-S) spectra at four temperatures according to the model parameters given in Table 1. (o-o-o), Spectra taken from Figure 2

improved the fits. This is especially true for the flank near 823 nm of the P-860 band, which at high temperatures is quite pronounced and has a non-negligible overlap with the 813 nm band shift, thereby partly reducing the 819 nm peak intensity. Furthermore, we have not attempted to fit the region around 795 nm and beyond to shorter wavelength, as the (T-S) spectra in that region are weak and not very detailed.

Discussion

The P-860 band (823–950 nm)

Our fits show that the total area of all the involved bands, except the appearing 809 nm band, are constant over the whole temperature range, both for the absorption spectra and the normalized (T-S) spectra. Heating from 24 K only causes band broadening for the 803 and 813 nm bands and a change in the intensity distribution between the two components of the P-860 band with maxima at 855 and 892 nm. The bandwidths of those components appear to be practically independent of temperature. This can be understood on the basis of recent hole-burning experiments [13], where it was shown that the width of the long wavelength absorption band of P-860 is largely determined by the lifetime of its excited state. The normalized surface areas of the 855 and 892 nm bands as a function of temperature do not show an Arrhenius behavior. Rather we find by plotting the areas of the 855 and 892 nm bands against $1/T$ (figure not shown) that the ratio of these areas changes abruptly between 100 and 150 K. Apparently, in this temperature region a conformational change of the structure of P-860 occurs leading to absorption at longer wavelength of the primary donor. The non-Arrhenius behavior of the intensity of the two bands of P-860 seems to exclude an assignment of the bands to two exciton components of the P-860 dimer.

The 800 nm region (800–823 nm)

For the 800 nm region of the ADMR-monitored (T-S) spectrum at 1.2 K a tentative interpretation was given earlier, based on the (T-S) spectrum of *Rps. viridis* [3]. Our present results substantiate our interpretation that the features of this spectral region are composed of a blue shift, a red shift and an appearing band. We demonstrated here that the absorption bands responsible for these shifts correspond exactly to the two components of the broad 800 nm band in the absorption spectrum. Note that the shifts in the (T-S) spectra of R-26 are qualitatively similar to the shifts in the (T-S) spectra of *Rps. viridis*, but are much smaller in amplitude. At first sight the temperature dependence of the shifting bands seems to be quite different for the two species. This, however, is a result of the lower spectral resolution in R-26. In fact, for both bacteria the change in spectral shape with temperature can be completely accounted for by broadening of the two shifting bands (at 803 and 813 nm in *Rps. sphaeroides* R-26 and at 834 and 851 nm in *Rps. viridis*) with increasing temperature. Note then that oscillator strengths of the 803 and 813 nm bands do not change upon ^{3}P-860 formation, i.e. we can explain all our spectra satisfactorily without invoking such a change. Thus, the 813 nm band cannot be an exciton band of P-860 because it would then bleach when a triplet is formed on P-860.

The correspondence of the two shifting bands in the (T-S) spectrum to the two components of the 800 nm band in the absorption spectrum strongly suggests that they are due to the two accessory BChls, which for simplicity we will label henceforth B-803 and B-813. It should be kept in mind, however, that the two bands are almost certainly not pure, and admixed with P-860 [17]. Our results suggest that such an admixture is not symmetric. First, the intensity of the B-803 and B-813 bands is unequal. Secondly, the difference in band position suggests that the B-813 band has a stronger interaction with P-860, as it is shifted further away from the in vitro Q_y band of BChl *a* which is at 780 nm. Thus, although by analogy to the RC structure of *Rps. viridis* as elucidated by X-ray diffraction [2] we may assume that the two accessory BChls are symmetrically placed with respect to P-860, their interaction with P-860 seems to be different. This is in line with the fact that only one branch of the symmetric pigment structure in the RC is photoactive. The difference in shift of B-803 and B-813 upon ^{3}P-860 formation may be caused by the asymmetry of ^{3}P-860 itself (localization of the triplet state on one BChl) and therefore need not reflect the difference in interaction between B-803 and B-813 with P-860.

The appearing band at 809 nm cannot be a triplet band of a BChl, because the in vitro triplet spectrum does not show narrow bands in this region [14]. We have previously [3] attributed this band to a monomeric absorption band of ^{3}P-860, the triplet being localized (at least on an optical time scale) on one of the two BChls which make up the dimer. At 24 K the oscillator strength of the 809 nm band is nearly equal to that of B-813, i.e. corresponding to about 0.3 BChl, while at room temperature its amplitude is about 3 times reduced compared to its value at 24 K, the width having not changed much. The low intensity of the 809 nm band may reflect coupling with the adjacent ^{3}BChl of P-860 which coupling would then become somewhat stronger at higher temperatures, i.e. the triplet state then may become somewhat delocalized. Presumably, the decrease in intensity of the 809 nm band with increasing temperature is not due to an increase in (internal) charge transfer (CT) character of ^{3}P-860 [8].

The 740–800 nm region

Recently Shuvalov and Parson [16] have suggested that the ratio $(A_{785} - A_{797})/A_{785}$, where A stands for absorbance, showed an Arrhenius behavior. They concluded that at higher temperatures a CT state composed of e.g. [P-860$^+$B-800$^-$] was admixed to ^{3}P-860. We do not find the purported Arrhenius behavior (Figure 2), most probably because our MODS-monitored (T-S) spectra are much better resolved than those recorded with flash spectroscopy [15, 16]. Thus, in agreement with [8] we conclude that there is no evidence of an 'external' CT state admixed to ^{3}P-860. In our opinion most of the bleaching at 798 nm that is seen at 290 K is caused by the shift of the 813 nm band, while the other changes in the 740–800 nm region are mostly due to absorption changes of the BPhs.

The 600 nm region

At 600 nm (Figure 3) we see a small bleaching, the extent of which is somewhat smaller than expected for one BChl. The reason for this is that ^{3}BChl, and by inference also ^{3}P-860, shows considerable absorption in the 600 nm region [14]. The bleaching becomes somewhat stronger at higher temperatures, possibly reflecting a lower absorption of the triplet state at higher temperatures. This would also explain the change in shape of the 600 nm bleaching with temperature. There is only a small change of the bandwidth with temperature, which agrees with the lack of temperature dependence of the width of the 855 and 892 nm bands. The shift with temperature of 600 and 603 nm is rather small. Apparently the Q_x transition of the 855 nm band is almost equal to the Q_x transition of the 892 nm band.

Table 1. Band parameters used for the simulations of Figure 5

	P-860 λ_0 (nm)	^{3}P-860 λ_1 (nm)	Bandwidth (nm)				Relative area of band*			
			290K	116K	66K	24K	290K	116K	66K	24K
B-803	803	802.6	23	15	13	12	0.76	0.76	0.76	0.76
B-813	813	816.5	15	12.5	10.5	10	0.21	0.21	0.21	0.21
P-860 high T band	855	—	45	45	45	45	0.73	0.41	0.32	0.26
low T band	892	—	40	40	40	40	0.32	0.62	0.73	0.81
monomer band	809	—	9.5	7.2	6.4	6.0	0.06	0.09	0.12	0.16

*The area between 823 and 950 nm of the long wavelength band of P-860 is used for normalization at the four temperatures.

Summarising we conclude that the (T-S) spectrum of RC of *Rps. sphaeroides* R-26 and its temperature dependence can be explained by assuming two conformations of P-860, two shifting bands due to the accessory BChls that broaden with increasing temperature, and one appearing band attributed to one monomeric BChl of ^{3}P-860. We emphasize that the aim of the present paper is to give a global analysis of the (T-S) spectrum and its temperature dependence with a minimum set of assumptions. More refined modeling may change somewhat the values of the various band parameters collected in Table 1 but we feel that the model as it stands is both reasonable and accurate and that there is presently no need for more complicated models to explain our experimental results.

Acknowledgements

We thank Ms. L. Nan for preparing the reaction centers, Dr. J. Schmidt of the Center for the Study of the Excited State of Molecules for his interest, the cryogenic facilities and the loan of several components of our MODS spectrophotometer, Mr. L. van As for carefully constructing the glass top of the flow cryostat, Mrs. J. van Egmond and W. Versluijs for making the additional mechanical parts and magnet and Dr. J.B. Pedersen for advice on the least squares fitting routine. This work was supported by the Netherlands Foundation for Chemical Research (SON), financed by the Netherlands Organization for the Advancement of Pure Research (ZWO).

References

1. Clayton RK and Yamamoto T (1976) Photochem Photobiol 24:67–70
2. Deisenhofer J, Epp K, Miki O, Huber R and Michel H (1984) J Mol Biol 180:385–395
3. Den Blanken HJ and Hoff AJ (1982) Biochim Biophys Acta 681:365–374
4. Den Blanken HJ and Hoff AJ (1983) Chem Phys Lett 98:255–262
5. Den Blanken HJ, Meiburg RF and Hoff AJ (1984) Chem Phys Lett 105:336–342
6. Feher G (1971) Photochem Photobiol 14:373–387
7. Feher G and Okamura MY (1978) In Clayton RK and Sistrom WR, eds. The Photosynthetic Bacteria, pp 349–386. New York: Plenum Press
8. Hoff AJ and Proskuryakov II (1985) Chem Phys Lett 115:303–310
9. Hoff AJ, den Blanken HJ, Vasmel H and Meiburg RF (1985) Biochim Biophys Acta 806:389–397
10. Hoff AJ, Lous EJ, Moehl KW and Dijkman JA (1985) Chem Phys Lett 114:39–43
11. Marquardt DW (1963) J Soc Indust Appl Math 7:431–441
12. Martin J-L, Breton J, Hoff AJ, Migus A and Antonetti A (1985) Proc Natl Acad Sci USA, in the press
13. Meech SR, Hoff AJ and Wiersma DA (1985) Chem Phys Lett, in the press
14. Pekkarinen L and Linschitz H (1960) J Am Chem Soc 82:2407–2411
15. Shuvalov VA and Parson WW (1981) Biochim Biophys Acta 638:50–59
16. Shuvalov VA and Parson WW (1981) Proc Natl Acad Sci USA 78:957–961
17. Zinth W, Knapp EW, Fischer SF, Kaiser W, Deisenhofer, J and Michel, H (1985) Chem Phys Lett 119:1–4

Photosynthesis Research 9, 103−112 (1986)
© 1986 Martinus Nijhoff/Dr. W. Junk Publishers, Dordrecht.

Manganese-histidine cluster as the functional center of the water oxidation complex in photosynthesis

SUBHASH PADHYE[1]*, TAKESHI KAMBARA[1]**, DAVID N. HENDRICKSON[1]+ and GOVINDJEE[2]+

[1]School of Chemical Sciences, University of Illinois at Urbana-Champaign, 352 Noyes Laboratory, 505 S. Mathews, Urbana, IL 61801, USA
[2]Department of Physiology and Biophysics, and Plant Biology, University of Illinois at Urbana-Champaign, 289 Morrill Hall, 505 S. Goodwin Avenue, Urbana, IL 61801, USA

(*Received 23 July 1985*)
Key words: oxygen evolution, model, manganese-histidine complex

Abstract. The recent model of Kambara and Govindjee for water oxidation [Kambara T. and Govindjee (1985) Proc. Natl. Acad. Sci. U.S.A., 82:6119−6123] has been extended in this paper by examining all the data in order to identify the most likely candidate for the 'redox-active ligand' (RAL), suggested to operate between the water oxidizing complex (WOC) and Z, the electron donor to the reaction center P680. We have concluded that a very suitable candidate for RAL is the imidazole moiety of a histidine residue. The electrochemical data available on imidazole derivatives play heavily in this identification of RAL. Thus, we suggest that histidine might play the role of an electron mediator between the WOC and Z. A model of S-states in terms of their plausible chemical identity is presented here.

Abbreviations

J, electronic spin of ion; P680, reaction center chlorophyll; RAL, Redox active ligand; S_n, state of the oxygen-evolving system; WOC, water oxidation complex; Z, electron donor to P680

Introduction

After examining all the data in the literature Kambara and Govindjee [18, 19] have recently proposed a new model for the molecular mechanism of water oxidation in photosynthesis. Various highlights of this new model, to be referred hereafter as the KG model, are schematically indicated in Figure 1. Two pools of manganese ions, each with two manganese ions, were proposed to exist, one of which is located in the hydrophobic cavity in the 'intrinsic' 34kD protein and the other on the hydrophilic surface of the 'extrinsic'

---

*Permanent address: Department of Chemistry, University of Poona, Poona-411 007, India
**Permanent address: Department of Engineering Physics, University of Electro-Communications, Chofu, Tokyo, 182, Japan
+Address for offprints: D.N. Hendrickson and Govindjee. Address see above

---

Dedicated to Prof. L.N.M. Duysens on the occasion of his retirement

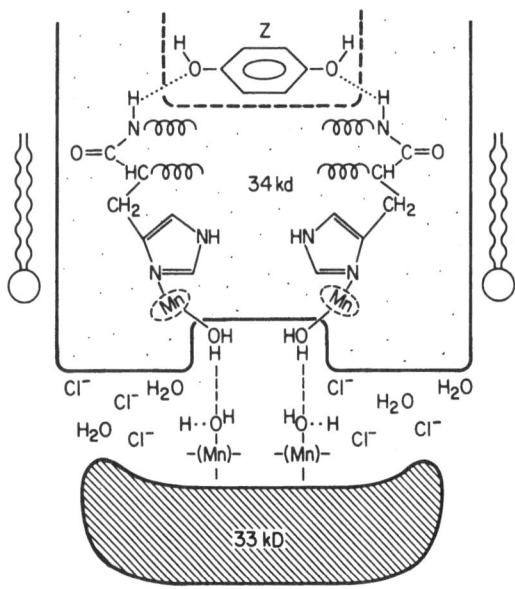

Figure 1. Schematic view of the modified Kambara-Govindjee [19] model for the molecular mechanism of water oxidation in photosynthesis. Two manganese ions, (Mn), are located in a region between the 'extrinsic' 33kD and 'intrinsic' 34kD proteins. Water oxidation occurs at the other two manganese ions [Mn] embedded in the 'intrinsic' protein. The water oxidation is assisted, in our hypothesis, by two redox-active ligands in the form of imidazole moieties.

33kD protein. These two pools of manganese ions are connected by hydrogen bonds through which protons and electrons can be transferred. The oxidation of two water molecules to give one dioxygen molecule is carried out by the two intrinsic manganese ions and protons are transferred from the intrinsic manganese ions to the extrinsic manganese ions. The oxidation state of the intrinsic manganese ions oscillate between Mn(III) and Mn(IV). It was further proposed that electron transfer occurs from this array of manganese ions in the water oxidizing complex (WOC) to Z^+ (Z is a plastoquinol and an electron donor to the reaction center chlorophyll P680). This transfer occurs via a redox active ligand (RAL), one of which is bound to each of two intrinsic manganese ions. Thus, it is two Mn(IV) ions and two RAL^+ that remove four electrons from two H_2O molecules to give one O_2.

One of the features of the KG model, as compared to earlier models [see 13,], is the inclusion of RAL between the manganese center in the WOC and Z. The existence of such an intermediate had also been suggested, among others, by Bouges-Bocquet [5] and by Boska and Sauer [4] on the basis of kinetic analyses. Each RAL serves as a one-electron sink that is intimately involved with one of the two intrinsic manganese ions. Recent work [22] on model manganese complexes has shown that ligands based on o-quinones can serve as one-electron sinks interacting with manganese ions. When the temperature of a solution containing a manganese complex with two such ligands coordinated to Mn was changed, it was found that each ligand reversibly changes from the 1- to 2-state with a concomitant change in the manganese ion from Mn(II) to Mn(IV). It is thus possible that the p-quinone plastoquinone functions as the RAL, shuttling between the semiquinone(1-) and hydroquinone(2-) forms. However, the reported number of plasto-quinone moieties in PSII is about 3 [27, 40] and in view of the other require-ments for plastoquinones in the electron-transport chain, there does not seem to be sufficient quantity of plastoquinone present for it to serve as the RAL [19].

In the development of the KG model [19] it was also suggested that aromatic amino acid residues such as histidine and tyrosine could function as the RAL. In this paper we examine this issue in detail and propose histidine as the likely candidate for the RAL. Although the amino acid analysis and sequence of the 'intrinsic' 34kD polypeptide is not yet available [13], the gene mapping data reported by Rasmussen et al. [30] for the membrane D_2 polypeptide of pea chloroplast genome have shown that it is part of the PSII and suggest that it is identical to the 'intrinsic' polypeptide associated with the electron donor of PSII [34]. The remarkable feature of the D_2 poly-peptide is that it is rich in histidine residues (7–8 histidines out of 300 residues) and very low in the lysine residues which are found to be abundant in the 'extrinsic' 33kD polypeptide [20].

The suggestion made in this paper that histidine may serve as the RAL moiety which mediates electron transfer between the WOC and Z by inter-acting with the intrinsic manganese ions is shown to be consistent with the redox-potential requirements of an intermediary between the WOC and Z, the potential ligand capabilities of RAL, and the pH dependence of O_2 evolution. A detailed scheme is proposed to show the functioning of histidine as the RAL as the intrinsic manganese ion pair cycles between the various redox states of the water oxidation complex (i.e., the so-called S_n states). Our expectations as to the EPR signals that should be seen for the different S_n states are discussed.

Histidine as the Redox Active Ligand

The electron mediator between WOC and Z is presently assumed to be a RAL. This RAL might either act as a ligand of the intrinsic manganese ions, or it might just be an one-electron sink that is in proximity to an intrinsic man-ganese ion, i.e., each RAL might not be directly bonded to a manganese ion. If it is just close to a manganese ion, effective electron flow between the intrinsic manganese ion and the RAL would be possible by a tunneling mechanism.

Catechols have *not* been detected in the PSII. As we indicated above, there also does *not* seem to be sufficient plastoquinone present for it to be the RAL. Furthermore, the mid-point potentials of various redox couples of catechols and quinones as given by Rich [32] do not support their identification as the RAL, since the potential of RAL/RAL$^+$ should be between that of H_2O/O_2 ($E_{m,7} = +0.8$ V) and Z/Z^+ ($E_{m,7} = +1.0$ V). The only one that has the appropriate potential for placement between the WOC and Z is the $QH_2/QH_2 \cdot^+$ couple at a potential of $\sim +0.9$ V. However, in this protonated form plastoquinol would not be expected to be a good ligand for a transition metal ion such as the manganese ion (see [29] and references cited therein). Furthermore, if Z is a bound plastoquinol PQH_2 [28] (perhaps, plastoquinol A [39], it would also not bind directly to the intrinsic manganese ions.

The mediation of electron flow between the WOC and Z by specific amino acid side chains is attractive in light of accumulating data indicating that the binding site of the intrinsic manganese ions is close to hydrophobic polypeptides [7, 25, 42]. It was pointed out by Isied [17] that low lying, empty π^* orbitals in tyrosine or the filled 2p orbitals in histidine can facilitate electron mediation. The formation of a radical cation, i.e., RAL$^+$, due to transfer of an electron to Z^+, is assumed for the electron mediation by RAL in the KG model [19]. The formation of a radical cation is indeed preferable for these aromatic amino acids, for their oxidation redox potentials are in the range of +0.8 to +1.0 V [26]. This is the desirable range for RAL an electron mediator between the WOC and Z. In this connection, we mention that an intermediate M, suggested by Renger (see e.g. ref. [31]) to have a redox potential in the range of +200 to +400 mV, would not fulfill our criterion for RAL.

There are two important considerations that lead to the choice of histidine as the RAL among all possible amino acid side chains. It has been shown by Brabec and Mornstein [6] that tyrosine and histidine both are electrochemically active. Tyrosine is oxidized at +0.7 V *vs.* SCE, while histidine shows a peak at +1.1 V *vs.* SCE at a graphite electrode. Furthermore, it was shown by coulometric measurements that oxidation of tyrosine involves a two-electron process in which the quinonoidal radical intermediates are not formed. The fact that tyrosine undergoes a two-electron oxidation process at +0.7 V may eliminate it from contention for RAL, because RAL is to serve as a one-electron sink. On the other hand there are precedents in the literature for the formation of radical cations from imidazole or substituted imidazoles [33, 35]. Eaton and Wilson [12] have shown that imidazole ligands can mediate the transfer of electrons to heme proteins by forming a transient radical under mild conditions.

If the RAL needs to be coordinated to the intrinsic manganese ions, then, in comparison to other amino acid residues the imidazole moiety of histidine is particularly well suited for this. For example, it has been shown by several workers (see references cited in [14]) that free tyrosine functions as a bidentate ligand without involving the aromatic hydroxyl group, while free histidine

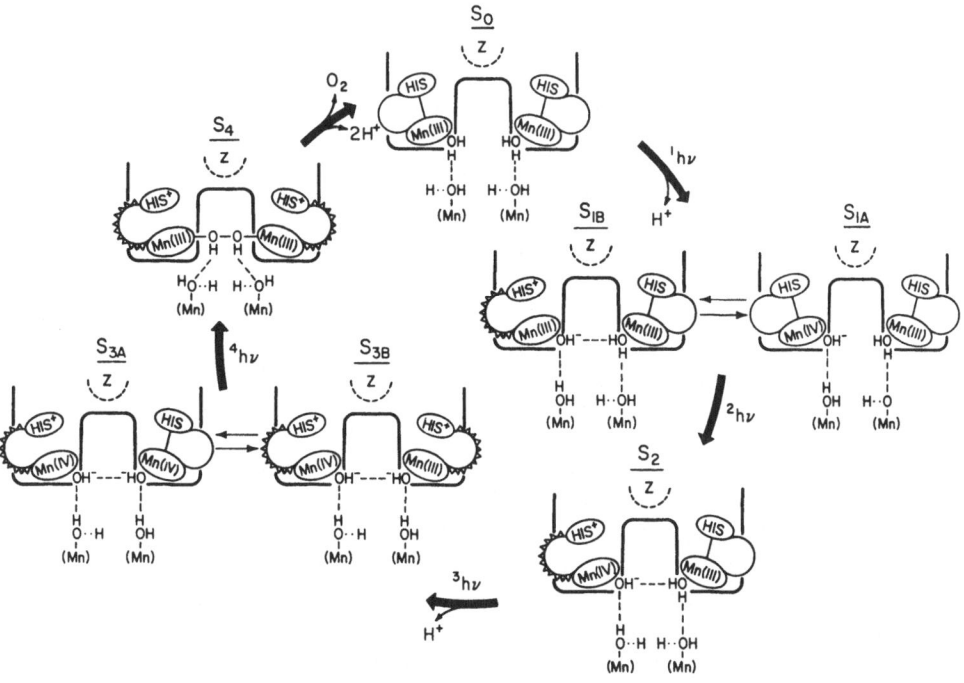

Figure 2. A diagram for the 'Histidine Model' for Kok's S_n states. Oxidation state changes are shown for the 'intrinsic' manganese ions and for the nearby histidine (His) residues. The wavy line connecting a Mn ion and its nearby His symbolically indicates a protein conformation different than is present for the state where a solid line is shown.

exhibits an ambivalent donor character depending upon the pH conditions [24]. At physiological pH in the thylakoid membrane, even though it is part of a polypeptide chain a histidine residue will still be able to interact through one nitrogen of the imidazole moiety. Seela et al. [38] have reported the preparation and characterization of a Mn(III) complex which has a coordinated imidazole ligand. Furthermore, conformational changes in the protein structure could modulate the nature of the interaction between the imidazole nitrogen atom of one RAL and a manganese ion.

Description of the 'Histidine Model'

The present model, which is called the 'histidine model' here, is based upon the KG model [19] for water oxidation in photosynthesis but utilizes the coordination and redox chemistry of histidine to explain the established facts relating to electron transfer from H_2O to Z. The microscopic mechanism of electron transfer and water oxidation in photosynthesis, consistent with Kok's four photon scheme, is shown in Figure 2. Only the two 'intrinsic' manganese ions are thought to be involved in water oxidation and they are shown as being bound into the 'intrinsic' 34kD protein. The two 'extrinsic' manganese ions are located nearby. These two ions are believed to be Mn(III) ions (each with spin $J = 2$) which are *not* involved in the redox processes and therefore do not change their oxidation states.

We shall first examine the reactions of the WOC as the latter passes through Kok's S_n states. Absorption of a photon leads to the oxidation of the reaction center P680 to P680$^+$; this is followed by electron flow from Z to P680$^+$ producing Z$^+$. During the transition of S_0 to S_1, Z$^+$ is reduced by an electron from one of the two histidine-Mn(III) pairs, and a proton, originating in the water molecule bound to this manganese ion, is released. The S_1 state is suggested to exist in two different forms which are in equilibrium. In one form (S_{1B}) the oxidized histidine-manganese pair is described as His$^+$-Mn(III), whereas in the other it is described as His-Mn(IV). Such an equilibrium involving the shuttling of an electron between a ligand and a manganese ion is well established [22] for manganese complexes with o-quinone-derived ligands. We suggest that the interconversion of S_{1A} and S_{1B} may be triggered by a change in the protein conformation. In fact, it is possible that the imidazole moiety is coordinated to the Mn(IV) ion in S_{1A} and that the conformational change moves the imidazole moiety such that His$^+$ is not coordinated to the Mn(III) ion in S_{1B}.

During the transition of S_1 to S_2, a second electron is removed from the *same* histidine-manganese pair as in the S_0 to S_1 transition. This produces a His$^+$-Mn(IV) pair.

During the S_2 to S_3 transition, an electron transfers to Z$^+$ from the second histidine-manganese pair, for it is not possible to oxidize further the first pair beyond His$^+$-Mn(IV). The S_{3B} state is produced and, in analogy with the $S_{1A} \rightleftharpoons S_{1B}$ equilibrium, there is an equilibrium between the S_{3B} and S_{3A} states. Furthermore, a proton originating in the H_2O bound to the manganese ion at the second histidine-manganese pair is released. Now both intrinsic manganese ions have a OH$^-$ ligand.

During the S_3 to S_4 transition, the second histidine-manganese pair is fully oxidized to His$^+$-Mn(IV). Although it is obviously difficult to be sure of the order of events, the net result is to form an O–O bond between the two coordinated 'OH$^-$' ligands. The resulting '$O_2H_2^{2-}$' moiety is then deprotonated, coupled with electrons flowing back to the two His$^+$-Mn(IV) pairs. With the elimination of one O_2 molecule and two protons, two His-Mn(III) pairs are regenerated. Two H_2O molecules are inserted, with each His-Mn(III) having one H_2O molecule coordinated to it.

Explanation of EPR signals

Finally, it is appropriate to comment on what the EPR signals would be expected to look like for the S states in the present model. It is known that

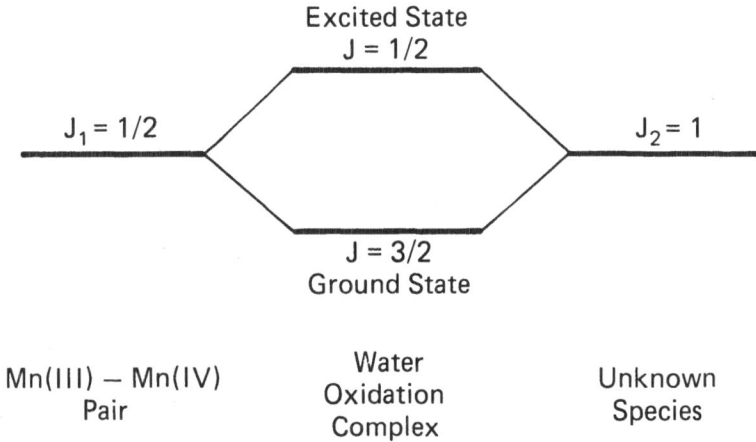

Figure 3. Energy level diagram for the low lying electronic states (J = 3/2 and 1/2) of the Mn(III)-Mn(IV)-(unknown species) complex which corresponds to the water oxidation complex, WOC (center). J_1 is an electronic spin of the antiferromagnetically coupled Mn(III)-Mn(IV) pair, suggested to be the intrinsic Mn pair, and J_2 is an electronic spin of the unknown species, suggested to be the extrinsic Mn(III)-Mn(III) pair.

the manganese complexes in the WOC do not give an EPR signals *in situ* at physiological temperatures [3]. A low-temperature EPR signal for the S_2 state of the WOC has been reported [11, 15], and, on the basis of the magnitude of observed hyperfine structure, the signal was attributed to a binuclear Mn(III)Mn(IV) site [11, 15] or a tetranuclear Mn_3(III)Mn(IV) site [10]. A caveat has been raised [23] about deciding which oxidation states of manganese are present if only the magnitudes of manganese hyperfine interactions are known from EPR simulations. In addition, de Paula and Brudvig [9] reported the continuous power saturation and temperature dependence of three EPR signals (*vide infra*) that are generated by low-temperature illumination of dark-adapted PSII membranes to give the S_2 state. de Paula and Brudvig [9] have concluded that the simplest model which could accommodate the observation of non-Curie type temperature dependence and the power saturation characteristics they observe would be a Mn(III) (electronic spin J = 2)-Mn(IV)-(J = 3/2) pair involved in a relatively strong *antiferromagnetic* exchange interaction. The interaction between Mn(III) and Mn(IV) would give a resultant J_1 = 1/2 for the pair (see Figure 3, left). If this Mn(III)-Mn(IV) pair would interact *ferromagnetically* with an unidentified species that has a J_2 = 1 state (Figure 3, right), we would obtain two low-energy electronic states of the Mn(III)-Mn(IV)-(unknown species) complex which corresponds to the WOC. One state, the ground state, is characterized by J = 3/2, which simply means that it has three unpaired electrons. The other is an excited electronic state with J = 1/2, which means it has one unpaired electron. If the energy separation between the J = 3/2 and $J = 1/2$ states of this Mn(III)-Mn(IV)-(unknown species) complex is comparable to thermal energies, then the $J = 1/2$ excited state will be populated, according to Boltzman distribution. It is believed [9] that the non-Curie law type EPR signal arises from complexes that are in this $J = 1/2$ state.

Even though the EPR signal observed for the S_2 state is clearly very useful and, in fact, is one of the few ways to directly monitor the electronic structure of the polymanganese WOC, it must be emphasized that very small interaction energies dramatically affect the appearance of an EPR signal [2]. A magnetic exchange interaction between two paramagnetic centers is always propagated by an orbital pathway; however, the distance between the two paramagnetic centers can be quite large, i.e., greater than 10–15Å [16].

The combination of manganese ions and His$^+$ indicated for the S_2 state in Figure 2 could lead to the type of EPR signals that have been reported seemingly only under one condition. If the His$^+$ radical was involved in a very weak magnetic exchange interaction with the Mn(IV) ion, then such an EPR signal could result. A diminishingly weak or non-existent exchange interaction could be the result of the His$^+$ being at an appreciable distance from the Mn(IV) ion, or could result from a poor orbital pathway for such an interaction. In this circumstance the two 'intrinsic' manganese ions could be the Mn(III)-Mn(IV) pair that is experiencing a relatively strong antiferromagnetic interaction to give a $J_1 = 1/2$ ground state. The two 'extrinsic' manganese ions form a Mn(III)-Mn(III) pair, which if they are also involved in a somewhat weaker antiferromagnetic interaction, would have a $J = 0$ ground state with a thermally populated (at liquid-helium temperatures) $J_2 = 1$ excited state. The weakly coupled extrinsic Mn(III)-Mn(III) pair corresponds to the unknown species in the model of de Paula and Brudvig [9]. A ferromagnetic interaction between the $J_2 = 1$ excited state of the 'extrinsic' Mn(III)-Mn(III) pair and the $J_1 = 1/2$ ground state of the Mn(III)-Mn(IV) pair would give a $J = 3/2$ ground and $J = 1/2$ excited states for the tetranuclear array as shown in Figure 3. This ferromagnetic interaction between the 'intrinsic' Mn(III)-Mn(IV) pair and the 'extrinsic' Mn(III)-Mn(III) pair could be propagated by the hydrogen bonds pictured in Figure 2. Weak magnetic exchange interactions have been reported [21] to exist between two manganese ions that are connected by hydrogen-bonding contacts.

Objections to the formulation of S_2 state in Figure 2 could be raised by noting that no EPR signal [8] or electronic absorption band [40] corresponding to His$^+$ have been reported for the S_2 state of the WOC. An organic radical such as His$^+$ might be expected to give a relatively sharp EPR signal close to $g = 2.0$. However, if His$^+$ is near to a paramagnetic transition metal center, then the EPR signal for His$^+$ can be distorted even to the point that it is quite difficult to detect. An example of this type of effect can be seen in the work [1, 36, 37] on horseradish peroxidase compound I (HRP I). HRP I is two oxidizing equivalents above the native Fe(III) state of this heme-protein. It has been characterized as having Fe(IV) ion bonded to a porphyrin π-cation radical. The early EPR studies [1] of HRP I revealed only minute fractions of an unpaired electron for the porphyrin radical. It has been shown that only if either dispersion-derivative rapid-passage techniques [36] or rapid adiabatic passage conditions [37] are employed to record the EPR can the value of one unpaired-electron per porphyrin be obtained. The influence of the Fe(IV) ion on the porphyrin radical results from a combination of through-space dipole–dipole interaction of the two unpaired-electron centers and a *weak* magnetic exchange interaction between the Fe(IV) ion and the porphyrin radical [37]. Furthermore, in order to simulate the line shape of the porphyrin radical EPR signal, it was necessary to suggest that there is a distribution of magnitude of magnetic exchange interaction present [37].

Thus, in this new model, it is not unreasonable to suggest that a EPR signal for His^+ cannot be easily seen for the S_2 state. In this new model, the EPR signal observed [9, 10, 11, 15] for the S_2 state could be explained by proposing that His^+ is only involved in a *very weak* magnetic exchange interaction with the intrinsic Mn(III)-Mn(IV) pair. This Mn(III)-Mn(IV) pair experiences a relatively strong antiferromagnetic exchange interaction to give a $S = 1/2$ state which is ferromagnetically coupled to the $S = 1$ state of the *extrinsic* Mn(III)-Mn(III) pair. If there is a distribution in the magnitude of the weak magnetic exchange interaction between His^+ and the intrinsic Mn(III)-Mn(IV) pair, then it would take particular care to see the EPR signal for His^+. The distribution in magnitude of exchange interaction could reflect a distribution in relative orientation of the His^+ imidazole moiety relative to the Mn(III)-Mn(IV) pair.

There does not seem to be adequate data available at this time to say why it is apparently not possible to identify an electronic absorption band assignable to His^+ [40, 41]. The present authors are unaware of a study in which the optical band for His^+ as a part of a protein structure has been assigned. Thus, with all the bands seen in the *difference* electronic absorption spectrum for the S_2 state, it is difficult to know whether or not a band for His^+ is seen. In the end, we emphasize that what we have presented here is a reasonable *working hypothesis* that remains to be tested.

Acknowledgements

SP acknowledges the award of a Fullbright fellowship under the CIES-UGC program. DNH is grateful for funding from the National Institutes of Health (grant HL 13652). G thanks the National Science Foundation (PCM 83-06061).

References

1. Aasa, R, Vanngård T and Dunford HB (1975) Biochim Biophys Acta 391:259–264

2. Abragam A and Bleaney B (1970) Electron Paramagnetic Resonance of Transition Ions. Oxford: Clarendon Press
3. Amesz J (1983) Biochim Biophys Acta 726:1–12
4. Boska M and Sauer K (1984) Biochim Biophys Acta 765:84–87
5. Bouges-Bocquet B (1980) Biochim Biophys Acta 594:85–103
6. Brabec V and Mornstein V (1980) Biophys Chem 12:159–165
7. Bricker TM, Metz JG, Miles D and Sherman LA (1983) Biochim Biophys Acta 724: 447–455
8. Brudvig G, personal communication
9. de Paula JC and Brudvig GW (1985) J Am Chem Soc 107:2643–2648
10. Dismukes GC, Ferris K and Watnick P (1982) Photobiochem Photobiophys 3:243–256
11. Dismukes GC and Siderer Y (1981) Proc Natl Acad Sci USA 78:274–278
12. Eaton DR and Wilson KM (1979) J Inorg Biochem 10:195–203
13. Govindjee, Kambara T and Coleman W (1985) Photochem Photobiol 42:187–210
14. Gregely A and Kiss T (1970) In Sigel H ed. Metal Ions in Biological System Vol. 9, pp 143–172. New York and Basel: Marcel Dekker Inc
15. Hansson Ö and Andreasson LE (1982) Biochim Biophys Acta 679:261–268
16. Hendrickson DN (1985) In Wilett RD, Gatteschi D, Kahn O, eds. Magneto Structural Correlations in Exchange Coupled Systems, pp 523–554. Dordrecht: Reidel Publishing Co
17. Isied SS (1984) Prog Inorg Chem 32:443–517
18. Kambara T and Govindjee (1985) Biophys J 47:419a
19. Kambara T and Govindjee (1985) Proc Natl Acad Sci USA 82:6119–6123
20. Kuwabara T and Murata N (1979) Biochim Biophys Acta 581:228–236
21. Laskowski EJ and Hendrickson DN (1978) Inorg Chem 17:457–470
22. Lynch MW, Hendrickson DN, Fitzgerald BJ and Pierpont CG (1984) J Am Chem Soc 106:2041–2049
23. Mabad B, Tuchagues J-P, Hwang YT and Hendrickson DN (1985) J Am Chem Soc 107:2801–2802
24. Martin RB (1979) In Sigel H ed. Metal Ions in Biological System vol. 9, pp 1–39, New York and Basel: Marcel Dekker
25. Metz JG and Seibert M (1984) Plant Physiol 76:829–832
26. Moore GR and Williams RJF (1976) Coord Chem Rev 18:125–197
27. Murata N, Miyao M, Omata T, Matsunami H and Kuwabara T (1984) Biochim Biophys Acta 765:363–369
28. O'Malley PJ and Babcock GT (1984) Biochim Biophys Acta 765:370–379
29. Pierpont CG and Buchanan RM (1981) Coord Chem Rev 38:45–87
30. Rasmussen OF, Bookjans G, Stummann BM and Henningsen KW (1984) Plant Molec Biol 3:191–199
31. Renger G (1978) In Metzner H ed. Photosynthetic Oxygen Evolution. pp 229–248. London, New York and San Francisco: Academic Press
32. Rich PR (1982) Faraday Discuss Chem Soc 74:349–364
33. Samuni A and Neta P (1973) J Phys Chem 77:1629–1635
34. Satoh K, Nakatani HY, Steinback KE and Arntzen CJ (1983) Biochim Biophys Acta 724:142–152
35. Schmidt J and Borg DC (1976) Radiat Res 65:220–237
36. Schulz CE, Chiang R and Debrunner PG (1979) J Phys Colloq (Orsay, France) 40: C2-534–C2-536
37. Schulz CE, Rutter R, Sage JT, Debrunner PG and Hager LP (1984) Biochem 23: 4743–4754
38. Seela JL, Huffman JC and Chrisou G (1985) J Chem Soc Chem Commun 58–60
39. Tabata K, Itoh S, Yamamoto Y, Okayama S and Nishimura M (1985) Plant and Cell Physiol 26:855–864
40. van Gorkom HJ, personal communication
41. Yamamoto Y, Tabata K, Isogai Y, Nishimura M, Okayama S, Matsuura K and Itoh S (1984) Biochim Biophys Acta 767:493–500
42. Yuasa M, Ono T and Inoue Y (1984) Photobiochem Photobiophys 7:257–266

Photosynthesis Research 9, 113–124 (1986)
© *1986 Martinus Nijhoff/Dr. W. Junk Publishers, Dordrecht.*

Mechanism of proton-pumping in the cytochrome b/f complex

PIERRE JOLIOT and ANNE JOLIOT

Institut de Biologie Physico-Chimique, 13, rue Pierre et Marie Curie. 75005 Paris, France

(*Received 1 August 1985*)

Key words: cytochrome b, cytochrome b/f complex, electron transfer, plastoquinone, Proton-pumping

Abstract. Several models have been proposed to interpret the mechanism of proton-pumping associated with the electron transfer reactions in the cytochrome b/f complex. Energetics considerations suggest that the proton pump is coupled to the oxidation of cytochrome b by plastoquinone. Experiments performed in living cells under anaerobic conditions suggest that proton-pumping can occur through two independent mechanisms. When the two b cytochromes are reduced prior to a flash illumination i.e. after a long dark anaerobic incubation (> 10 minutes), proton-pumping is very likely associated with the reduction of a semiquinone by cyt b which occurs at a site close to the inner face of the membrane. The electrogenic phase is associated with the transfer of protons *via* a transmembrane channel. This process is not inhibited by 2-n-nonyl-4-hydroxyquinoline *N*-oxide (NQNO). Under repetitive-flash or under aerobic conditions, proton-pumping occurs according to a modified Q-cycle mechanism, which is inhibited by NQNO.

Introduction

Cytochrome (cyt) b/f and cyt b/c complexes are known to catalyze the oxidation of a lipid-soluble quinone (plastoquinone or ubiquinone) and the reduction of a water-soluble electron carrier (plastocyanine or cyt c). Electron transfer within these complexes is associated with a proton-pumping process, the mechanism of which is a matter of controversy. We will essentially discuss here the structural and functional properties of the cyt b/f complex, which very likely are common to the class of b/f and b/c complexes.

Cyt b/f complex is formed of 4 or 5 polypeptides [3, 10]. Three of them are associated with the two cyt b, the FeS center and cyt f, respectively. The FeS center and cyt f are very likely located close to the inner face of the membrane, as no electrogenic phase is associated with the electron transfer to these carriers [16]. The position of the two cyt b in the membrane has been determined on the basis of the amino-acid sequences of the corresponding polypeptides [24, 29]. The two heams are located in a position rather symmetric with respect to the two faces of the membrane and the distance between them is about 20Å. Less information is available on the number and the location of the site(s) able to fix plastoquinone. Photoaffinity labelling

---

Dedicated to Prof. L.N.M. Duysens on the occasion of his retirement

suggests that the site where plastoquinone is oxidized is located on the cyt b polypeptide at a short distance from the FeS polypeptide [21, 22].

Proton-pumping associated with the electron transfer from plastoquinone to Photosystem (PS) can be followed by measuring light-induced pH changes occurring on both faces of the membrane using a glass electrode [7] or indicator dyes [26, 28]. Fowler and Kok [7] suggested that more than 1 proton were pumped through the membrane for each electron transferred from PSII acceptors to PSI donors. This result was later confirmed by Velthuys [28]. On the contrary, under highly oxidizing conditions in the presence of ferricyanide, Hope and Matthews [9] did not observe any proton uptake on the outer face of the membrane which could be associated with the intersystem electron transport. Under similar experimental conditions, we observed [14] that the electron transfer reactions which are generally associated with proton uptake (cyt b oxidation) are slow ($t_{1/2} \sim 1s$). We cannot exclude that in oxidizing conditions the 2 electrons of plastoquinone are sequentially transferred to the FeS protein without involvement of a proton-pumping mechanism. A proton-pumping process can also be characterized by the increase of the membrane potential with which it is obligatorily associated. In living cells, a flash induces a fast increase of the membrane potential (phase a, $t_{1/2} < 100\,ns$), which corresponds to the transmembrane photochemical charge separation, followed by a slow increase (phase b, $t_{1/2}$, few ms) during which one extra electron is transferred for each positive charge formed by the PSI reaction [13]. The same phenomenon has been observed by Velthuys [26] in isolated chloroplasts when a fraction of the plastoquinone pool is reduced prior to the illumination.

Energetics of the electron transfer reactions

The coupling of a proton pump with an electron transfer reaction requires that enough redox energy is available at this step to allow proton-pumping against a transmembrane electrochemical potential (100 mV to 200 mV). The knowledge of the energetics of the electron transfer reactions in the b/f complex could give us some insight on the possible steps where a proton pump could be coupled (scheme 1).

A characteristic feature of the cyt b/f complex is to include 2 electron transfer chains, the redox potentials of which widely differ. There is a fair agreement in the literature on the values of the potentials of FeS and cyt f, i.e. 280 mV and 340 mV respectively. On the contrary, some controversy is raised concerning the value of the redox potential of cyt b: in the isolated cyt b/f complex. Hurt and Hauska [11] and Clark and Hind [4] distinguish two forms of cyt b, a high potential form cyt b_h around −30 mV and a low potential form cyt b_l around −140 mV. In chloroplasts, Girvin and Cramer [8] find a common value for both cyt b, around −30 mV. The redox potentials in solution at pH 7 of the couples PQH_2/PQ^- and PQ^-/PQ are

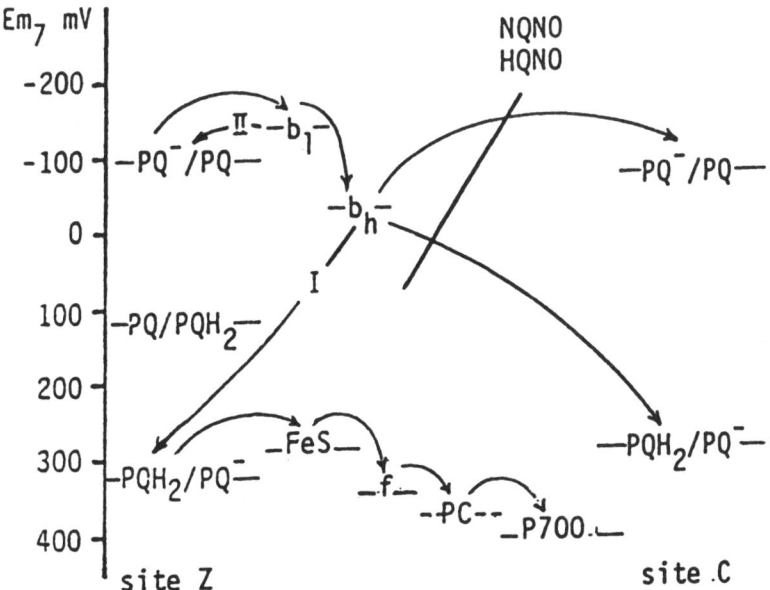

Scheme 1

380 mV and −160 mV respectively [23]; when the plastoquinone molecules are attached to the cyt b/f complex, the redox potentials of these two couples will also depend upon the relative affinity of the oxidized, semiquinone and reduced forms of plastoquinone for the binding site. If the affinity of PQ^- is higher than the affinity of PQ and PQH_2, the redox potential of the PQH_2/PQ^- couple is shifted toward lower values and the one of the PQ^-/PQ couple toward higher values. In scheme 1, in order to match the potentials of the two quinone couples with the redox potentials of the high and low potential chains respectively, we arbitrarily suppose that the affinity of PQ^- for the quinol-oxidase site Z is about 10 times higher than the affinity of PQ and PHQ_2. These energetic considerations allow us to understand the process of oxidation of a plastoquinol molecule attached to the cyt b/f complex. A first electron is transferred to the high potential chain while the second one is transferred to the low potential one. This concerted process has been first demonstrated in vitro on the isolated mitochondrial b/c complex by Baum et al. [1] and by Velthuys [27] in the cyt b/f complex contained in functional chloroplast membranes.

On the basis of scheme 1, we conclude that little energy is available in the electron transfer reactions occurring between PQH_2 and cyt f *via* the high potential electron transfer chain. A proton pump coupled to this fraction of the electron transfer chain could only work against a low electrochemical gradient of protons. In order to increase the available energy on the high potential chain, one can assume a higher relative affinity of PQ^- for its binding site, but in this case the reduction of cyt b by the PQ^-/PQ couple becomes thermodynamically less favorable.

The more probable hypothesis remains that the proton pump is coupled to the oxidation of cyt b by plastoquinone. The reduction of PQ in PQH_2 by two cyt b corresponds to a redox potential drop of about 200 mV, which would permit to pump protons against an equivalent transmembrane electrochemical potential. It is generally assumed that reduction of plastoquinone by cyt b occurs at a site C different from the quinol-oxidase site Z; the redox potentials of the couples PQH_2/PQ^- and PQ^-/PQ then can differ from one site to the other. Experiments we performed in isolated chloroplasts in the presence of ferricyanide give some information on the relative redox potentials of cyt b and of the couple PQ^-/PQ at site C [14]. When a group of two flashes is given to dark-adapted material, the plastoquinol molecule formed by PSII reduces through the concerted process one molecule of cyt b ($t_{1/2} \sim$ 10 ms). Then, the re-oxidation of cyt b is a very slow process ($t_{1/2} \sim 1$ s), which shows that the electron of cyt b cannot be transferred to PQ. Therefore, we have to assume that the redox potential of the couple PQ^-/PQ at site C is significantly lower than the one of cyt b_h (see scheme 1). A faster re-oxidation of cyt b is observed when more than one plastoquinol has been formed by PSII (illumination by 4 flashes) and that 2 electrons could be sequentially transferred at site C to reduce PQ into PQH_2.

In any case, scheme 1 only gives a rough description of the energetics of the electron transfer reactions in the cyt b/f complex. We have to keep in mind that when several electron carriers are included in the same protein complex, the apparent redox potential of one carrier can depend upon the redox states of the other carriers either by direct electrostatic interactions or through conformational changes of the protein. We suggested [15] that anti-cooperative interactions occur between cyt f and cyt b.

Models

Two classes of models have been proposed to explain the mechanism of proton-pumping by the cyt b/f or b/c complexes. A first class of models is derived from the original Q-cycle proposed by Mitchell [20]. The essential features of the modified Q-cycle [5] are the following (scheme 2): (1) The cyt b/f complex includes two sites for plastoquinone fixation; the concerted process of plastoquinol oxidation occurs at a site Z close to the inner face of the membrane while plastoquinone reduction by the two cyt b occurs at a site C close to the outer face. (2) The electrogenic phase is mainly associated with electron transfer from site Z to site C *via* the two cyt b. (3) There is no proton or ion movement between sites C and Z.

In a second class of models, a pumping device involving a transmembrane proton flux is coupled to the redox changes of one of the electron carriers included in the b/f or b/c complex. The basic features of this class of models are the following: (1) The electrogenic phase is mainly associated with proton movement from the outer face to the inner face of the membrane *via* a

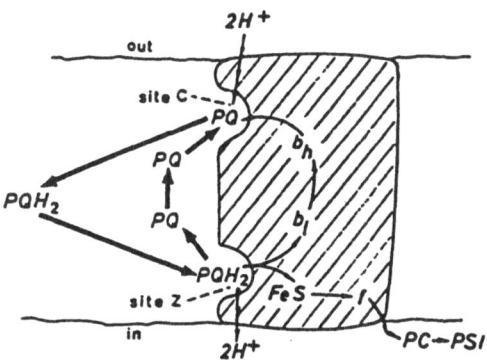

Scheme 2

proton channel. (2) The proton channel includes a gate coupled to the redox state of an electron carrier which orients the flux of protons from the outside to the inside of the thylakoid; Wikström and Krab [30], in the original form of the b-cycle, proposed that the proton pump was coupled to the redox changes of cyt b. In this model there is a single site for plastoquinone fixation. More recently, Girvin and Cramer [8] and we ourselves [16] assumed that the coupling occurs with a carrier of the high potential chain (FeS). An alternate hypothesis which we presently favor, is that a proton-pumping process could be coupled to the redox changes of plastoquinol at site Z. Recently, Wikström and Saraste [31] proposed a model somewhat intermediary between a b-cycle and a Q-cycle, which will be discussed later.

Two classes of inhibitors are known to block the electron transfer and the proton-pumping associated with the cyt b/f or b/c complex. In chloroplasts, 2,5-dibromomethylisopropyl-p-benzoquinone (DBMIB) is known to block both the concerted reduction of FeS and cyt b by plastoquinol and the electrogenic phase [2, 12]. DBMIB very likely competes with plastoquinol at site Z. 2-n-heptyl-4-hydroxyquinoline N-oxide (HQNO) [25] and 2-n-nonyl-4-hydroxyquinoline N-oxide (NQNO) [18] specifically block cyt b oxidation but not its reduction. It is then very likely that these inhibitors compete with plastoquinone at site C where plastoquinone reduction occurs. NQNO is about 10 times more efficient that HQNO and we observed that these inhibitors block with about the same efficiency the secondary quinone acceptor site of PSII. Site C and the binding site of the secondary quinone acceptor of PSII should have some common structural properties. The effect of inhibitors which demonstrates that there are two independent sites for plastoquinone fixation is a strong argument in favor of a Q-cycle model.

Experiments performed in highly reducing conditions are, however, difficult to interpret in terms of a Q-cycle model. Diner and Delosme [6] titrated the slow electrogenic phase in isolated chloroplasts in the presence of anthraquinone-2-sulfonate as mediator and found a midpoint potential of around −250 mV. Girvin and Cramer [8] recently demonstrated that the slow electrogenic phase remains essentially unchanged in amplitude and kinetics for potentials as low as −200 mV; this phase is totally inhibited by DBMIB. In these conditions, Girvin and Cramer did not observe significant photoinduced changes in the redox state of cyt b. Using spinach chloroplasts in the presence of dithionite but without any mediator, we do observe a large electrogenic phase but, contrary to Girvin and Cramer, also a large photo-oxidation of cyt b. This discrepancy can perhaps be explained by the use by Girvin and Cramer of a mediator which could rapidly re-reduce cyt b. We also observed a large flash-induced phase b and oxidation of cyt b in living algae (*Chlorella sorokiniana*) that were dark adapted for several minutes in anaerobic conditions [17]. All these results show that a phase b of normal amplitude is observed when both cyt b and the plastoquinone pool are in their reduced form prior to the flash illumination. Different hypotheses can be proposed to interpret these data: (1) An electrogenic phase and proton-pumping process can be associated with the electron transfer through the high potential chain of the cyt b/f complex. In this case, cyt b turnover is not obligatorily involved in the proton-pumping process (scheme 3A). (2) Oxidation of quinol at site Z by the oxidized FeS center leads to the formation of a semiquinone PQ$^-$ which can move to site C. Then, one electron is transferred from cyt b to PQ$^-$ which is then re-reduced to PQH$_2$ with uptake of 2 protons from the outer phase of the membrane. It is excluded that a semiquinone can be released in the membrane without being immediately destroyed by a dismutation reaction. Recently, Wikström and Saraste [31] proposed a modified version of the b-cycle in which a semiquinone anion is transferred from site Z to site C *via* a specialized structure, termed the 'Q-pocket', in the trans-membrane part of the protein (scheme 3B). (3) The reduction of the semi-quinone by cyt b occurs at site Z. In this case, a proton pump should be coupled to the changes in the redox states of plastoquinone molecules at site Z. The protons involved in the reduction of PQ$^-$ are picked up from the outer phase of the thylakoid *via* a transmembrane proton channel (scheme 3C).

Experimental

Experiments we recently performed on mutant strains of *Chlorella sorokiniana* which lack PSII give some insight on the mechanism of proton-pumping by cyt b/f complex under reducing conditions [17]. Incubation of living algae

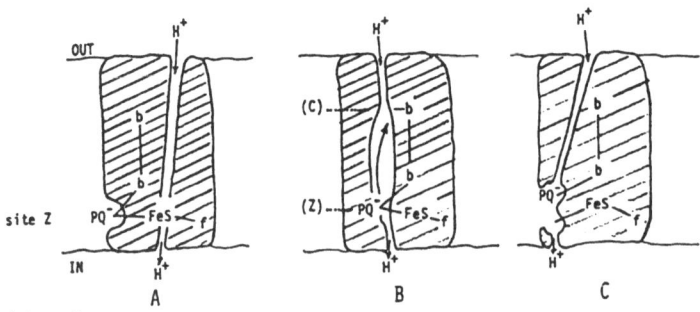

Scheme 3

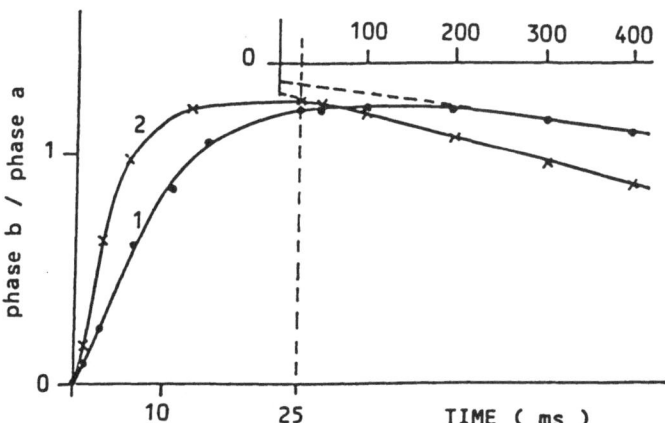

Figure 1. Phase *b* induced by a non-saturating flash (hitting 30% of the PSI centers). S8 mutant strain of *Chlorella sorokiniana* in anaerobic conditions. Curve 1, single-flash illumination after 20 minutes dark-adaptation. Curve 2, repetitive flashes 6 s apart. The amplitude of phase *a* was identical for both curves

for more than 20 minutes in the dark in anaerobiosis reduces a major fraction of both cyt b. Illumination by a strong continuous light oxidizes more than 1 cyt b molecule per cyt b/f complex [16]. In a subsequent dark period, the re-reduction of cyt b is biphasic; a fast phase is completed in less than 15 s while the second phase is completed in more than 10 minutes. The reduced *minus* oxidized spectrum of cyt b differs slightly for the fast and the slow phases of reduction; these results suggest that there are 2 forms of cyt b, a high potential form cyt b_h rapidly reduced and a low potential form cyt b_l slowly reduced under anaerobic conditions. The following experiments have been formed using non-saturating flashes. In these conditions, the cyt b/f complex never receives more than one positive charge among the high potential carriers (FeS and cyt f), which simplifies the interpretation of the kinetics of the electron transfer reactions. In a first series of experiments, algae dark adapted for more than 10 minutes are illuminated by a single flash, i.e. when a major fraction of the two cyt b is reduced. A phase *b* of large amplitude is observed (up to 1.5 charges transferred through the membrane). This phase *b* is slower and generally larger than the one observed in repetitive-flash illumination or under more oxidizing conditions (Figure 1). A large oxidation of cyt b occurs in the same time range (0–100 ms). Surprisingly, a saturating concentration of NQNO does not inhibit but even slightly accelerates cyt b oxidation (Figure 2); in the same experimental conditions, the slow electrogenic phase is not inhibited by NQNO. It is important to stress that a better correlation is observed between the kinetics of cyt b to oxidation and of the slow electrogenic phase when using non-saturating flashes rather than saturating ones, with which more than one positive charge could be transferred to a cyt b/f complex [17].

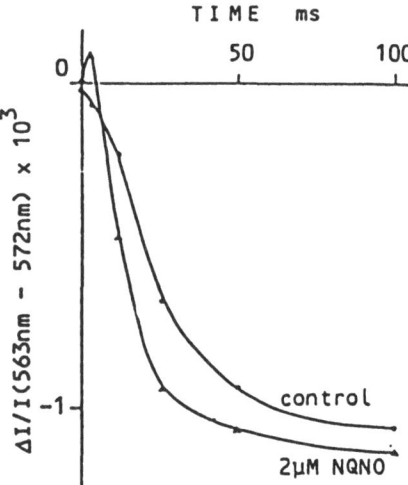

Figure 2. Time-course of cyt b redox changes induced by a single-flash illumination given to S52 mutant strain of *Chlorella sorokiniana* after 30 minutes dark-adaption to anaerobic conditions. Non-saturating flash hitting 45% of PSI centers

Among the three hypotheses proposed above to explain the proton-pumping mechanism in reducing conditions, only hypothesis (3) is consistent with our data. The fact that the electrogenic phase correlates with cyt b oxidation argues against hypothesis (1). We observed that in reducing conditions, NQNO induces a small red shift of the α-band of cyt b. Such a shift has already been observed in isolated b/c complex [19]; one can then conclude that NQNO is actually bound to the cyt b/f complex in dark-adapted material. Therefore, hypothesis (2) is ruled out as proton-pumping can occur when site C is occupied by the inhibitor. Interestingly enough, our data in repetitive-flash illumination are better interpreted by a classic Q-cycle process. We thus favor a model in which two different mechanisms of proton-pumping are involved depending upon the experimental conditions.

After dark-adaption (scheme 4), the 2 cyt b are in their reduced form (b_l b_h state). Removal of the positive charge located among the high potential carriers leads to the formation of a semiquinone form, associated with the release of two protons on the inner face of the membrane. Then cyt b_l transfers an electron to the semiquinone at site Z which is re-reduced to plastoquinol. The essential feature of this model is that we assume that the two protons involved in this reaction are picked up from the outer face of the membrane. This model implies a gate mechanism to prevent an uptake of protons from the inner face of the membrane. In scheme 3C, we suggest that site Z is divided into 2 subsites Z1 and Z2. At site Z1, plastoquinone exchanges electrons with the FeS center and protons *via* a channel connected

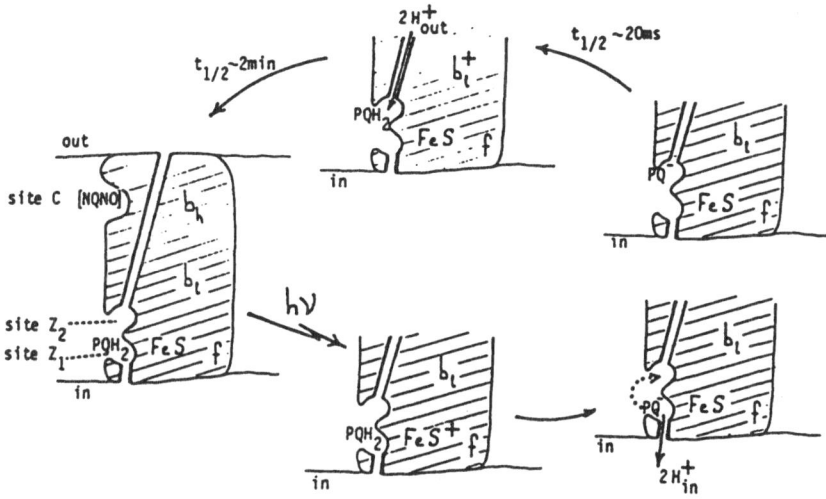

Scheme 4

with the internal aqueous phase of the thylakoid. At site Z2, plastoquinone exchanges electrons with cyt b_1 and protons *via* a channel connected with the external aqueous phase. In this model, the gating mechanism is associated with a small displacement of plastosemiquinone from site Z1 to site Z2.

A different behavior is observed when algae are submitted to a series of non-saturating flashes 8.6 s apart (Figure 3). Each flash induces a large slow electrogenic phase ($\sim$1.2 electrons transferred through the membrane per positive charge formed by PSI) and only small absorption changes in the α-band of cyt b. It is very likely that in these conditions reduction and oxidation of cyt b occur in the same time-range and widely overlap. Contrary to the single-flash experiment, addition of increasing concentrations of NQNO progressively slows down the rate of cyt b oxidation without effect on the reduction phase. In the presence of non-saturating concentrations of NQNO, phase *b* becomes clearly biphasic; a first phase, completed in about 15 ms, correlates with the reduction of cyt b; the second phase which lasts for several hundred ms correlates with the oxidation of cyt b. For saturating concentrations of NQNO, the first phase is unaffected and corresponds to the transfer of 0.4 charges across the membrane per positive charge formed by PSI. The second phase is too slow to be accurately deconvoluted from the membrane potential decay.

These experiments can be easily interpreted in terms of a modified Q-cycle mechanism (scheme 5): When a positive charge is transferred from PSI to the cyt b/f complex, a semiquinone is formed at site Z whatever is the state of the cyt b/f complex (reaction 1); then, we assume that under repetitive-flash illumination, about half of the complexes are in the $(b_l^+ \, b_h)$ state and the other half in the state $(b_l^+ \, b_h^+)$. For the complexes in the $(b_l^+ \, b_h)$

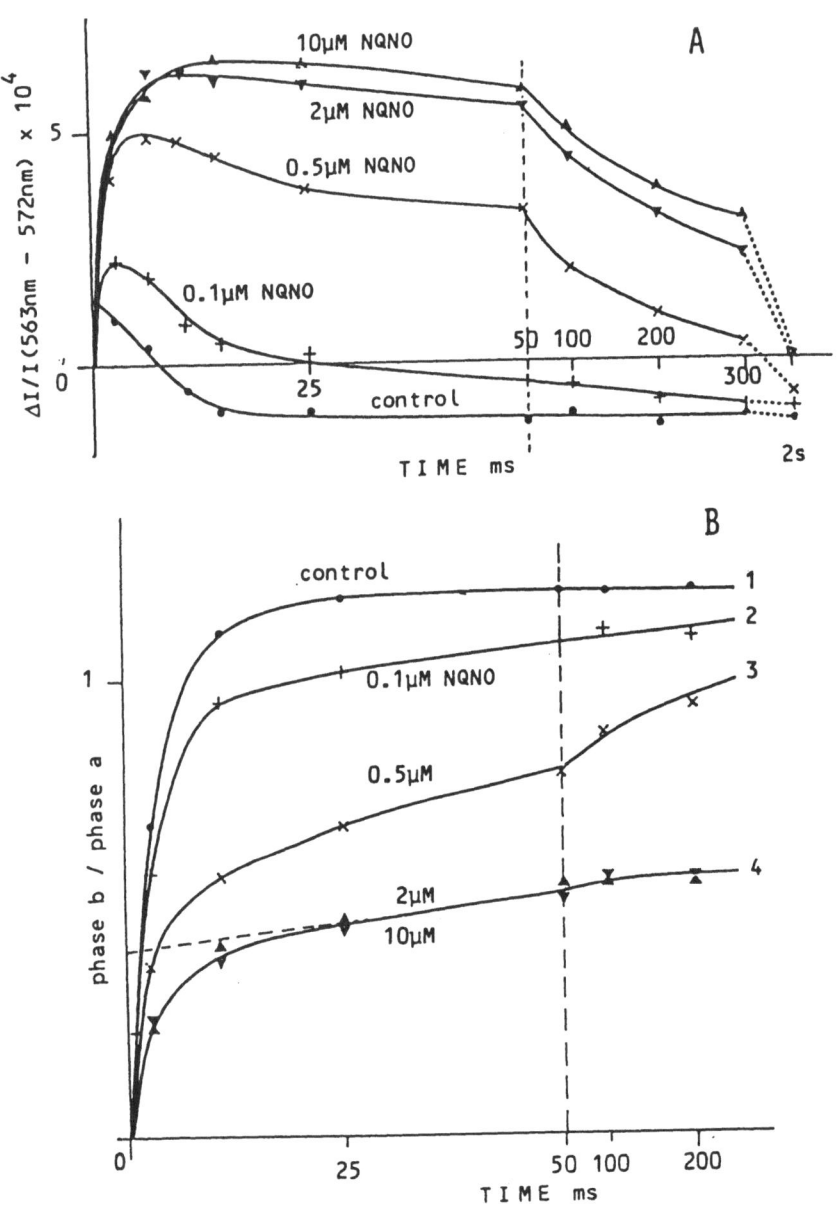

Figure 3. Time-course of the flash-induced redox changes of cyt b (Figure 3A) and of phase *b* (Figure 3B) in the presence of various concentrations of NQNO. S8 mutant strain. Repetitive non-saturating flashes 8.6 s apart hitting 39% of PSI centers. Phase *b* has been deconvoluted from the decay of the membrane potential

state, a flash induces the formation of PQ at site Z and of the (b₁ bₕ) state (reaction 2A). Then, PQ – which is an uncharged species – jumps from the Z to the C site and a conventional Q-cycle is operating (reaction 2B). This process is obviously blocked when site C is occupied by NQNO. For the

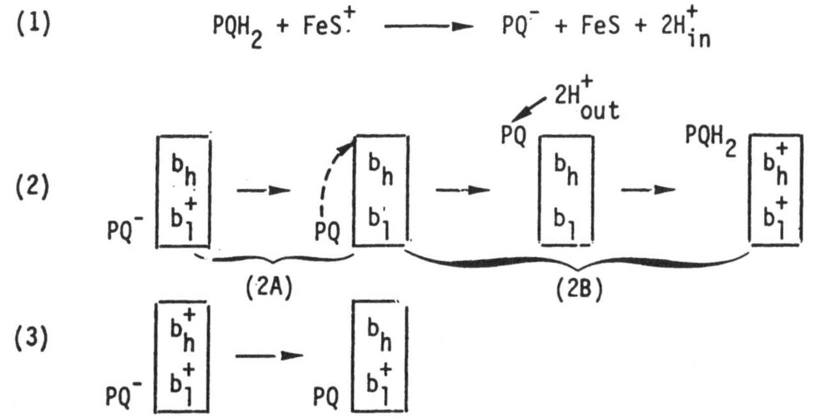

(1) $\qquad PQH_2 + FeS^{+} \longrightarrow PQ^{-} + FeS + 2H^{+}_{in}$

(2)

$$PQ^{-}\begin{bmatrix}b_h\\b_l^{+}\end{bmatrix} \longrightarrow PQ\begin{bmatrix}b_h\\b_l'\end{bmatrix} \xrightarrow[\;2H^{+}_{out}\;]{PQ} \begin{bmatrix}b_h\\b_l\end{bmatrix} \xrightarrow{PQH_2} \begin{bmatrix}b_h^{+}\\b_l^{+}\end{bmatrix}$$

$$\underbrace{\qquad\qquad}_{(2A)} \qquad \underbrace{\qquad\qquad\qquad\qquad}_{(2B)}$$

(3)

$$PQ^{-}\begin{bmatrix}b_h^{+}\\b_l^{+}\end{bmatrix} \longrightarrow PQ\begin{bmatrix}b_h\\b_l^{+}\end{bmatrix}$$

Scheme 5

complexes in state ($b_l^{+}\, b_h^{+}$), the flash induces the formation of the ($b_l^{+}\, b_h$) state and of PQ (reaction 3) which is rapidly replaced by PQH_2 molecules from the pool. Our model correctly predicts that the part of phase b associated with cyt b reduction is NQNO-insensitive. For half of the complexes, a flash induces the reduction of cyt b_h^{+} located close to the outer face of the membrane and for the other half the reduction of cyt b_l^{+} located close to the inner face. In the presence of NQNO, the electrogenic phase corresponds on the average to the transfer of 1 electron within the half of the membrane thickness, i.e. 0.5 charges transferred through the membrane as compared with the measured value of 0.4. Our model also explains why a slow oxidation of cyt b is still observed in repetitive-flash illumination in the presence of a saturating concentration of NQNO. When site C is occupied by the inhibitor, the oxidation of the two cyt b could occur at site Z. This reaction (step II, scheme 1) is thermodynamically less favorable and slower than the transfer of electrons to semiquinone (step I, scheme 1) which occurs in a single-flash experiment.

An open question is the physiological use of the proton-pumping mechanism associated with the redox changes of plastoquinol at site Z. A possible hypothesis would be that this process is involved in cyclic electron transfer around PSI.

References

1. Baum H, Rieske JS, Silman HI and Lipton SH (1967) Proc Natl Acad Sci USA 57: 798–805
2. Bouges-Bocquet B (1977) Biochim Biophys Acta 462:371–379
3. Clark RC and Hind G (1983) J. Biol Chem 258:10348–10354
4. Clark RC and Hind G (1983) Proc Natl Acad Sci USA 80:6249–6253
5. Crofts AR, Meinhardt SW, Jones KR and Snozzi M (1983) Biochim Biophys Acta 723:202–218
6. Diner BA and Delosme R (1983) Biochim Biophys Acta 722:443–451

7. Fowler CF and Kok B (1976) Biochim Biophys Acta 423:510–533
8. Girvin ME and Cramer WA (1984) Biochim Biophys Acta 767:29–38
9. Hope AB and Matthews DB (1983) Aust J Plant Physiol 10:363–372
10. Hurt EC and Hauska G (1981) Eur J Bioch 591–599
11. Hurt EC and Hauska G (1983) FEBS Lett 153:415–419
12. Isawa S and Pan RL (1978) Biochem Biophys Res Commun 83:1171–1177
13. Joliot P and Delosme R (1974) Biochim Biophys Acta 357:267–284
14. Joliot P and Joliot A (1984) Biochim Biophys Acta 765:210–218
15. Joliot P and Joliot A (1984) Biochim Biophys Acta 765:219–226
16. Joliot P and Joliot A (1985) Biochim Biophys Acta 806:398–409
17. Joliot P and Joliot A Submitted
18. Jones RW and Whitmarsh J (1985) Photobiochem Photobiophys 9:119–127
19. Kamensky Y, Konstantinov AA, Kunz WS and Surkov S (1985) FEBS Lett 181: 95–99
20. Mitchell P (1976) J Theor Biol 62:327–367
21. Oettmeier W, Masson K and Olshewski E (1983) FEBS Lett 155:241–244
22. Oettmeier W, Masson K, Soll HJ, Hurt EC and Hauska G (1982) FEBS Lett 144: 313–317
23. Rich P and Bendall DS (1980) Biochim Biophys Acta 592:506–518
24. Saraste M (1984) FEBS Lett 166:367–372
25. Selak MA and Whitmarsh J (1982) FEBS Lett 150:286–292
26. Velthuys BR (1978) Proc Natl Acad Sci USA 75:6031–6034
27. Velthuys BR (1979) Proc Natl Acad Sci USA 76:2765–2769
28. Velthuys BR (1980) FEBS Lett 115:167–170
29. Widger WR, Cramer WA, Hermann RG and Trebst A (1984) Proc Natl Acad Sci USA 81:674–678
30. Wikström M and Krab K (1980) Curr Top Bioenerg 10:51–101
31. Wikström M and Saraste M (1984) in Bioenergetics (Ernster L. ed) Elsevier Science Publishing BV pp 49–94

Photosynthesis Research 9, 125–134 (1986)
© *1986 Martinus Nijhoff/Dr. W. Junk Publishers, Dordrecht.*

ESR spectroscopy demonstrates that cytochrome b_{559} remains low potential in Ca^{2+}-reactivated, salt-washed PSII particles

DEMETRIOS F. GHANOTAKIS[1], CHARLES F. YOCUM[1] and
GERALD T. BABCOCK[2]

[1] Division of Biological Sciences, The University of Michigan, Ann Arbor,
MI 48109–1048, USA
[2] Department of Chemistry, Michigan State University, East Lansing,
MI 48824–1322, USA

(*Received 26 August 1985*)
Key words: photosystem II, cytochrome b_{559}, polypeptide, calcium

Abstract. Cytochrome b_{559} in various Photosystem II preparations was studied by using low temperature ESR spectroscopy. This technique was used because it is able to distinguish high from low potential forms of the cytochrome owing to the g-value differences between these species. Moreover, by using low temperature irradiation to oxidize cyt b_{559} we have avoided the use of redox mediators. Previous work (Ghanotakis DF., Topper J.N. and Yocum, C.F. (1984) *Biochim. Biophys. Acta* 767, 524–531) demonstrated that reduction and extraction of manganese of the oxygen evolving complex, which might be expected to alter the redox properties of cyt b_{559}, occurs when certain PSII preparations are exposed to reductants. The ESR data presented here show that a mixture of high potential and lower potential cyt b_{559} species is observed in the oxygen evolving Photosystem II complex. Treatment of PSII membranes with 0.8 M Tris converts the high potential form(s) to those of lower potential. Exposure of the membranes to 2 M NaCl shifts a significant amount of high potential cyt b_{559} to lower potential form(s); addition of $CaCl_2$ reconstituted oxygen evolution activity but did not restore cyt b_{559} to its high potential form(s).

Abbreviations

Chl, chlorophyll; cyt, cytochrome; DCBQ, 2,5-dichloro-benzoquinone; DDQ, 2,3-dichloro-5,6-dicyano-1,4-benzoquinone; ESR, electron spin resonance; OEC, oxygen evolving complex; PS, photosystem

Introduction

Despite considerable effort to elucidate the function of cytochrome b_{559}, its role in photosynthesis remains unclear [6]. The involvement of cyt b_{559} in the electron transport chain between the two photosystems was proposed early on by Cramer and Butler [5]. More recently, an accumulation of circumstantial evidence suggests a close association of cyt b_{559} with the oxidizing side of Photosystem II. It has been shown that at $-196\,°C$ cyt b_{559} is photooxidized by Photosystem II [16]; in addition, treatments which inhibit oxygen evolution invariably cause a marked decrease in the midpoint potential of cyt b_{559} [10, 18]. A series of experiments carried out by

Dedicated to Prof. L.N.M. Duysens on the occasion of his retirement

Matsuda and Butler [20, 21] has clearly demonstrated that high potential cyt b_{559} requires the structural integrity of the photosynthetic membrane and that disruption of that integrity causes cyt b_{559} to be modified to lower potential forms.

Butler has pointed out that much of the confusion about the function of cyt b_{559} derives from the unstable nature of its redox properties [4]. Even though it has been common in the literature to speak of the high potential and low potential forms of cyt b_{559} as if they were well-defined chemical species, cyt b_{559} exists as a heterogeneous population which assumes a range of midpoint potentials [10]. Recent molecular biological, biochemical and spectroscopic evidence has revealed the likely basis for the ease with which the redox properties of cyt b_{559} can be modified and restored [1, 15, 28]. The heme iron of cyt b_{559} is apparently ligated in its axial positions by histidine nitrogens from two distinct polypeptide chains. The dihedral angle between the two histidines imidazole planes controls the ESR properties of the cytochrome. These had been shown previously to be diagnostic of the heme redox potential [2, 19, 24]. In lower potential forms the dihedral angle is close to $0°$ (coplanar imidazole rings) and g_z is close to 2.94; as the redox potential increases, the dihedral angle also increases and g_z approaches its high potential value of 3.08 [1]. Thus, simple shifts in the histidine geometry, presumably controlled by the relative orientation of the two different polypeptides involved, can be related to the redox properties of cyt b_{559}.

In this communication we have studied the effect of various inhibitory treatments on the state of cyt b_{559} in O_2-evolving PSII preparations. We have used ESR spectroscopy to characterize the redox properties of the protein, both because of the relationship between g-value and redox potential noted above and because the use of redox mediators may be avoided by using this approach. Our results show that treatment of PSII membranes with either high concentrations of Tris-buffer or 2 M NaCl converts high potential cyt b_{559} to a low potential form. Although addition of high concentrations of $CaCl_2$ to high-salt treated PSII membranes reconstituted oxygen evolution activity, conversion of the low potential form(s) to the high potential form(s) of cyt b_{559} was not observed.

Materials and methods

Photosystem II membranes were prepared and treated with 0.8 M NaCl as described in [12]. ESR measurements were carried out by using a Bruker ER 200D spectrometer. An Oxford ESR-9 liquid helium cryostat was used to maintain sample temperature; frequencies were measured with a Hewlett-Packard frequency counter and magnetic fields were determined with a Bruker gaussmeter accessory. Low temperature illumination was carried out by focusing the heat filtered output of a 300 W slide projector onto the ESR sample tube which was contained in a liquid nitrogen filled clear glass dewar.

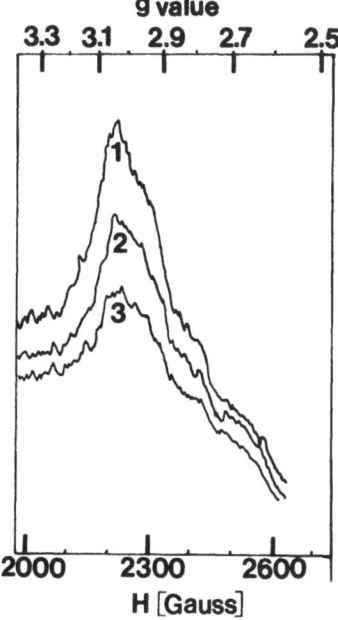

Figure 1. ESR spectra of intact Photosystem II membranes oxidized with 1 mM DDQ. The following conditions were used: Modulation amplitude, 20 G; time constant, 0.1 sec; sweeptime, 200 sec; gain 1.6×10^6; temperature, 18 K; microwave power, (1) 50 mW, (2) 20 mW and (3) 10 mW; each spectrum was obtained as the average of three scans. Chlorophyll concentration, 6.5 mg/ml

Results

Cyt b_{559} is usually described as being in a high or low potential form on the basis of whether it is reduced by hydroquinone. Optical spectroscopy in the α-band region of the cytochrome absorption spectrum is commonly used in this assay. As noted above, the state of cyt b_{559} can also be examined by use of ESR spectroscopy. High potential cyt b_{559} has characteristic g values ($g_x = 1.36$, $g_y = 2.16$ and $g_z = 3.08$) which are shifted upon conversion to the low potential form(s). The latter species have a range of g-values which have as asymptotes the following: $g_x = 1.50$, $g_y = 2.26$ and $g_z = 2.94$ [2, 19, 24]. The g_z component is the most convenient to monitor; g_y appears in a more congested region of the ESR spectrum and g_x, which occurs as a broad, weak feature at high field, has been observed only for the isolated protein [1].

As shown in Figure 1, after oxidation of cyt b_{559} by addition of 1 mM DDQ the ESR spectrum of the cytochrome in oxygen evolving PSII membranes shows a main peak around $g = 3.08$ along with a shoulder at $g = 2.94$; this spectrum is indicative of a high/low potential cyt b_{559} mixture. If we compare the ESR signal of cyt b_{559} shown in Figure 1 to that reported earlier

Table 1. Effect of various inhibitory treatments on oxygen evolution activity of photosystem II membranes

Treatment	Activity (μMol O$_2$/mg Chl·hr)*	
	−CaCl$_2$	+CaCl$_2$
None	720	730
0.8 M Tris, pH 8.0	0	0
2 M NaCl, pH 6.0	200	580

*Photosystem II membranes were assayed for oxygen evolution activity in a medium containing 50 mM MES, pH 6.0, 15 mM NaCl, ±15 mM CaCl$_2$. The chlorophyll concentration was 20 μg/ml and DCBQ (300 μM) served as an exogenous acceptor

by Bergström and Vanngård [2] for intact thylakoids, we observe an increased amount of low potential cyt b$_{559}$ in the isolated PSII complex; this is probably due to the exposure of the photosynthetic membrane to the detergent Triton X-100 (see Discussion section). A titration of cyt b$_{559}$ in an O$_2$-evolving PSII preparation has demonstrated that about 50% of the cyt b$_{559}$ was in the high potential form(s) (hydroquinone reducible), while another 50% was reduced by ascorbate [12]. However, various values of the high potential/low potential ratio in untreated PSII preparations have been reported [11, 17, 20, 21, 22] (see below), and there is uncertainty in the overall cyt b$_{559}$/PSII stoichiometry in the resolved preparations. Several groups report two cyt b$_{559}$ per P680 [e.g., 17, 22, 23, 26], whereas others report only one [8, 11, 12]. Figure 2a shows that upon inhibition of oxygen evolution activity by incubation with 0.8 M Tris-buffer (Table 1), cyt b$_{559}$ is converted to the low potential form(s) (g$_z$ = 2.94) as expected.

As indicated in Table 1, another treatment which inhibits oxygen evolution activity is exposure of the PSII complex to high ionic strength; this treatment is known to release two water soluble polypeptides (17 and 23 kDa) from the oxidizing side of Photosystem II without any release of functional manganese [13]. Previously, we have shown that the 17, 23 kDa-depleted PSII compex is very susceptible to exogenous reductants which reduce and destroy the Mn-complex [14]. Therefore, it is not possible to examine the state of cyt b$_{559}$ in salt treated membranes reliably by use of hydroquinone because such an addition would result in release of functional manganese with concomitant conversion of any high potential cyt b$_{559}$ to the low potential form(s) [10, 18]. To avoid this complication, we studied the state of cyt b$_{559}$ in high salt-treated PSII membranes by use of ESR spectroscopy. As shown in Figure 2c, after exposure of the PSII complex 2 M NaCl a significant amount of high potential cyt b$_{559}$ has been converted to the low potential form(s). Compared to the signal observed after Tris treatment (Figure 2a) however, we notice that after salt wash a fraction of the cytochrome still remains as the high potential form(s).

Since addition of high concentrations of CaCl$_2$ to salt washed PSII membranes is known to reconstitute high rates of oxygen evolution activity (Table

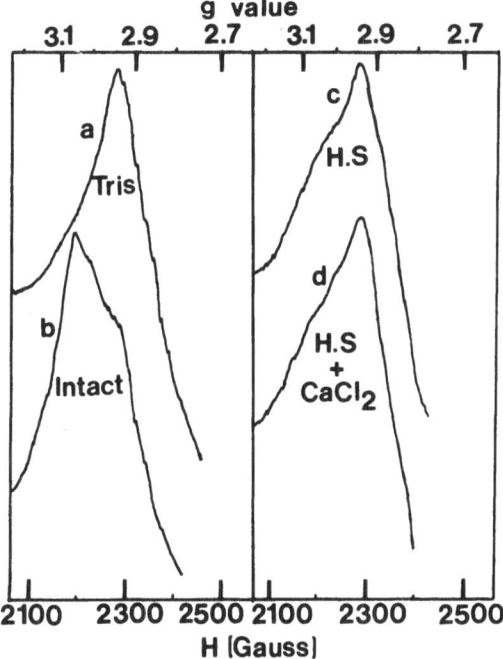

Figure 2. ESR spectra of (a) Tris treated; (b) intact; (c) 2 M NaCl-treated (H.S.), and (d) 2 M NaCl-treated plus 40 mM CaCl₂ PSII preparations oxidized with 1 mM DDQ. The following conditions were used: Microwave power, 30 mW; modulation, 20 G; time constant, 0.1 sec; sweeptime, 200 sec; gain, 1.6×10^6; temperature, 18 K; each spectrum was obtained as the average of six scans. The chlorophyll concentration was 6.5 mg/ml

1), we studied cyt b_{559} in 2 M NaCl-treated PSII membranes which had been reactivated by addition of 40 mM CaCl₂. In a separate series of time-resolved, room temperature measurements (not shown), we observed that addition of 40 mM Ca²⁺ to concentrated, salt-washed PSII membranes (∼4 mg chl/ml) was sufficient to restore the rapid Z^+ reduction kinetics which accompany the reactivation of oxygen evolution [3, 13]. As shown in Figure 2d, even though CaCl₂ restores oxygen evolution activity, cyt b_{559} remains in its low potential form(s).

Since we cannot exclude the possibility that the reduced form of DDQ, which is present in the medium after addition of 1 mM of the oxidant as revealed by a strong ESR signal around $g = 2.00$ (data not shown), reduced and destroyed the Mn-complex [14] in the experiments in Figure 2, we repeated the experiment but this time we oxidized cyt b_{559} by illumination at $-196\,^{\circ}C$ [16]. Figure 3b shows that in PSII membranes frozen in the dark, low potential form(s) of cyt b_{559} are oxidized while high potential form(s) are in the reduced state and thus ESR silent. Illumination of the PSII

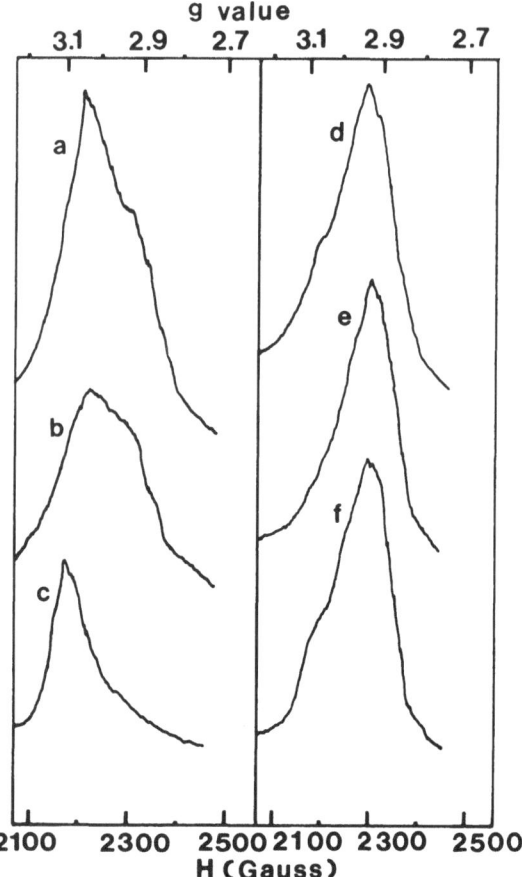

Figure 3. ESR spectra of (a) intact PSII membranes illuminated for 4 min at −196 °C; (b) dark-adapted PSII membranes; (c) spectrum (a) − spectrum (b); (d) salt-washed PSII membranes plus 40 mM CaCl$_2$ illuminated for 4 min at −196 °C; (e) dark-adapted, salt-washed PSII membranes plus 40 mM CaCl$_2$; (f) salt-washed PSII membranes plus 40 mM CaCl$_2$ illuminated for 4 min at 4 °C and then frozen and illuminated for another 4 min at −196 °C. Experimental conditions the same as those in Figure 2

complex at −196 °C results in photooxidation of the high potential form(s) (Figures 3a, 3c). Examination of cyt b$_{559}$ in CaCl$_2$-reactivated salt-washed PSII membranes by the same method showed that only a small fraction of the cytochrome is in the high potential form(s) (Figures 3d, 3e). An attempt to expose the higher S-states to CaCl$_2$ before freezing, by illumination at 4 °C in the presence of CaCl$_2$, was also not successful in restoring cyt b$_{559}$ to its high potential form(s) (Figure 3f).

Discussion

Although there are data in the literature which provide circumstantial links between high potential cyt b_{559} and the water-splitting capacity of PSII, elucidation of its function is problematic [4, 7]. Treatments which inhibit oxygen evolution activity (e.g. exposure to 0.8 M Tris, heptane extraction, etc.) cause a decrease in the midpoint potential of cyt b_{559}. In the case of heptane extraction, Okayama and Butler [25] showed that low potential cyt b_{559} was restored to the high potential form(s) upon reconstitution of oxygen evolution activity by addition of plastoquinone A and β-carotene. It has been suggested that high potential cyt b_{559} requires the structural integrity of the photosynthetic membrane and that disruption of that integrity, by use of treatments such as incubation with chaotropic reagents or detergents, causes modification of cyt b_{559} to the low potential form(s) and loss of oxygen evolution. Results which support this contention were provided by Matsuda and Butler [20, 21] who showed that low potential forms of cyt b_{559} in PSII preparations could be restored to the high potential form by incubation with digalactosyldiacylglycerol (DGDG)-containing liposomes. Accompanying the restoration of high potential cyt b_{559} in this system was in increase in rate of oxygen evolution [21]. Similarly, a comparison of the rates of O_2 evolution in PSII particles prepared in three different laboratories with the high potential/low potential ratio measured on the same preparations in the same laboratories shows that the higher the ratio the greater the O_2 rate [11, 12, 17].

Whether the link between high potential cyt b_{559} and O_2 evolution is causal, however, has been difficult to establish [6]. The results presented here provide some insights into this question. Exposure of the PSII complex to 2 M NaCl releases two water soluble polypeptides (17, 23 kDa) and decreases oxygen evolution activity to 20–30% of the control. The low temperature ESR data on salt-washed PSII membranes show that in 17, 23 kDa polypeptide-depleted membranes a significant fraction of high potential cyt b_{559} has been converted to the low potential form(s). The fraction of the cytochrome which remains in the high potential form(s) may be associated with the residual activity of the PSII complex (see Table 1). When we reconstituted high rates of oxygen evolution activity by addition of external calcium, however, we observed no restoration of the low potential cyt b_{559} to the high potential form(s). This observation suggests that high potential cyt b_{559} *per se* is not necessary for high O_2 evolution rates. It also rationalizes the correlation noted above between high potential cyt b_{559} and oxygen rates in PSII preparation as it is quite likely that the 17 and 23 kDa polypeptides may be removed to a greater or smaller extent during isolation. Loss of these polypeptides would decrease O_2 evolution and increase the concentration of low potential forms of the cytochrome. Evidence supporting this idea can be taken from the recent report by Preston and Critchley, who noted that the low rates of O_2 evolution in their PSII preparations could be enhanced by addition of Ca^{2+} [27].

The situation is more complex than the above analysis indicates, however, as it is clear that Ca^{2+} reactivation of oxygen evolution in polypeptide-depleted PSII preparation does not reconstitute the native system. Dekker et al. have shown that even though $CaCl_2$ restores high rates of oxygen evolution activity the reactivated system differs from the intact system in that the $S_3-->S_4-->S_0$ transition is 2–3 times slower in the salt-washed system [9]. Similarly, the observations of Ghanotakis et al. [14] on the accessibility of the oxidizing side of PSII to exogenous reductants have demonstrated that the structural organization of PSII in calcium-reconstituted salt-washed membranes is different from that of the intact system. To these observations we now add our results on the differences between the properties of cyt b_{559} in untreated and polypeptide-depleted, Ca^{2+} reactivated preparations.

Figure 4 shows a model which summarizes the findings presented here and links them to recent suggestions as to the membrane and protein structural factors which control the properties of cyt b_{559} [1]. In both high and low potential cyt b_{559} the axial ligands are histidines, each one of which is contained in a separate and distinct polypeptide chain. The high potential form is achieved by structural factors which twist the plane of one of the two histidine imidazole rings out of a parallel orientation with respect to the second ring, i.e., the dihedral angle between the imidazole planes is not zero. In Figure 4a these structural forces are suggested to be the nonparallel orientation of the two histidine bearing polypeptide chains. This transmembrane organization is stabilized by the interaction of the two membrane polypeptides with the peripheral 17 and 23 kDa. Upon removal of the two extrinsic polypeptides, (Figure 4b), the transmembrane polypeptides relax to a more parallel orientation which allows the imidazole ring to assume the low energy, parallel (i.e., zero dihedral angle) conformation which is typical of low potential cyt b_{559} [1]. At the same time, the release of the peripheral polypeptides depletes the Ca^{2+} binding site and the OEC relaxes to an inactive configuration. Upon addition of high concentrations of Ca^{2+} in Figure 4c, the oxygen evolving complex assumes an active conformation, which is different from that of the untreated system; cyt b_{559} remains in low potential conformation.

Such a model is also able to rationalize the early observations of Butler and coworkers [4]. These data showed that cyt b_{559} existed as a heterogeneous population with a range of midpoint potentials after inhibition of O_2 evolution. In the model of Figure 4, this suggests that the dihedral angle between the two imidazole planes is somewhat labile and can assume a range of values all of which, however, should be fairly close to zero. Because of the link established between imidazole geometry and principal g-values for various forms of cyt b_{559}, this variability in imidazole orientation may be expected to be manifested in the ESR properties of the cytochrome. An

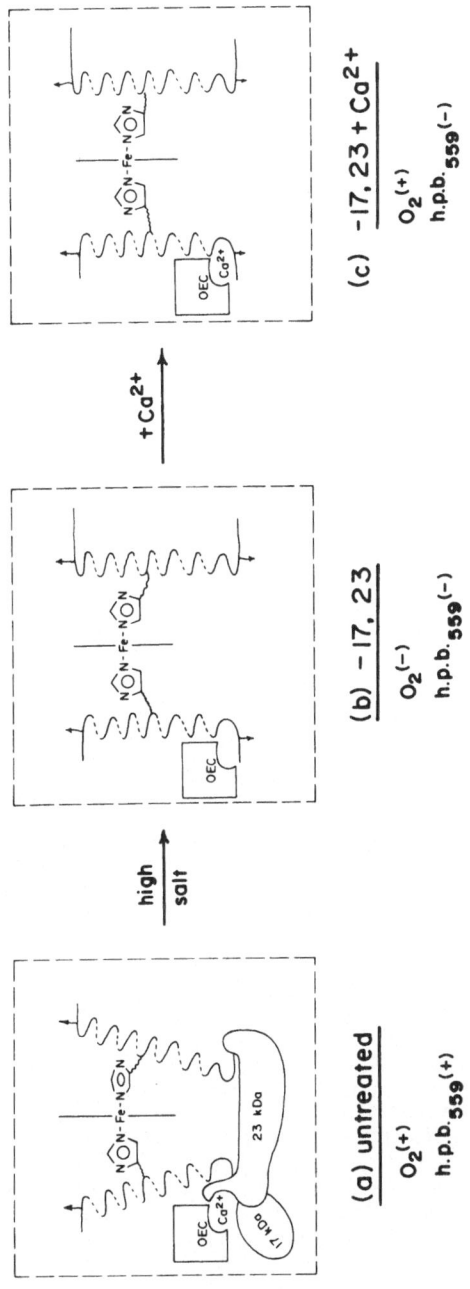

Figure 4. Model for the states of the oxygen evolving complex and cytochrome b_{559} in (a) untreated, (b) 17, 23 kDa polypeptide-depleted and (c) Ca^{2+} reactivated preparations. See text for details

example of this may be found in the recent work in inhibited systems by Matsuura and Itoh [22] which showed that g_z was 2.98 for Tris or heat-treated PSII preparations but 2.95 for the same preparation following a 1 M $CaCl_2$ wash.

Acknowledgements

This research was supported by grants from NSF (PCM-8214240) and USDA/ CRGO (G-82-CRCR-1-1127) to CFY and by USDA/CRGO (5901-04010-9-0344) to GTB.

References

1. Babcock GT, Widger WR, Cramer WA, Oertling WA and Metz JG (1985) Biochemistry 24:3638–3645
2. Bergström J and Vanngärd T (1982) Biochim Biophys Acta 682:452–456
3. Boska M, Blough NV and Sauer K (1985) Biochim Biophys Acta 808:132–139
4. Butler WL (1978) FEBS Lett 95:19–25
5. Cramer WA and Butler WL (1967) Biochim Biophys Acta 143:332–339
6. Cramer WA and Whitmarsh J (1977) Annu Rev Plant Physiol 28:133–172
7. Cramer WA, Whitmarsh J and Widger WR (1981) in 'Photosynthesis Electron Transport and Phosphorylation' (Akoyounoglou G, ed.) pp 509–521, Balaban Int Sci Serv, Philadelphia
8. Dekker JP (1984) Ph.D. Thesis, University of Leiden
9. Dekker JP, Ghanotakis DF, Plijter JJ, Van Gorkom HJ and Babcock GT (1984) Biochim Biophys Acta 767:515–523
10. Erixon K, Lozier R and Butler WL (1972) Biochim Biophys Acta 267:375–382
11. Ford RC and Evans MCW (1983) FEBS Lett 160:159–164
12. Ghanotakis DF, Babcock GT and Yocum CF (1984) Biochim Biophys Acta 765: 388–398
13. Ghanotakis DF, Babcock GT and Yocum CF (1984) FEBS Lett 167:127–130
14. Ghanotakis DF, Topper JN and Yocum CF (1984) Biochim Biophys Acta 767: 524–531
15. Herrmann RG, Alt J, Schiller D, Cramer WA and Widger WR (1984) FEBS Lett 179:239–244
16. Knaff DB and Arnon DI (1969) Proc Natl Acad Sci USA 63:956–962
17. Lam E, Baltimore B, Ortiz W, Chollars S, Melis A and Malkin R (1983) Biochim Biophys Acta 724:201–211
18. Lozier R, Baginsky M and Butler WL (1971) Photochem Photobiol 14:323–328
19. Malkin R and Vanngärd T (1980) FEBS Lett 111:1–4
20. Matsuda H and Butler WL (1983) Biochim Biophys Acta 724:123–127
21. Matsuda H and Butler WL (1983) Biochim Biophys Acta 725:320–324
22. Matsuura K and Itoh S (1985) Plant Cell Physiol 26:537–542
23. Murata N, Miyao M, Omata H, Matsunami H and Kuwabara T (1984) Biochim Biophys Acta 765:363–369
24. Nugent JHA amd Evans MCW (1980) FEBS Lett 112:1–4
25. Okayama S and Butler WL (1972) Plant Physiol 69:769–774
26. de Paula JC, Innes JB and Brudvig GW (1985) Biochemistry, in press
27. Preston C and Critchley C (1985) FEBS Lett 184:318–322
28. Widger WR, Cramer WA, Hermodson M, Meyer D and Gullifor M (1984) J Biol Chem 259:3870–3876

Photosynthesis Research 9, 135–147 (1986)
© *1986 Martinus Nijhoff/Dr. W. Junk Publishers, Dordrecht.*

Photosynthetic and respiratory electron transport in a cyanobacterium

JACK MYERS

Department of Botany and Zoology, The University of Texas, Austin, Texas 78712, USA

(Received 3 September, 1985)

Key words: blue-green algae, cyanobacterium, cytochrome $C553$, electron transport, photosynthesis, respiration

Abstract. In the cyanobacterium *Agmenellum quadruplicatum* steady-state redox conditions were monitored in vivo for cytochrome $(f + c553)$ and P700 versus intensities of an actinic light 1 or light 2 (mainly absorbed by photosystems, and 2, respectively). Parallel measurements of O_2 evolution were used to calibrate intensities for rates of electron transfer. Results show that the quality of actinic light (as light 1 or light 2) depends on intensity as well as wavelength. The contribution of electron flow from respiration is confirmed by observations of relative rate of photoreaction 1 estimated from Ip (intensity × fraction of P700 reduced). With 3,- (3,4-dichlorophenyl-1, 1-dimethylurea (DCMU) the rate of photoreaction 1 depends upon, and is sensitive to small changes in, the rate of dark respiration. Very slow transient dark reductions of Cyt $(f + c553)$ and P700 following any low intensity actinic light 1 are attributed to respiratory electron flow. Cyclic electron flow around photoreaction 1 cannot be large compared to dark respiration and cannot vary significantly with light intensity.

Introduction

The common link between photosynthesis and respiration, inferred for *Anacystis nidulans* from measurements of O_2 evolution [11] has now been described in detail for several cyanobacteria (blue-green algae) [e.g., 9, 20, 21]. Thylakoid membrane components, including plastoquinone and cytochrome *f*, serve electron transfer between the two photoreactions of photosynthesis and also serve the respiratory pathway from NAD(P)H to cytochrome oxidase. Since cytochrome *f* and P700 thereby become reduced in darkness, changes in their redox conditions in light describe the opposing effects of photoreactions 1 or 2. This kind of observation was used to obtain the first evidence for the Z-scheme [8, 12]. I have applied and extended this approach in a systematic study on whole cells of two cyanobacteria. The objective simply was to see whether the cyanobacterial system in real life behaves as expected from in vitro studies.

This paper is contributed in honor of my longtime friend, L.N.M. Duysens, who has carried still further the eminence of the Dutch tradition in biophysics.

Materials and methods

All measurements reported below were made on *Agmenellum quadruplicatum*, strain PR6 (ATCC 29404) [19] obtained from C. Van Baalen. It was grown in continuous culture in ASP-2 medium [24] at 30 °C, aerated with 2% CO_2 in air, and under tungsten illumination to give a specific growth rate about 1.0 day^{-1}. *Anacystis nidulans*, strain T × 20, grown at 39 °C as previously described [16] was also used in comparative experiments.

Steady-state redox conditions of P700 (703 vs 735 nm) and of cytochrome $(f + c553)$ (554 vs 540 nm) were monitored by a dual wavelength spectrophotometer [15] which presented a beam of 3.3 nm half-band width (HBW) chopped at 120 Hz. Time constant of the output amplifier was usually set as 0.2 s but was reduced to 0.05 s for observation of transients with a compatible Brush Mark 220 recorder. Harvested cell suspension was diluted in growth medium to about 6 μM Chlorophyll (Chl), aerated with 2% CO_2 in air, and held in darkness at 25 °C until use. 1.0 ml sample was transferred to a cuvette which presented 10 mm to the measuring beam and 5 mm to an actinic beam entering at right angles. The photomultiplier, about 3 cm behind the cuvette, was protected by a filter complementary to the actinic beam and the exit measuring beam was limited by a 4 mm wide diaphragm.

The actinic beam, which completely illuminated the 10 × 20 mm cross section of the cuvette, was provided by a General Electric DEK/DFW projection lamp operated usually at 80.0 ± 0.05 volts DC. The projection system included an electrical shutter (3 ms transit time), calibrated screens for intensity control, a 1.0 cm water filter and a Balzers 6143 infra-red filter. Wavelength definition was provided by one of the following filter combinations: (a) Baird 4432 plus Ditric 500C or 580C cut-off filters giving light mainly absorbed by photosystem I (light 1) at 440 nm (HBW 65 nm); (b) Ditric 680 and Schott RG5 giving light 1 at 680 nm (HBW 12 nm); (c) Ditric 620 and Corning 2418 giving light 2 at 620 (HBW 11 nm). The filter combinations were checked for stray infra-red transmission out to 12 μm and found clean to 3 OD. Intensities were measured by a YSI Radiometer calibrated against a standard lamp.

Rates of electron transport were measured in terms of net O_2 exchange using a Clark-type electrode immersed in a thermostatted (25 °C) cuvette of 10 mm path length and 2 cc volume. Illumination was provided by a projection system similar to that used for the actinic beam of the spectrophotometer and limited by the same sets of filters. Equivalence of intensities in the O_2-electrode and spectrophotometer cuvettes was provided by measurements with an inserted 5 × 5 mm silicon cell. An alternative source of illumination was an EG & G F × 76 flash tube [16].

The steady-state redox conditions of Cyt $(f + c553)$ and P700 were estimated from the light-minus-dark difference signals observed at 1×10^{-3} absorbance full scale and with time periods sufficiently long (usually 10 s) that the result was independent of time period.

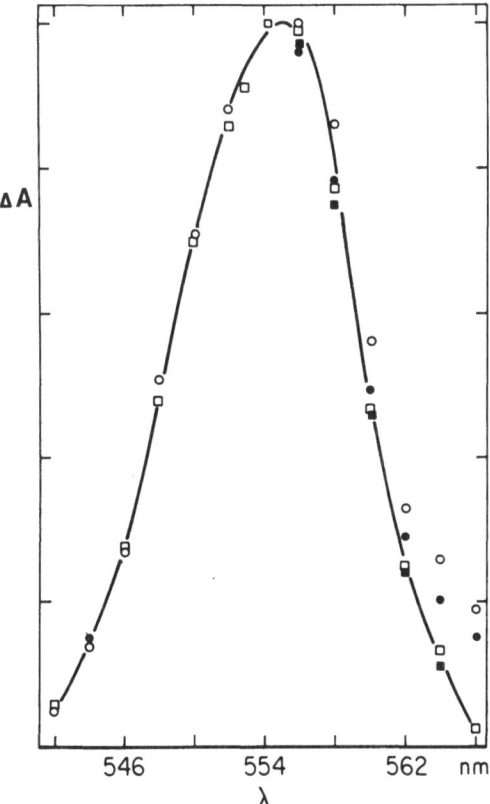

Figure 1. The light-dark difference in absorption observed in *Ag. quadruplicatum* for the wavelength-pairs ($\lambda - 540$) nm, with wavelength resolution of 1.6 nm (HBW), and with actinic lights of the following wavelengths and intensities (nEcm^{-2} s^{-1}) ○, 680 nm at 33; ●, 620 nm at 24; ■, 440 nm at 2.4; □, 440 nm at 11

Results

The 554/540 nm in vivo cytochrome signal

As an indicator of cytochrome f there is a choice between 421 and 554 nm. After exploring both regions I chose 554/540 nm, partly for convenience, although effects of varied intensity of 680 or 620 nm (cf. Figure 3) were substantially identical whether monitored at 554/540 or at 421/410 nm. The choice of 554/540 nm and the generally used 3.3 nm HBW implies that the signal lumps together cytochrome f and cytochrome c-553 which seem to be closely coupled in time [17]. The spectrum of the light-dark difference signal (Figure 1) was independent of actinic wavelength and intensity except for small variations beyond 560 nm which were not explored.

The maximum light-dark absorption change at 554/540 nm was about 0.3 of the change at 703/735 nm. This is about the same as the ratio of

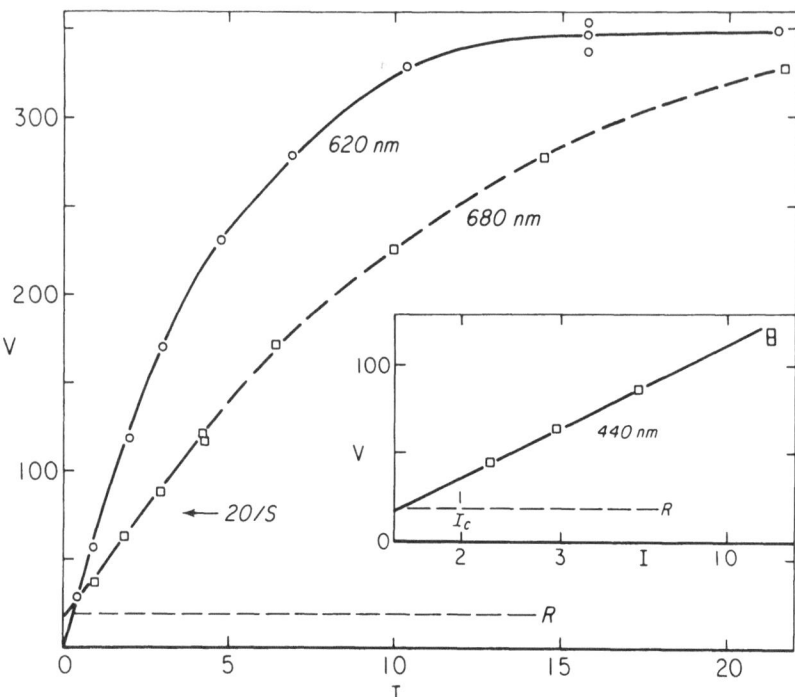

Figure 2. Rates of net O_2 evolution V in $O_2Chl^{-1} h^{-1}$ versus intensity I in $nE\,cm^{-1}\,s^{-1}$ for lights of 620, 680, and 440 nm. V is taken as the rate observed in light minus the rate observed in the immediately following dark period. For 620 nm V extrapolates to zero at zero I; at 680 and 440 nm V extrapolates to a value of V approximately equal to the rate of dark respiration R = 18. R was measured between 2 and 5 minutes after a light period. The rate V = 74 observed with 20 Hz saturating flashes is given for reference

absorption coefficients usually assigned to cytochrome c-553 and P700 (20/70) and implies that in *Ag. quadruplicatum* the pool sizes of cytochrome $(f + c$-553$)$ and P700 are about equal, as observed more precisely in a *Synechococcus* [9]. (However, in *A. nidulans* the 554/540 nm signal was more nearly 0.15 of that at 703/735 nm.) In *Aq. quadruplicatum* the reaction center concentrations were estimated as 1 reaction center (RC2)/250 Chl(flash yield of oxygen evolution) and 1 P700/120 Chl.

Rates of electron transfer verus light intensities

Figure 2 shows light intensity curves for oxygen evolution at 620, 680, and 440 nm. All data describe rate in light minus rate observed in an immediately following dark period, i.e., they are conventionally corrected for dark respiration. In light 2 (620 nm) the curve extrapolates to zero rate at zero I. In light 1 (680 or 440 nm) the curve extrapolates to a rate approximately equal to that of dark respiration. Corresponding rates of electron transfer may be rationalized as follows.

Let q designate the fraction of system 2 reaction centers (Q) in the oxidized (open) condition; let p designate the fraction of P700 in the reduced (open) condition. For any incident quantum flux I let a_1 and a_2 designate the fractions absorbed and partitioned to P700 and Q, respectively. Let k_1 and k_2 be probabilities that an excitation reaching an open reaction center will cause it to transfer an electron. Then the rate for photoreaction 1 is $v_1 = k_1 a_1$ Ip and for photoreaction 2, $v_2 = k_2 a_2$ Iq. At low intensities of 620 nm the rates of photoreactions 1 and 2 are equal:

$$v_1 = k_1 a_1 Ip = v_2 = k_2 a_2 Iq$$

It is necessary to suppose that under low intensity light 2, P700 is so completely reduced that it does not compete as an acceptor with cytochrome oxidase which is the presumed rate-limiting step in respiration. For a low intensity light 1 (680 or 440 nm), P700 is oxidized and serves as an acceptor for electron flow both from respiration (R) and from v_2:

$$v_1 = k_1 a_1 Ip = R + k_2 a_2 Iq$$

An additional condition occurs with 3-(3,4-dichlorophenyl-1,1-dimethylurea (DCMU) inhibition. Then both q and v_2 become negligible and

$$v_1 = k_1 a_1 Ip = R$$

Two predicted special consequences of the above will be tested below. (a) Under intensity, I_c (defined as in the inset of Figure 2), $v_2 = R$ and $v_1 = 2R$; at intensity I_c addition of DCMU should change the rate v_1 from 2R to R and should lower p by a factor of 0.5.

Translating O_2 exchange rates into electron transfer rates requires an assumption that there are only two significant O_2 exchange reactions: evolution via the oxidizing side of photoreaction 2 and uptake via the terminal oxidiase(s) of respiration. An additional uptake resulting from reduction via the superoxide anion [2, 5, 18] may occur at the beginning of an illumination. In practice the steady-state rates described in Figure 2 were reached only after about 3 to 5 minutes. In contrast, steady-state concentrations of Cyt $(f + c\text{-}553)$ and P700 were reached within seconds. Evidently, just as previously envisioned [18], electron transfer quickly reaches a steady-state rate and the apparent O_2 induction period represents a transition from O_2 to CO_2 reduction.

Redox condition of Cyt (f + c-553) versus light intensity

Figure 3 shows responses to 680 and 620 nm. At low (photochemically limiting) intensities classically expected effects are observed. The light 1 (680 nm) drives Cyt $(f + c\text{-}553)$ oxidized whereas the light 2 (620 nm) maintains it close to the reduced condition as in darkness. At high intensities both wavelengths drive Cyt $(f + c\text{-}553)$ oxidized. This second result also is expected because the light-saturated rate is determined only by the

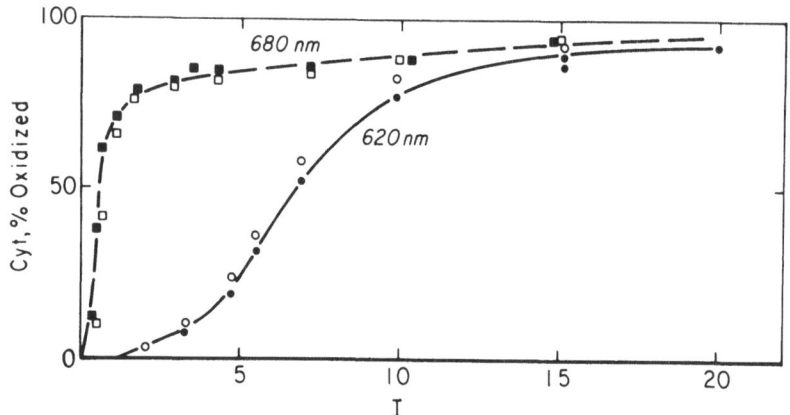

Figure 3. The redox state of Cytochrome $(f + c\text{-}553)$ as % oxidized versus intensity $(nEcm^{-2} s^{-1})$ for actinic 680 and 620 nm. % oxidized was taken from the light-dark absorption difference for 554/540 nm with 100% as the maximum signal obtained at $30 \, nEcm^{-2} \, s^{-1}$ of 680 nm

rate-limiting step, the oxidation of plastoquinol. The result has been observed previously [cf. 6] but the very different profiles for Cyt $(f + c\text{-}553)$ oxidation have not previously been made explicit. As judged by the original criterion [8], the redox state of Cyt $(f + c\text{-}553)$, 620 nm is a light 2 only at photochemically limiting intensities.

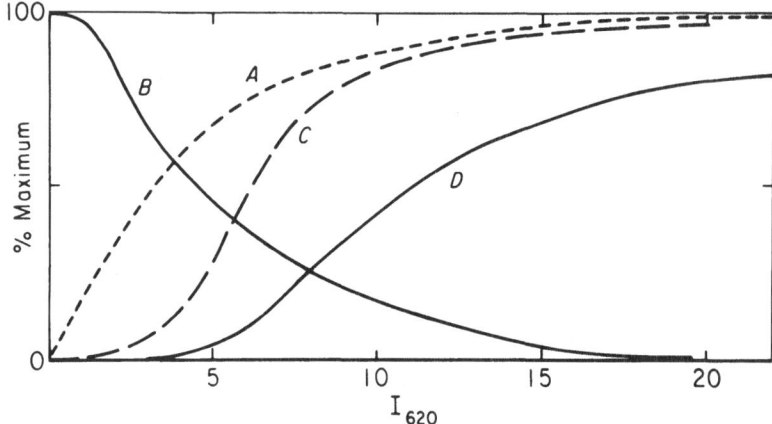

Figure 4. Behavior of components versus intensity $(nEcm^{-2} s^{-1})$ of 620 nm. Reference curve A is taken from the curve for 620 nm of Figure 2, normalized to V_{max}. Curve B, $\Delta V/\Delta I$, taken from the point-to-point slopes of curve A, shows the differential quantum yield for O_2 evolution as % of its maximum value. Curve C is the % of cyt $(f + c\text{-}553)$ oxidized. Curve D is the % P700 oxidized. Curve B is drawn as a best fit for two data sets; curves and D are drawn as best fits for five data sets

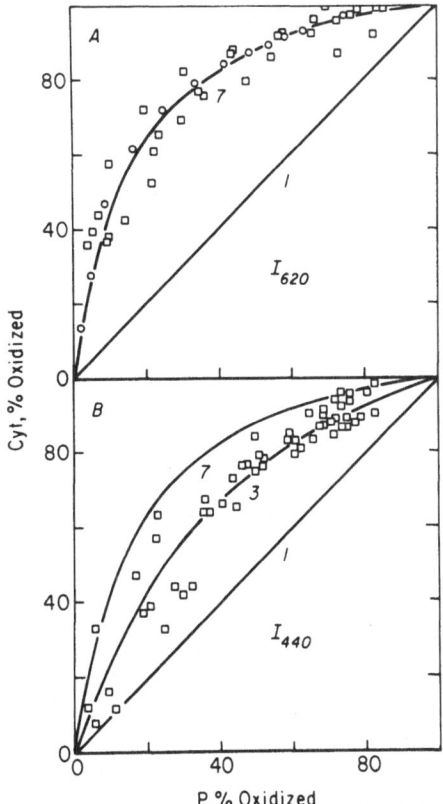

Figure 5. Redox state of Cytochrome (f + c-553) versus redox state of P700 at steady-state as controlled by intensity. Both the cytochromes and P700 were assumed to be completely reduced in darkness. 100% oxidation of P700 was estimated as the light-dark difference signal observed with 18 nEcm^{-2} s^{-1} of 440 nm and 2 μM DCMU. 100% oxidation of Cyt (f + c-553) was estimated from the light-dark difference signal with 30 nEcm^{-2} s^{-1} which was not further increased by added DCMU. A, as controlled by I at 620 nm. The curves are drawn for K = 1 and K = 7(□), all data from five data sets; ○, taken from averaged curves C and D of Figure 4. B, as controlled by I at 440 nm. The curves are drawn for K's of 1, 3, and 7. Points are all data from six data sets

Redox conditions controlled by light 2 intensity

Figure 4 presents a comparison of redox conditions of three components versus intensity of 620 nm. For reference curve A is replotted from Figure 3 as % of maximum rate of O$_2$ evolution. Curve B shows the first derivative of A estimated as Δv/ΔI from experimental points. Curve C shows % oxidation of Cyt (f + c-553) and D shows % oxidation of P700.

If the rate of O$_2$ evolution by photoreaction 2 is $v_2 = k_2 a_2 I q$, then curve B estimates the decay in q with increasing light intensity. It is evident and expected that q is not simply related to Cyt (f + c-553)$_{ox}$ or to P700$_{ox}$. However, the Cyt (f + c-553) and P700 components are closely related as

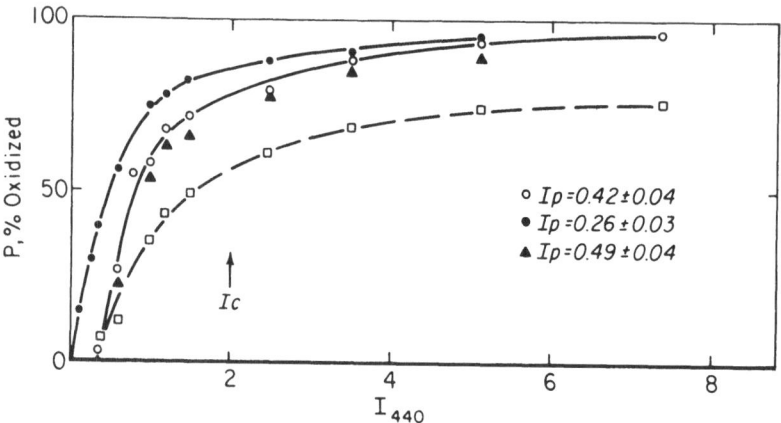

Figure 6. Redox state of P700 versus intensity ($nEcm^{-2} s^{-1}$) of actinic 440 nm in four different preparations. □, at 2.2 hour in darkness after harvest; ○, at 2.4 hour in darkness after harvest + $2 \mu M$ DCMU; ●, at 26 hour in darkness + $2 \mu M$ DCMU; ▲ at 26.5 hour in darkness + $2 \mu M$ DCMU + 50 mM glycerol. The product Ip is shown for each of the three data sets with added DCMU; standard deviations are for values of p (usually 5 or 6) between 15 and 85% reduced

shown more formally in Figure 5A. A best fit is obtained for K = 7 as the apparent equilibrium constant, $K = Cyt_{ox} \times P_{red}/Cyt_{red} \times P_{ox}$.

Figure 5B shows corresponding data under control by light 1 at 440 nm. The data appear to give a better fit to K = 3 but their wide scatter allows only the suggestion that K may be lower in light 1. The only reason to consider the difference is that the apparent variability and uncertainty in K seen in chloroplasts [10, 14] seems to extend to cyanobacteria.

Redox condition of P700 versus intensity of light 1

Figure 6 shows the effect of I_{440} for different preparations. In effect these are redox titration curves in which P700 serves as an internal indicator. I_{440} provides an adjustable oxidizing rate opposing respiratory electron flow as a steady reducing rate. Addition of DCMU makes I_{440} a "perfect" light 1, a stronger oxidant, and displaces the titration curve to the left. Long dark incubation lowers the reducing rate provided by respiration and moves the titration curve still further to the left. Then addition of glycerol, which increases respiration rate (see below), moves the titration curve back toward the right. Similar displacements of the titration curve (toward lower values of I for lower R) occurred also in the absence of DCMU (data not shown).

Figure 6 provides two tests of predictions made previously in discussion of Figure 2. At $I_c = 2$ nE cm^{-2} s^{-1} (cf. Figure 2 insert) addition of DCMU is expected to lower the rate of photoreaction 1, $v_1 = k_1 a_1 Ip$, by a factor of 0.5. The effect of added DCMU seen in Figure 6 is a lowering of p $(100-\%P_{ox})$ from 44 to 22. (The coincidence of predicted and observed

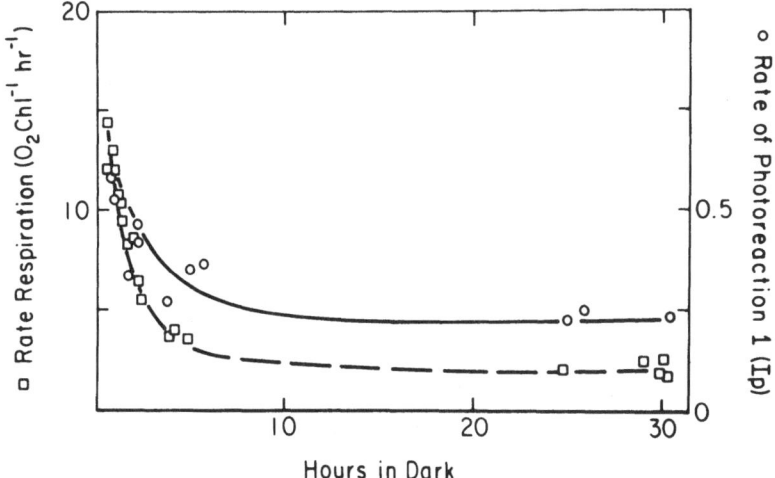

Figure 7. Decays in rates of dark respiration and photoreaction 1 with dark incubation. Relative rate of photoreaction 1 was estimated as Ip where p is the steady-state fraction of P700 reduced under incident intensity I of 440 nm actinic light in the presence of $2\,\mu$M DCMU. Dark respiration was measured without any preceding light exposure

values probably is better than should be expected because of some necessary uncertainty in absolute match of intensity scales for Figures 2 and 6.) A second prediction was that in the presence of DCMU, $R = k_1 a_1 Ip$ so that the product Ip should be constant for any one value of R. As noted on Figure 6 the Ip product was, in fact, sensibly constant for each preparation with DCMU. A corollary prediction is that Ip should vary with change in R.

Rate of photoreaction 1 versus rate of respiration

Search for methods to vary respiration in *Ag. quadruplicatum* met with only limited success. Dark starvation gave a rapidly decreasing respiration rate, especially during the first two hours, as shown in Figure 7. Concomitantly there was a decrease in relative rate of photoreaction 1 under I_{440} as estimated from the Ip product. However, the predicted simple proportion between Ip and R was not found and may not be experimentally achievable. R was measured in continuous darkness while p was measured under repetitive light and dark periods.

Search for exogenous substrates yielded one, glycerol, which increases the rate of respiration within minutes after addition. Effects of glycerol were studied in spite of an unfortunate characteristic that it gives a severe inhibition of net O_2 evolution after prolonged incubation. The inhibition may result from some conversion to glyceraldehyde which is a known inhibitor of photosynthesis in *A. nidulans* [13] and was found to completely suppress growth of *Ag. quadruplicatum* at 1 mM concentration (data not shown). As seen in Table 1, addition of glycerol increases rate of dark O_2 uptake up to

Table 1. Effects of dark starvation and various additions on dark O_2 uptake and photoreaction 1 in *Ag. quadruplicatum*. O_2 uptake was measured without any addition, with 50 mM glycerol, with 1.0 mM benzyl viologen, or with 50 mM glycerol + 1.0 mM benzyl viologen. Relative rates of photoreaction 1 were measured as Ip where p is the steady-state fraction of P700 reduced under incident intensity I ($nE\,cm^{-2}\,s^{-1}$) of 440 nm actinic light

Hours dark	No addition	+ Glycerol	+ Benzyl viologen	+ Glycerol + Benzyl viologen
Rate of dark O_2 uptake ($O_2 Chl^{-1}\,h^{-1}$)				
1	9.4	13.1	41	38
3	7.6	10.7	34	34
5	4.7	13.8	20	19
24	3.8	9.8	23	22
30	2.6	7.1	–	–
Relative rate photoreaction 1 (Ip)				
1	0.52	–	0.19	–
2.5	0.33	0.43	0.24	0.28
4	0.27	–	0.15	0.25
26	0.25	0.49	0.17	0.22

a factor of about 3 in dark-starved cells and also increases the rate of photoreaction 1 in terms of the Ip product observed with added DCMU.

Table 1 also shows effects of added benzyl viologen (chosen in preference to methyl viologen because in whole cells of *Ag. quadruplicatum* it was more effective). At intensity high light added benzyl viologen gave the expected behavior of close to zero net O_2 exchange which changed to rapid O_2 uptake on addition of 1 mM KCN (data not shown). In the dark benzyl viologen gave a more than four-fold increase in O_2 uptake which was not increased further by added glycerol. Benzyl viologen also lowered the rate of photoreaction 1 as observed for the Ip produce under DCMU inhibition. A reasonable explanation of the effects of benzyl viologen is provided by a study on spheroplasts of *Aphanocapsa* [20] : viologen served as an alternative pathway to O_2 for oxidation of NADH and NADPH, presumably decreasing electron flow through Cyt $(f + c\text{-}553)$ to P700.

Transient dark reduction of Cyt (f + c-553) and P700

In spite of the rather long instrumental time constant, large differences were easily observed in transient dark reductions depending upon the character of the preceding illumination. Summary data are presented in Table 2. I consider first the more complete measurements for Cyt $(f + c\text{-}553)$ at 554/540 nm. A dramatic and consistent result was that the transient rate of dark reduction was rapid ($t_{1/2} \leqslant 0.1$ s) following a saturating light intensity (either 620 or 680 nm) but much slower ($t_{1/2} \geqslant 0.3$ s) following a low intensity light 1 (440 or 680 nm). Explanation depends on differences in the redox state of the plastoquinone pool [1] as can be inferred from Figure 3 for Cyt $(f + c\text{-}553)$. At light saturating intensity the plastoquinone pool must be almost completely

Table 2. Effects of previous actinic lights and various treatments on the half-times or dark reduction of Cyt $(f + c\text{-}553)$ and P700

Treatment and preceding illumination	$t_{1/2}$ dark reduction, s, 554/540 nm	measured at 703/735 nm
No Addition		
High light 1 or light 2	$\leqslant 0.1$	$\leqslant 0.1$
Low light 1	0.3	0.2
24 hour dark starvation		
High light 1 or light 2	$\leqslant 0.1$	–
Low light 1	0.5	0.3
Low light 1 + 50 mM glycerol	0.3	0.2
+ DCMU, 2 μM		
High light 1	0.5	0.3
Low light 1	0.8	0.5
+ DAD, 50 μM		
low light 1	$\leqslant 0.1$	–

High light 1 means 15 nE cm^{-2} s^{-1}, 680 nm
High light 2 means 15 nE cm^{-2} s^{-1}, 620 nm
Low light 1 means 4 nE cm^{-2} s^{-1}, 680 or 11 nE cm^{-2} s^{-1}, 440 nm

reduced while Cyt $(f + c\text{-}553)$ is almost completely oxidized. Transfer to darkness allows observation of reduction of Cyt $(f + c\text{-}553)$ at the rate-limiting step and a $t_{1/2}$ of the order of 10 ms (faster than my instrument can detect). At a low intensity of light 1 the entire electron transfer chain is mostly oxidized. Dark reduction then can proceed only at the much slower rate provided by respiratory electron flow, about 3 e/s or $t_{1/2} = 0.2$ s as viewed by Cyt $(f + c\text{-}553)$. Dark reduction following a low light 1 becomes even slower after dark starvation but is restored by added glycerol, as expected from observed respiration rates (Table 1). The effect of added DCMU (which I cannot explain) is to give still slower dark reduction. Addition of diaminodurene (DAD) gave a very rapid dark reduction consistent with a marked increase in Ip under light 1 (data not shown).

As viewed by P700 rates of dark reduction were always faster than for Cyt $(f + c\text{-}553)$, as expected if P700 is the final sink and Cyt $(f + c\text{-}553)$ an intermediate transfer agent. However, all the differences seen for Cyt $(f + c\text{-}553)$ were also seen for P700.

Discussion

The results provide a related and consistent series of observations on in vivo electron transport in a cyanobacterium. They show explicitly that electron transport depends on intensity as well as wavelength of actinic light. They show that respiratory electron flow to photoreaction 1 is (a) predicted from the light 1 intensity curve for O_2 evolution, (b) confirmed by observations of P700 oxidation versus intensity which show an expected effect of DCMU on the Ip product and dependence of Ip on rate of respiration, and (c) confirmed by the very slow rates of transient dark reduction observed after a low light 1 when the plastoquinone pool is oxidized.

One unexpected result remains to be considered. In *Ag. quadruplicatum* at 25 °C the rate of electron transport contributed by respiration is only about 3% of the rate contributed by photosynthesis at light saturation. Nevertheless, changes in this small and limited rate, as by dark starvation or by addition of glycerol, made significant changes in the relative rate of photoreaction 1 observed as the Ip product under DCMU inhibition. I find no way to escape two important conclusions: any cyclic electron flow occurring around photoreaction 1 cannot be large compared to dark respiration and cannot vary significantly with light intensity. A corollary is that cyclic phosphorylation, in the sense originally defined [3, 4], cannot make significant contribution to the in vivo photosynthesis of this cyanobacterium growing at ordinary light intensities.

The above supports the conclusion of Biggins [6] from time-resolved measurements of dark reduction of Cyt f in *Porphyridium*. Two components of the reduction could be distinguished: a fast component of about 25 ms ($t_{1/2}$) and a slow component of 150 ms or longer. The relative magnitude of the two components varied with actinic wavelength and intensity in the same way observed more crudely in the present study with *Ag. quadruplicatum*. With increasing intensity of an actinic light 2 the slow component became a smaller fraction of the total decay, approaching a limit of less than 10%. Since the slow component was attributed to cyclic flow, it was a necessary conclusion that cyclic flow must become negligible at high light intensity. Instead of the ad hoc assignment of the slow component to cyclic flow, I prefer to assign it to the respiratory electron flow from NAD(P)H which also occurs in the dark. However, the net conclusion demanded is the same: any rate of cyclic flow is of the order of magnitude of dark respiration and becomes progressively less significant at increasing rates of non-cyclic electron flow.

There is an extensive literature on cyclic phosphorylation in cyanobacterium [7, 22] and implicitly accepted proposals that cyclic phosphorylation is a fundamental feature of all photosyntheses [3, 23]. I have set forth a contradictory notion only because the observations fit no other conclusion. I would be pleased to have personal communications on shortcomings perceived in my argument or on hard evidence for significant participation of cyclic phosphorylation in photosynthesis of cyanobacterium.

References

1. Amesz J (1973) Biochim Biophys Acta 301:35−51
2. Anderson JW, Foyer CH and Walker DA (1983) Biochim Biophys Acta 724: 69−74
3. Arnon DI (1959) Nature 184:10−21
4. Arnon DI, Whatley FR and Allen MB (1958) Science 127:1026−1034

5. Asada K and Badger MR (1984) Plant and Cell Physiol 25:1169−1179
6. Biggins J (1973) Biochemistry 12:1165−1170
7. Bottomley PJ and Stewart WDP (1976) Arch Microbiol 108:246−258
8. Duysens LNM, Amesz J and Kamp BM (1961) Nature 190:510−511
9. Hirano M, Satoh K and Katoh S (1980) Photosynthesis Res 1:149−162
10. Joliot P and Joliot A (1984) In Sybesma C ed, Advances in Photosynthesis Research vol I pp 399−406, Nijhoff/Junk, The Hague
11. Jones LW and Myers J (1963) Nature 199:670−672
12. Kok B and Hoch G (1961) In McElroy WD and Glass B ed, Light and Life pp 397−423, Johns Hopkins Press, Baltimore
13. Lara C, Romero JM, Flores E, Guerrero MG and Losada M (1984) In Sybesma C ed, Advances in Photosynthesis Research vol II pp 715−718, Nijhoff/Junk, The Hague
14. Marsho TV and Kok B (1970) Biochim Biophys Acta 223:240−250
15. Myers J, Graham JR and Wang R (1980) Plant Physiol 66:1144−1149
16. Myers J. Graham JR and Wang R (1983) Biochim Biophys Acta 722:281−290
17. Nanba M and Katoh S (1983) Biochim Biophys Acta 725:272−279
18. Radmer RJ and Kok B (1976) Plant Physiol 58:336−340
19. Rippka R, Derulles J, Waterbury JB, Herdman M and Stanier RY (1979) J Gen Microbiol 111:1−69
20. Sandmann G and Malkin R (1983) Biochim Biophys Acta 725:221−224
21. Scherer S and Boger P (1982) Arch Microbiol 132:329−332
22. Simonis W and Urbach W (1973) Annu Rev Plant Physiol 24:89−114
23. Stanier RY (1961) Bacteriol Rev 25:1−17
24. Van Baalen C (1962) Bot Mar 4:129−139

Photosynthesis Research 9, 149–158 (1986)
© 1986 Martinus Nijhoff/Dr. W. Junk Publishers, Dordrecht.

Anoxygenic photosynthetic hydrogen production and electron transport in the cyanobacterium *oscillatoria limnetica*

CHRISTIAAN SYBESMA, DIEDERIK SCHOWANEK, LUIT SLOOTEN and NADIA WALRAVENS

Biophysics Laboratory, Vrije Universiteit Brussel, Pleinlaan 2, B-1050 Brussels, Belgium

(Received 3 September 1985)

Key words: anoxygenic photosynthesis, cyanobacteria, hydrogen production, oscillatoria limnetica

Abstract. The induction of anoxygenic photosynthesis in the cyanobacterium *Oscillatoria limnetica* by sulfide was shown to involve the synthesis of a "sulfide oxidizing factor"; this factor, partly adsorbed on the thylakoid membrane, can be recovered in the soluble phase and is active also on membranes from oxygenically grown cells. The factor is required for sulfide dependent light-induced hydrogen evolution. It accelerates electron transport from sulfide to the electron donor of photosystem I, P700, in membranes from cells in which anoxygenic photosynthesis is induced. The plastiquinone analogue DBMIB does not inhibit electron transport to P700 but accelerates it. The analogue might promote cyclic electron transport involving P700, thus preventing electrons to reach hydrogenase.

Abbreviations

chl, chlorophyll; DBMIB 3,5-dibromo-3-methyl-6-isopropyl-8-benzoquinone; DCMU, 3-(3,4-dichlorophenyl)-1,1-dimethylurea; EDTA, ethylenediamine tetraacetate; FAD, flavin adenine dinucleotide; HEPES, N-2-hydroxyethyl-piperazine N'-2-ethane sulfonic acid; PS, photosystem

Introduction

Cyanobacteria from a group of primitive organisms in which plant-type photosynthesis evolved, involving the oxidation of water and the evolution of oxygen. This is remarkable since morphologically these organisms are more akin to photosynthetic bacteria. It now appears that at least a number of species of cyanobacteria can carry out a facultative anoxygenic form of photosynthesis involving only photosystem (PS) I and using sulfide as an electron donor for CO_2 fixation [2, 6, 7, 10, 12]. Up to now, this phenomenon was found to be most manifest in *Oscillatoria limnetica*.

The anoxygenic mode of photosynthetic growth is induced, after a lag phase of about two hours in about 48 hrs in the presence of sulfide [6, 7, 12].

Dedicated to Prof. L.N.M. Duysens on the occasion of his retirement

The presence of dithionite reduces the time of the completion of the induction considerably (to about 12 hrs) and abolishes almost completely the lag phase [4]. The absence of any effect of DCMU on growth, the lack of red drop and enhancement effect and the action spectrum of CO_2 fixation [10] show convincingly that only PS I is involved. If sulfide is excluded, *O. limnetica* shifts immediately back to oxygenic photosynthesis, involving PS II [10].

During anoxygenic growth, in the absence of CO_2 (and also in the presence of CO_2 at sulfide concentrations >4 mM) hydrogen is evolved in the light at a rate of about one-tenth of that of CO_2 fixation. This hydrogen evolution is also sulfide dependent [3]. It is of considerable interest to gain more knowledge about the anoxygenic light-driven electron transport from sulfide to the hydrogen evolving system as well as to the CO_2 fixation system, in view of the evolutionary links of these organisms with on one hand the chloroplast and on the other hand the photosynthetic bacteria. Such knowledge may also help to assess the feasibility to use these organisms to remove sulfides from waste and at the same time produce some storable energy.

Materials and methods

Oscillatoria limnetica, a gift from Dr. E. Padan, The Hebrew University, Jerusalem, Israel, was grown in a medium described by Cohen et al. [6] for both oxygenic and anoxygenic growth. The growth temperature was about $35\,^{\circ}$C. Whole cells were harvested by low speed centrifugation and resuspended in fresh growth medium. Spheroplasts were prepared from harvested whole cells resuspended in 8% glycin, 20 mM KH_2PO_4 and 25 mM EDTA (pH $= 6.8$) by 20 to 30 min incubation in 2‰ lysozyme and 1% NaCl. Debris was removed by low speed centrifugation. The spheroplasts were then precipitated by 10 min centrifugation at 600 × g and resuspended in the above mentioned buffer solution to which about 2% (w/v) NaCl was added. Thylakoid membranes were prepared from spheroplasts by mild osmotic shock (sometimes followed by 15 to 30 s sonication) and centrifugation at 2000 × g at low temperature ($4\,^{\circ}$C) to remove debris.

The supernatant containing the membrane fractions was centrifuged for 20 min at 28,000 × g. The pellet then was resuspended in a buffer solution containing 8% glycin, 2.5 mM HEPES, 2 mM KH_2PO_4 and 0.5 mM $MgCl_2$ at pH 7. When such preparations were made from anoxygenically grown cells, care was taken to keep the suspension anaerobic. *Anacystis nidulans*, strain UTEX 625, also a gift from Dr. Padan, was grown in the medium described in [6] but without Turks Island Salts at about $25\,^{\circ}$C. Only whole cells were used in the experiments.

Chlorophyll concentrations were measured according to the method given in [16].

Hydrogen production (and uptake) was measured with an Intersmat 10C gas chromatograph with katharometer detection and a molecular sieve column of 180 cm, 60–80 m/i in 20 ml test tubes stoppered with a rubber septum. The carrier gas was Ar. Oxygen evolution and uptake were measured with a Clark-type oxygen electrode.

Light-induced electron transport was investigated by measuring absorbance changes induced by actinic flashes obtained from an eximer laser-pumped dye laser (EMG 102 + FL 2000, Lamda Physik, Göttingen). The dye used was oxazine 4, which allowed tuning between 660 and 720 nm. Unless mentioned otherwise, the dye laser was tuned to 680 nm. The intensity of the flashes was saturating.

The absorbance changes were measured with a laboratory-build single beam spectrophotometer, controlled by a Tektronix 4051 microcomputer which was used also for data processing. The signals were averaged over 25 to 100 flashes, 12 s apart. The flash duration was 10 ns. Most of the chemicals used were purchased from Sigma. DBMIB was a gift from Dr. A. Trebst, Ruhr Universität, Bochum, FRG.

Results

Hydrogen production

Since light-induced hydrogen production with sulfide as electron donor occurs through a hydrogenase (there is no ATP requirement and no inhibition by uncouplers [3]) it seems important to compare hydrogenase activity in *O. limnetica* with that in other cyanobacteria in which no anoxygenic photosynthesis could be detected (e.g. *A. nidulans*). Hydrogenase activity was measured in whole cells in the dark using methylviologen as a mediator which was kept reduced by dithionite under anaerobic conditions [1]. The results are shown in Table 1, column 2. In *O. limnetica* the rate shown was reached almost immediately after the start of the experiment and remained constant over a considerable time. The rate measured in *A. nidulans* was substantially lower. This could indicate a low intracellular hydrogenase concentration in this organism but it could also be due to a hydrogenase which is predominantly suited for (recycling) hydrogen uptake. This would confirm findings of Peschek [15]. We found some support for an uptake hydrogenase by measuring high rates of H_2 uptake under anaerobic conditions in the presence of CO_2 in *A. nidulans* (results not shown).

Hydrogen production in the light with sulfide as an electron donor was tested also for both organisms. It was shown previously [5, 11] that anoxygenic photosynthesis as well as photoproduction of hydrogen from sulfide is induced in *O. limnetica* by anaerobic incubation with sulfide. The presence of dithionite accelerates the induction period by a factor of about 4. PS II is rendered inoperative; DCMU affected neither the photoproduction of hydrogen, nor growth in the presence of CO_2. In the experiment, whole

Table 1. Comparison of hydrogenase activity (column 2) and of sulfide dependent hydrogen production in the light (columns 3 and 5) in *O. limnetica* and *A. nidulans*. The hydrogenase activity is given as the rate of hydrogen evolution in the dark in the presence of 20 mM dithionite and 5 mM methylviologen [1]. Sulfide-dependent hydrogen production in the light was determined by transferring the cells to growth medium containing in addition 4 mM (*O. limnetica*) or 1 mM (*A. nidulans*) Na_2S, 10 μM DCMU, 10 μM HEPES and 10 mM dithionite and measuring the appearance of hydrogen in the gas phase. The rates given are those determined at the times indicated (columns 4 and 6) after the first appearance of hydrogen. Although the measurements of hydrogenase activity did not involve photosynthetic electron transport, rates are given in μM/H_2/mg chl *a* · h; the concentration of chlorophyll *a* in the original cell-samples is used in all cases as a reference to calculate rates of hydrogen evolution

| | Temp. °C | Hydrogenase activity μM H_2/mg chl *a* · h | Photoproduction of H_2 from S^{2-} | | | |
			μM/mg chl *a* · h	measured after	μM/mg chl *a* · h	measured after
O. limnetica	35	26.5	5.5	3 h	-20	30 h
A. nidulans	28	1.3	0.04	22 h		

cells of *O. limnetica* (grown oxygenically) as well as of *A. nidulans* strain UTEX 625 were suspended anaerobically in growth medium containing in addition Na_2S, dithionite and DCMU. Hydrogen production was measured chromatographically. Table 1, columns 3 and 5, gives the rates of hydrogen production in the light determined at various times (columns 4 and 6) after the first appearance of hydrogen in the gasphase. In the presence of dithionite this occurred about 15 min after the addition of sulfide. In the absence of dithionite this took about 2 h. Confirming the results of Oren and Padan [11], these results show that in *O. limnetica*, there is an induction period of about 12 hrs (48 hrs in the absence of dithionite) before the maximum rate was reached.

In *A. nidulans*, no appreciable rate of photoproduction of hydrogen could be detected even after very long induction periods. Peschek [14] showed that in *A. nidulans* (strain L 1402-1) in the presence of DCMU there is a maximum rate of CO_2 fixation at a concentration of 0.1 mM sulfide and no CO_2 fixation at concentrations >0.5 mM. If in *A. nidulans* (strain UTEX 625), as in *O. limnetica*, the rate of CO_2 fixation in the presence of sulfide and DCMU is about 10 times the rate of hydrogen production [3], the rate of CO_2 fixation would be of the same order of magnitude as the rate measured by Pescheck [13] in *A. nidulans* (strain L1402-1). In order to investigate the induction process in more detail, the hydrogenase activity was also measured in cell-free preparations from both oxygenic cells and cells induced to anoxygenic photosynthesis (by sulfide in the presence of dithionite). Membrane fractions were prepared from both type of cells as described in MATERIALS AND METHODS. The supernatants of the 28,000 × g centrifugation were both saved and hydrogenase activity was measured in the dark as before in all four preparations. The results are given in Table 2. From this experiment it it clear that not only in the induced (anoxygenic) cells but also in the "normal", oxygenic cells hydrogenase is present. The hydrogenase seems to be partially

Table 2. Hydrogenase activity in $\mu M\, H_2$/mg chl $a \cdot$ h in thylakoid membrane fractions and supernatant from *O. limnetica* measured in the dark in the presence of 20 mM dithionite and 5 mM methylbiologen.

	Oxygenic cells	Anoxygenic cells
Membranes	4.7	5.7
Supernatant	5.8	75.0
Total	11.7	80.7

soluble. In induced (anoxygenic) cells there is about 8 times more hydrogenase activity, most of it appearing in the supernatant after lysis of the cells. In the same preparations also sulfide-dependent hydrogen production was measured. In spite of the fact that a considerable amount of hydrogenase activity could be detected in these preparations (Table 2) no sulfide-dependent hydrogen production was found in membranes from oxygenically grown cells, even after addition of (native) supernatant. In none of the supernatants hydrogen production in the presence of sulfide was found either. Only in membranes from induced (anoxygenic) cells could sulfide-dependent hydrogen production be detected.

In order to investigate this further, the following experiment was set up. Membrane fractions from both oxygenic cells and induced (anoxygenic) cells were prepared as described above and the supernatants saved. Then four suspensions were made: oxygenic membranes in "oxygenic supernatant", oxygenic membranes in "anoxygenic supernatant", anoxygenic membranes in "oxygenic supernatant" and anoxygenic membranes in "anoxygenic supernatant". In these four preparations hydrogen photoproduction in the presence of 4 mM sulfide was measured. The results are shown in Table 3. These results show that the hydrogenase activity (max. rate of hydrogen production in the dark by dithionite-methylviologen, as shown in Table 2) cannot be the rate limiting step of the sulfide-dependent hydrogen production in the light. The sulfide-dependent hydrogen production requires, therefore, in addition to a functional photosynthetic electron transport system (present in all membrane fractions) and hydrogenase, a "sulfide oxidizing factor" that is present only in cells induced to anoxygenic photosynthesis. The fact is easily solubilized and can, in soluble form, mediate electron transport from sulfide by hydrogenase even in membranes from oxygenically grown cells (Table 3, second row, first column).

Electron transport

Light-induced electron transport in *O. limnetica* was examined by measuring absorbance changes at 435 nm induced by a single turn-over flash. At this wavelength, the absorbance changes probably are due predominantly to P700 oxidation [9]. Thylakoid membrane fractions were used. These membranes were devoid of PS II activity, as was determined by measuring oxygen

Table 3. Sulfide-dependent hydrogen production in the light in μM/mg chl $a \cdot$ h in membranes (columns) suspended in supernatants (rows) in the presence of $10\,\mu$M DCMU, 10 mM dithionite and 5 mM Na_2S

	Membranes from oxygenic cells	Membranes from anoxygenic cells
In supernatant from oxygenic cells	0	3.8
In supernatant from anoxygenic cells	1.5	6.7

evolution with a Clark-type oxygen electrode in the presence of $10\,\mu$M methylviologen. In addition, no ferricyanide reduction could be detected in illuminated thylakoid fractions in the presence of hydroxylamine. To determine the effect of sulfide in membrane fractions of oxygenically grown cells and membrane fractions from induced (to anoxygenic photosynthesis) cells, flash-induced kinetics of absorbance changes were measured in preparations similar to those used in the experiments on hydrogen evolution. The results are illustrated in Figure 1. Four preparations were made: oxygenic membranes suspended in "oxygenic supernatant" (A), anoxygenic membranes suspended in "oxygenic supernatant" (B), oxygenic membranes suspended in "anoxygenic supernatant" (C) and anoxygenic membranes suspended in "anoxygenic supernatant" (D). Curves 1 are those without sulfide, curves 2 are those obtained after addition of 4 mM Na_2S. Since addition of Na_2S to the suspensions increases the pH by about 0.8 unit, an appropriate amount of NaOH was added to the suspensions of the control curves 1. From the experiments it seemed that sulfide has a stimulating effect on the absorbance change, even in thylakoid fractions from oxygenic cells. In thylakoid fractions from oxygenic cells suspended in supernatant from the oxygenic preparation (Figure 1A) addition of sulfide does not seem to change drastically the kinetic behaviour. In both curves there is small relatively rapid phase and a larger much slower phase. If oxygenic thylakoid fractions are suspended in "anoxygenic supernatant", (Figure 1C), sulfide seems to increase the rapid phase to a much larger extent than the slow phase. In thylakoid preparations from anoxygenic cells sulfide had a more dramatic effect. The absorbance change, already being substantially larger in the absence of sulfide, decays much more rapidly when sulfide was added; the increase in the extent of the change seemed to be predominantly due to an increase of the rapid phase.

Figure 2 shows the effect of the plastoquinone analogue DBMIB on the flash-induced absorbance change at 435 nm in the presence of sulfide. Belkin and Padan showed [4] that DBMIB at a concentration of $20\,\mu$M inhibits about 70% of the sulfide dependent hydrogen photoproduction, concluding that sulfide donates electrons at the level of plastoquinone. The somewhat surprising result shown in Figure 2 does not support that conclusion; instead

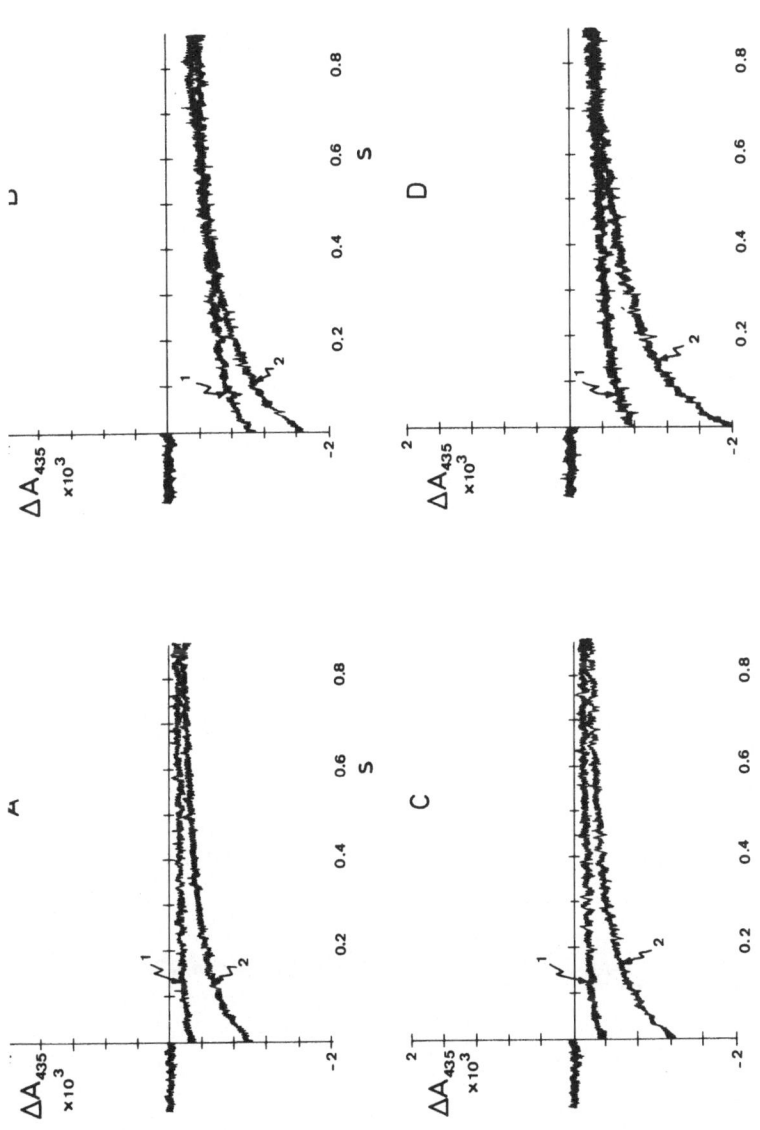

Figure 1. Flash-induced absorbance changes at 435 nm of (A) thylakoid membranes from oxygenic cells in "oxygenic supernatant", (B) thylakoid membranes from oxygenic cells in "anoxygenic supernatant", (C) thylakoid membranes from anoxygenic cells in "oxygenic supernatant", and (D) thylakoid membranes from anoxygenic cells in "anoxygenic supernatant". Curves 1 were measured in the presence of NaOH at a concentration to bring the pH to about 7.8. Curves 1 in the absence, curves 2 in the presence of 4 mM Na_2S. The intensity of the flash was saturating. The chlorophyll a concentration in the samples was 4 to 5 μg/ml. 25 Signals were averaged. Flashes were fired every 12 s

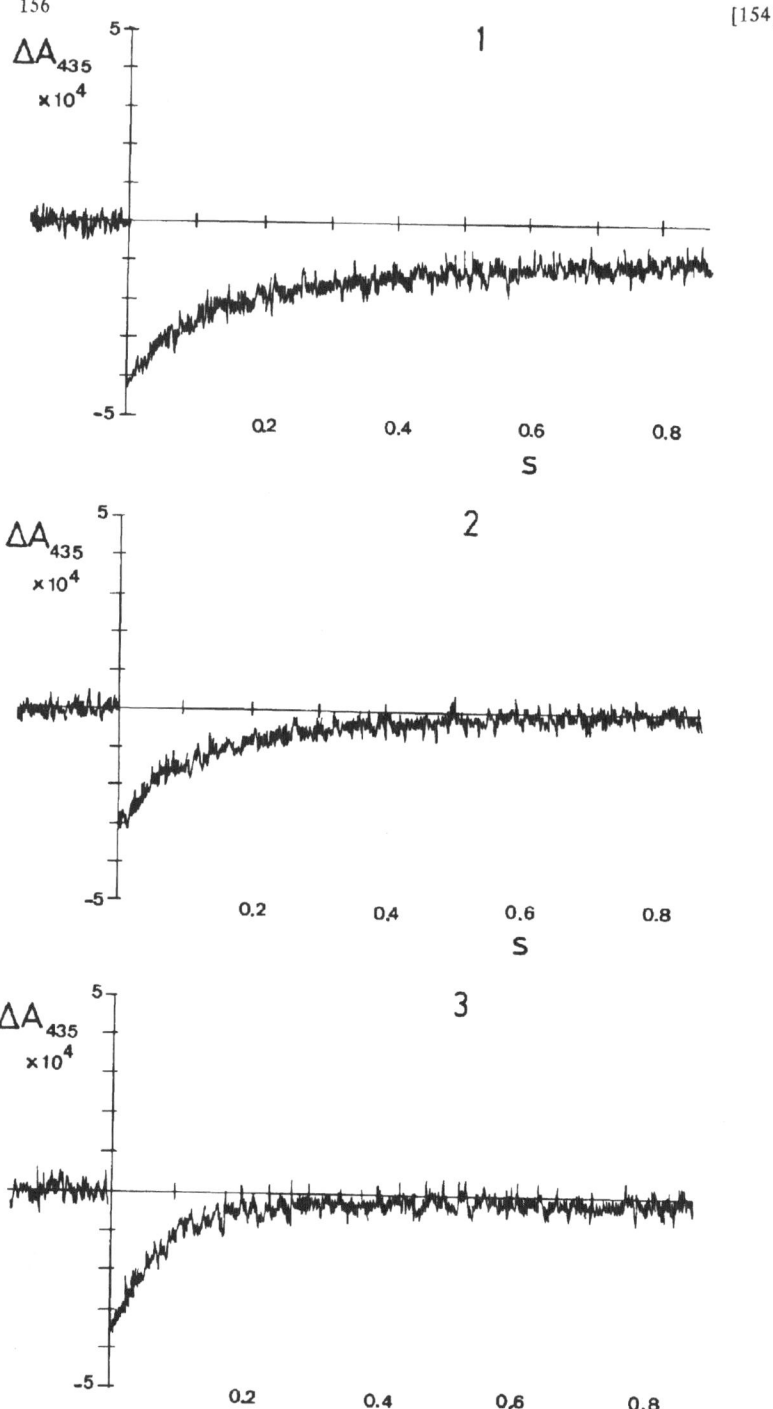

Figure 2. The effect of DBMIB on the flash-induced absorbance change at 435 nm in spheroplasts from induced cells (to anoxygenic photosynthesis) in the presence of 4 mM Na$_2$S and 1 μM gramicidin; curve 1, no DBMIB; curve 2, 10 μM DBMIB; curve 3, 40 μM DBMIB. The chlorophyll concentration was about 7 μg/ml. 25 Signals were averaged. Flashes were fired every 12 s

of inhibiting electron transport to P700, DBMIB seem to accelerate electron transport to P700.

Discussion

The experimental results described above show that a "sulfide oxidizing factor", synthesized during the induction process to anoxygenic photo-synthesis, is required to produce hydrogen in the light from electrons coming from sulfide. Without this factor sulfide-dependent hydrogen production occurs neither in oxygenic cells nor in membrane preparations from such cells, in spite of the presence of a considerable amount of hydrogenase in these cells. Addition of the factor to membrane preparations from oxygenic cells can make such preparation produce hydrogen in the light in the presence of sulfide (first column, second row of Table 3). However, the presence of sulfide alone in thylakoid membrane suspensions from oxygenic cells, even after several hours incubation time, never results in sulfide-dependent hydro-gen production in the light. This observation shows that synthesis of the "sulfide oxidizing factor" requires the integrety of the cell, i.e. the functioning of the genetic apparatus. This is in complete agreement with the finding of Belkin and Padan [5] that chloramphenicol, a protein synthesis inhibiting antibiotic, inhibits the induction to anoxygenic photosynthesis. The factor does not seem firmly bound to the thylakoid membrane of anoxygenic cells; apparently it is easily solubilized. Work is in progress in our laboratory to isolate and characterize the factor. It is tempting to compare it to the FAD-cytochrome c complex that has been reported [8] to mediate oxidation of sulfide in *Chromatium vinosum*.

The presence of the sulfide oxidizing factor seems to accelerate electron transport from sulfide to P700 (Figure 1). This is, at least qualitatively, in agreement with the experiments on hydrogen evolution. It provides for an indication (although no proof) that the electrons from sulfide to hydrogen in the reaction

$$Na_2S + 2 H^+ \rightarrow 2 Na^+ + H_2 + S$$

go through P700. Quantitatively there still are some difficulties with this interpretation; some electron transport from sulfide to P700 does not seem to require the factor (Figure 1A). Comparing Table 3 with Figure 1, there is about half of the amount of hydrogen evolved from oxygenic membranes in a suspension medium that contains the factor than there is in anoxygenic membranes in a medium without it, and there is a higher rate and extent of electron transport from sulfide to P700 in the latter case than in the former.

The result of the DBMIB experiment (Figure 2) is somewhat surprising. If the electrons from sulfide to the hydrogenase are moved by P700, one would expect, in view of the results from Belkin and Padan [4], inhibition of electron transport. Instead, an increase of the decay rate of the 435 nm absorbance change (P700 re-reduction ?) is found at concentrations equal or twice those used by Belkin and Padan. Ten times smaller concentrations had very little effect. It seemed as if DBMIB is shuttling electrons from sulfide to P700, or providing for an artificial cofactor for cyclic electron transport involving P700. The latter interpretation could explain both our observations and the result of Belkin and Padan [4]. More evidence is required, however, to support this interpretation.

References

1. Adams NWW, Mortenson CE and Chen JS (1981) Biochim Biophys Acta 594: 105–106
2. Belkin S and Padan E (1978) Arch Microbiol 116: 109–111
3. Belkin S and Padan E (1978) FEBS Lett 94: 291–293
4. Belkin S and Padan E (1983) J Gen Microbiol 129: 3091–3098
5. Belkin S and Padan E (1983) Plant Physiol 72: 825–828
6. Cohen Y, Padan E and Shilo M (1975) J Bacteriol 123: 855–861
7. Garlick S, Oren A and Padan E (1977) J Bacteriol 129: 623–629
8. Gray GO and Knaff DB (1982) Biochim Biophys Acta 680: 290–296
9. Masaru N and Katoh S (1983) Biochim Biophys Acta 725: 272–279
10. Oren A, Padan E and Avron M (1977) Proc Natl Acad Sci USA 74: 2152–2156
11. Oren A and Padan E (1978) J Bacteriol 133: 558–563
12. Padan E (1979) Annu Rev Plant Physiol 30: 27–40
13. Peschek GA (1978) Arch Microbiol 119: 313–322
14. Peschek GA (1979) Arch Microbiol 123: 81–92
15. Peschek GA (1984) Subcellular Biochemistry 10: 85–191
16. Porra RJ and Grimme LH (1974) Anal Biochem 57: 255–267

Photosynthesis Research 9, 159–166 (1986)
© 1986 Martinus Nijhoff/Dr. W. Junk Publishers, Dordrecht.

Oxidation of cytochrome c_2 by photosynthetic reaction centers of *Rhodospirillum rubrum* and *Rhodopseudomonas sphaeroides* in vivo. Effect of viscosity on the rate of reaction

H.N. VAN DER WAL[1], P.Y. GORTER[1] and R. VAN GRONDELLE[2]

[1]Department of Biophysics, Huygens Laboratory of the State University, P.O. Box 9504, 2300 RA Leiden and [2]Department of Biophysics, Physics Laboratory of the Free University, De Boelelaan 1081, 1081 HV Amsterdam, The Netherlands

(*Received 9 September 1985*)

Key words: cytochrome c_2, rhodospirillum rubrum, rhodopseudomonas sphaeroides, reaction center, diffusion

Abstract. In *Rhodospirillum rubrum* and *Rhodopseudomonas sphaeroides* it is shown that the oxidation of cytochrome c_2 involves a diffusion limited process. From analysis of the results it follows that the electron transfer probability must be very low. This is corroborated by in vitro studies using the isolated components.

Introduction

The first discovery in 1954 of the involvement of a cytochrome in photosynthesis was the observation of a photooxidation of cytochrome c-428 in whole cells of the photosynthetic bacterium *Rhodospirillum rubrum* by Duysens [3]. The cytochrome reaction which is by far the most extensively studied in bacterial photosynthesis is the electron transfer from the soluble cytochrome c_2 to the oxidized reaction center bacteriochlorophyll P-870$^+$. In whole cells and chromatophores of *R. rubrum* this reaction proceeds with a halftime of about 250 μs [11, 12]. It was concluded that the cytochrome is mobile on the timescale of electron transfer reactions (ca. 10 ms) with respect to the reaction centers. Assuming that at least 8 cytochrome molecules could rapidly diffuse among 15–20 reaction centers the flash intensity dependence of the amount of cytochrome c_2 oxidized could be explained [11]. In vitro the reaction has an apparent rate constant of $2.8 \times 10^7 \, M^{-1} \, s^{-1}$ at an ionic strength of 30 mM [9].

In whole cells and chromatophores of *Rhodopseudomonas sphaeroides* the oxidation of cytochrome c_2 occurs in two phases: a fast phase with a half time of about 3 μs and a slower one with a half time of 200–400 μs [7]; (The latter depends on the ionic strength of the medium [2, 8]). In vitro one also observes two phases in the cytochrome c_2 oxidation kinetics. The fast phase appears to arise from cytochrome c_2 bound to the reaction center and has a first order rate constant of $10^6 \, s^{-1}$ [7] and the slow phase has a pseudo first order rate constant of $3.10^3 \, s^{-1}$ at high cytochrome c_2 concentrations [7]. It was proposed that the slow phase reflects a different bound state [7].

Table 1. Viscosity in cP at different concentrations of ethylene glycol and at two temperatures. The poise is related to the SI units system by $1\,cP = 10^{-3}\,kg\,m^{-1}\,s^{-1}$

EG % v/v	15.5°C	22.0°C
0	1.21	1.02
10	1.59	1.31
20	2.01	1.67
30	2.78	2.23
40	3.72	2.91

Experiments with variable flash intensity did not indicate a mobility of cytochrome c_2 with respect to the reaction centers both in vivo [8]) and in vitro [7]. In a study in which reaction centers were incorporated in negatively charged artificial membranes a mobility of externally added cytochrome c_2 was observed in some cases [5, 6].

In this paper we describe a number of experiments with whole cells of R. rubrum and Rps. sphaeroides in which the viscosity of the medium and the temperature were varied in order to relate the resulting change in rate constant of electron transfer from cytochrome c_2 to P-870$^+$ to a simple diffusion model. From the obtained kinetic data second order rate constants are calculated based on assumed concentrations of cytochrome c_2 in the cells, and these rates will be compared to the rate constants from in vitro experiments.

Materials and methods

Cells of R. rubrum and Rps. sphaeroides were grown anaerobically at temperatures of 30°C in a medium described in [10], illuminated by two 150 W incandescent lamps. The cells were harvested after a growth of one day, washed once with fresh medium and suspended at an optical density of $5\,cm^{-1}$ at $880-960\,nm$ (R. rubrum) and $5\,cm^{-1}$ at $870-960\,nm$ (Rps. sphaeroides) in a mixture of growth medium and ethylene glycol. The optical pathlength in all the experiments was 1 mm.

Flash-induced absorption changes were measured using a single beam spectrometer described before [12]. The viscosity of mixtures of growth medium and ethylene glycol were measured with the gravitational method on a Schott AVS/G automatical viscosity system and the results are given in Table 1.

The extinction coefficients used were $20\,mM^{-1}\,cm^{-1}$ for P-870 reduced minus oxidized at 604 nm and $60\,mM^{-1}\,cm^{-1}$ for cytochrome c_2 reduced minus oxidized at 420 nm.

Results

Figure 1 shows the kinetics of flash-induced oxidation and reduction of P-870$^+$ measured at 604 nm in R. rubrum whole cells. The fastest trace,

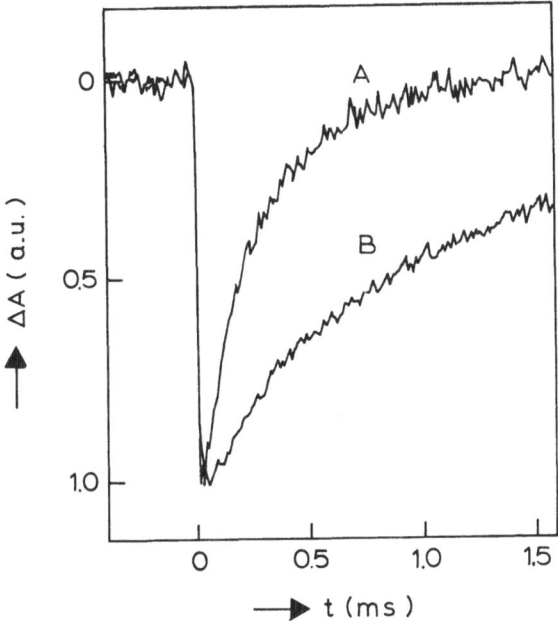

Figure 1. Oxidation reduction kinetics of P-870 in *R. rubrum*. Whole cells of *R. rubrum* were suspended in a mixture of growth medium and ethylene glycol to an optical density of 5 cm^{-1} at 880–960 nm. The traces were measured at 604 nm: (A) 0% ethylene glycol, (B) 40% ethylene glycol. The amounts of P-870 oxidized in the flash were 1 μM (A) and 0.7 μM (B)

measured in the absence of ethylene glycol, has a halftime of 200 μs, which is also the halftime for oxidation of cytochrome c_2 as measured at 420 nm (data not shown). This halftime is less than reported in [12] where the reduction of P-870$^+$ proceeded in two phases with halftimes of 300 μs and 15 ms. We ascribe this difference to different light intensities during growth of the cells. A high light intensity has the effect of increasing the amount of cytochrome c_2, in this case from 0.5 to close 1 molecule per reaction center.

The increase in viscosity (η) of the medium caused a decrease of the electron transfer rate (Figure 1); in 40% ethylene glycol the halftime was 800 – 900 μs. From the slope of the kinetics immediately after the flash the initial rate of P-870$^+$ reduction was determined as

$$k = - \frac{d[\text{P-870}^+]}{dt} \bigg/ [\text{P-870}^+]_{\text{max}}.$$

Two methods were used to find k, i.e.: (1) drawing a tangent at t = 50 μs after smoothing the kinetics by hand and (2) fitting the kinetics by computer with two exponentials. Both methods gave the same results. These results are shown in Figure 2 as k versus $1/\eta$ giving a fairly linear plot. To rule out the

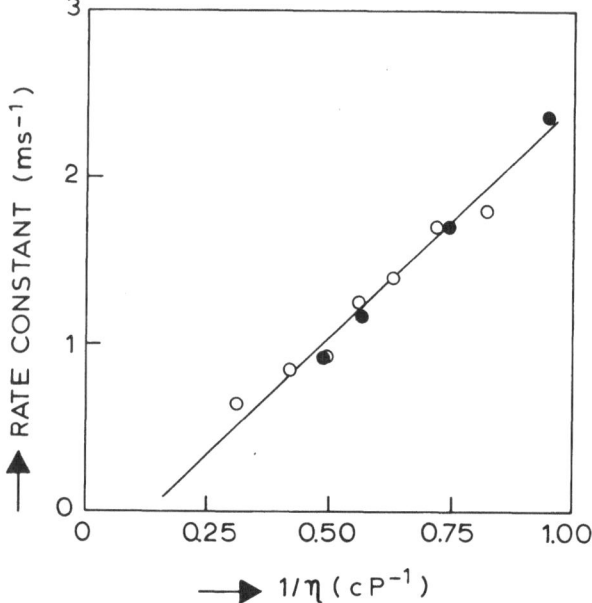

Figure 2. The rate of reduction of P-870$^+$ by cytochrome c_2 in whole cells of *R. rubrum* at different viscosities. The temperature was held constant at 15.5 °C (o) and 21 °C (●). The drawn line is a least squares linear approximation to the points. 1 cP = 10^{-3} kg m^{-1} s^{-1}

possibility that k was affected by ethylene glycol other than through its viscosity the experiments were performed at two different temperatures. This gave the opportunity to obtain the same viscosity at very different concentrations of ethylene glycol and Figure 2 shows that the same results are obtained proving that other effects of ethylene glycol are negligible.

We tried to calculate theoretical values of the rate constant with two models. From standard reaction theory the total rate k_d of a diffusion controlled second order reaction between species A and B is given by

$$k_d = f4\pi r_0(D_A + D_B)N^2 C_A C_B$$

where f is the fraction of collisions of A and B resulting in a reaction, r_0 is the distance of closest approach, C_A and C_B are the concentrations of species A and B in mol/m^3, D_A and D_B the diffusion coefficients, and N is Avogadro's number. We identify A with the membrane-bound reaction center and B with the cytochrome c_2 molecule, and assume that D_A equals zero. D_B is given by the Stokes-Einstein relation as $D_B = kT/6\pi r\eta$, where r is the radius of cytochrome c_2 (3.4 Å) [1] and the other parameters have their usual meaning. Assuming that $r_0 = r$, the normalized rate of P-870$^+$ reduction is then given by

$$\frac{d[\text{P-870}^+]}{dt} \bigg/ [\text{P-870}^+]_{max} = f\tfrac{2}{3} RTC_B/\eta,$$

thence plotting the normalized rate at 294 K versus $1/\eta$ yields a straight line with slope $1.6\ 10^3\ f\ C_B$. Assuming there are 15 molecules of cytochrome c_2 in a volume of 1 chromatophore, viz. a sphere with a radius of 25 nm, the concentration, C_B, of cytochrome c_2 equals $0.4\ \text{mol/m}^3$ ($= 0.4\ \text{mM}$). This gives a slope of $650\ f\ J \cdot m^{-3}$.

The least squares linear fit of the experimental results given in Figure 2 yields a slope of $3.0 \pm 0.2\ J\,m^{-3}$. Thus the theoretical models based on free diffusion of all the cytochrome c_2 present give values that are about 200 times too large. Several explanations can be suggested to interpret this difference. Not all cytochrome may be involved in the diffusion process because a large fraction of it is in compartments without reaction center; if the diffusion is along the membrane it may be severely hampered, or the probability of electron transfer 'f' may be low. We have an observation that seems to refute the first two possibilities. Experiments with isolated reaction centers and cytochrome c_2 from *R. rubrum* yield a second order rate constant for the reduction of P-870$^+$ of $1.5\ 10^7\,M^{-1}\,s^{-1}$ at an ionic strength of 100 mM. In this case no membranes were present. This should be compared with the rate constant in whole cells which is $0.5\ 10^7\,M^{-1}\,s^{-1}$. Therefore, neither an overestimation of cytochrome c_2 nor the presence of the membrane can be invoked to explain the large discrepancy between theory and experiment and we conclude that the probability of reaction 'f' is in the order of $10^{-2} - 10^{-3}$.

The above experiments were repeated with *Rps. sphaeroides*. In this species the flash-induced oxidation of cytochrome c_2 is biphasic [7] with a fast phase that was not resolved by our apparatus ($t_f \lesssim 5\ \mu s$ in [7]) and a slow phase of $t_s = 100\ \mu s$. The cytochrome c_2 oxidation kinetics of Figure 3 were obtained by correction of the signal for P-870$^+$ contribution. Figure 3 shows that the effect of high viscosity on the kinetics was to slow down t_s from about 100 μs to more than 400 μs at 40% ethylene glycol while the extent of the fast phase diminished. In experiments with different concentrations of ethylene glycol in which the viscosity was equalized by adjusting the temperature it appeared that t_s was the same but the extent of the fast phase decreased with increasing ethylene glycol concentrations. This indicates, in our opinion, that ethylene glycol in some way inhibits the fast cytochrome oxidation while the slow phase is only affected by viscosity as is the case in *R. rubrum*.

In Figure 4, t_s^{-1} is plotted versus $1/\eta$. We did not attempt to calculate an initial rate from the traces as given in Figure 3 because of the presence of the fast phase. The value of t_s was obtained taking the kinetics between 20 μs and the point where the kinetics run flat. For *R. rubrum* such a procedure yielded essentially the same plot at Figure 2.

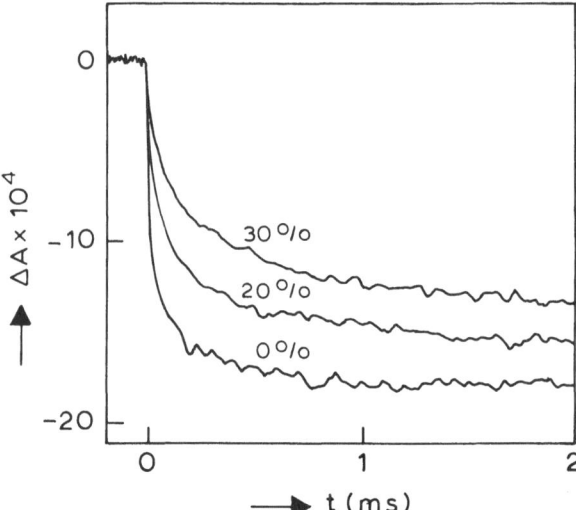

Figure 3. Oxidation kinetics of cytochrome c_2 in *Rps. sphaeroides*. Whole cells were suspended in a mixture of growth medium at 21°C and the amounts of ethylene glycol given in the figure

From the linear relationship between t_s^{-1} and $1/\eta$ we conclude that the slow phase in the cytochrome c_2 oxidation is also a diffusion controlled reaction.

Discussion

In all purple bacteria investigated the reduction pathway of P-870⁺ involves the oxidation of a soluble cytochrome. The rate constant of oxidation is determined by the rate of diffusion and the probability of electron transfer on collision. If the viscosity of the solvent is increased, only the first factor is affected in a manner expected for a diffusion reaction. For *R. rubrum* the probability of reaction upon collision is low: of the order of 10^{-2} to 10^{-3}. That the reaction between isolated cytochrome c_2 and *R. rubrum* reaction centers at physiological ionic strength has a slightly increased reaction probability compared to the rate in vivo may be explained by the presence of the membrane. In the latter case cytochrome c_2 diffusion is hampered similarly to the situation encountered with cytochrome c_2 added to negatively charged membranes [8]. From the two sets of points in Figure 2 we can conclude that the activation energy for the reaction of P-870⁺ and cytochrome c_2 is less than about 2 kcal/mol. A larger activation energy would have been visible as a shift of the two sets of points relative to each other. In [5, 6] a value of 8 kcal/mol was reported for the reaction in artificial membranes, but in this case the effect of the temperature on viscosity was disregarded.

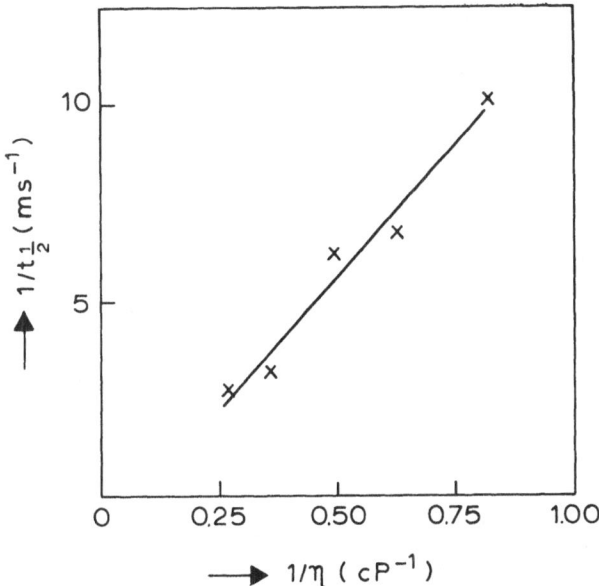

Figure 4. The rate of oxidation of cytochrome c_2 by P-870$^+$ in whole cells of *Rps. sphaeroides* at different viscosities. The drawn line is a least squares linear approximation to the points

For *Rps. sphaeroides* the reaction probability upon collision is higher than in *R. rubrum*. This difference is also apparent in vitro: with isolated cytochrome c_2 and reaction centers from *Rps. sphaeroides* (strain R-26) at 100 mM ionic strength the reaction rate for electron transfer is about 5.10^7 to 1.10^8 M^{-1} S^{-1} [unpublished observations] i.e. about five times higher than in *R. rubrum*.

In *Chromatium vinosum* soluble cytochrome c-551 is oxidized in a flash with a half time of about 400 μs. However, a direct comparison with *R. rubrum* and *Rps. sphaeroides* is difficult because in this case the electron is transferred from soluble c-551 to bound cytochrome c-555, but from the observed reaction rate [13] it may be surmised that the reaction probability is likewise small.

In all these systems, despite the low reaction probability, the oxidation of soluble cytochrome is still sufficiently fast under physiological conditions to compete efficiently with the rate-limiting steps in the ubiquinol cytochrome c_2 oxidoreductase which is in the order of 10 ms.

Many details of the soluble cytochrome reactions gradually become apparent, which allows a better understanding of the function of this important protein in energy conserving processes. This is in striking contrast to the persistance of our ignorance about the function of cytochrome c-428.

References

1. Canto CR and Schimmel PR (1980) Biophysical Chemistry, part II. San Francisco: Freeman WH and Comp
2. Dutton PL and Prince RC (1978) in Clayton RK and Sistrom WR, eds. The Photosynthetic Bacteria, pp. 525–570. New York: Plenum Press
3. Duysens LNM (1954) Nature 173: 692–693
4. Montroll EW (1969) J Math Phys 10: 753–765
5. Overfield RE and Wraight CA (1980) Biochemistry 19: 3322–3327
6. Overfield RE and Wraight CA (1980) Biochemistry 19: 3328–3334
7. Overfield RE, Wraight CA and Devault D (1979) FEBS Lett 105: 137–142
8. Prince RC, Cogdell RJ and Crofts AR (1974) Biochim Biophys Acta 347: 1–13
9. Rickle GK and Gusanovich MA (1979) Arch Biochem Biophys 197: 589–598
10. Slooten L (1972) Biochim Biophys Acta 256: 452–466
11. van Grondelle R (1978) Doctoral thesis, University of Leiden
12. van Grondelle R, Duysens LNM and van der Wal HN (1976) Biochim Biophys Acta 449: 169–187
13. van Grondelle R, Duysens LNM, van der Wel JA and van der Wal HN (1977) Biochim Biophys Acta 461: 188–201

Photosynthesis Research 9, 167–179 (1986)
© *1986 Martinus Nijhoff/Dr. W. Junk Publishers, Dordrecht.*

Photooxidation of mitochondrial cytochrome c by isolated bacterial reaction centers: Evidence for tight-binding and diffusional pathways

R.E. OVERFIELD and C.A. WRAIGHT

Department of Physiology and Biophysics and Department of Plant Biology, University of Illinois, Urbana, Illinois 61801, USA

(Received 29 September 1985)

Abstract. The binding of horse heart mitochondrial cytochrome c to isolated reaction centers from *Rhodopseudomonas sphaeroides* is described. The kinetics of photooxidation of cytochrome c following a short actinic flash is compared to the expected binding state of the cytochrome at various concentrations and at different ionic strengths. At low ionic strength a very tight binding site ($K_D \leq 10^{-8}$ M) is apparent which is non-functional with respect to electron donation to the bound reaction center. This tightly bound cytochrome can react with another reaction center in a diffusion limited, second order process. A weaker binding site ($K_D \simeq 0.3 \cdot 10^{-6}$ M) is also observed which is associated with rapid, first order electron transfer from cytochrome to reaction center. Both binding processes are weakened in the presence of salt and there is no detectable binding in 100 mM NaCl. Under such conditions cytochrome oxidation is entirely a diffusional, second order process. However, analysis of the flash intensity dependence of the extent of cytochrome oxidation, by the method of van Grondelle (van Grondelle, R. (1978) Ph.D. Thesis, State University, Leiden) indicated that the cytochrome was not freely mobile even in 100 mM NaCl, at least in the sense that reduced cytochrome only slowly dissociates from unactivated reaction centers. An overall kinetic/equilibrium scheme for cytochrome c binding and photooxidation by reaction centers is presented. This is very similar to that described earlier for cytochrome c_2 (Overfield, R.E., Wraight, C.A. and DeVault, D. (1979) FEBS Lett. *105*, 137–142), but the tight binding site and associated diffusion controlled oxidation is unique to cytochrome c.

Introduction

In chromatophores from *Rhodopseudomonas sphaeroides*, the kinetics of oxidation of cytochrome c_2 by the reaction center have been interpreted as indicating a tight binding between the cytochrome and reaction center [3]. On the other hand, van Grondelle and coworkers showed that functionally similar cytochromes in *Rhodospirillum rubrum* (cytochrome c_{420}) [17] and *Chromatium vinosum* (cytochrome c_{551}) [18] were mobile, and this was supported by the light saturation characteristics of cytochrome oxidation in a short flash [19]. A similar analysis in *Rp. sphaeroides*, however, supported the previous notion of a long-lived complex between cytochrome c_2 and the reaction center [11]. However, we subsequently showed that for cytochrome c_2 and isolated reaction centers, lack of mobility, as indicated by

---

Dedicated to Prof. L.N.M. Duysens on the occasion of his retirement.

the flash-saturation analysis, is not necessarily correlated with tight binding [9]. Furthermore, it was pointed out that the method is limited to detecting the mobility of the reduced cytochrome with respect to unactivated reaction centers. Most recently, Matsuura has shown that the kinetics of cytochrome c_2 oxidation in chromatophores and whole cells of *Rp. sphaeroides* are not responsive to changes in ionic strength, apparently supporting the non-diffusional nature of this process [7]. The problem is clearly not yet resolved.

We report here a cautionary note on the lack of correlation between binding characteristics and the kinetic order of associated processes. The oxidation of mitochondrial cytochrome c by isolated reaction centers from *Rp. sphaeroides* can show second order kinetics associated with very tight binding, and apparent lack of mobility not correlated with any detectable binding.

Methods

Reaction centers were prepared from *Rp. sphaeroides*, strain R26, essentially as described earlier [9]. Lauryl dimethylamine N-oxide (LDAO or Ammonyx-LO) was a gift of the Onyx Corporation, Division of Millimaster, Jersey City, N.J. 07023. The purified reaction centers were dialyzed for at least 24 hours against several changes of 10 mM Tris-HCl, pH 8.0 to remove $(NH_4)_2SO_4$ and NaCl used in the purification. The ratio of absorbance at 280 nm and 800 nm was 1.2–1.3 except for the data of Figure 1a.

Incorporation of the reaction centers into unilamellar bilayer vesicles was accomplished by a modification of the technique of Brunner et al. [2] as described earlier [8]. Phosphatidylcholine was purified from egg yolk by the method of Singleton et al. [15].

Horse heart cytochrome c (Sigma Type III) was prepared in Tris buffer and reduced to about 95% by minute additions of sodium dithionite.

The equilibrium binding determinations were made with Biogel P-300 columns. A column 40 × 1 cm was pre-equilibrated with cytochrome c in the reduced form in 10 mM Tris at pH 8.0 with or without 100 mM NaCl. The reaction centers were introduced in 1 ml at 1–10 μM, sometimes with extra cytochrome c. Fractions were collected, reduced with dithionite and measured for absorbance in the range 500–650 nm. The free level of cytochrome c was determined in the fractions before and after the reaction center peak from the absorbance at 550 nm. The reaction center concentration was determined from the absorbance peak at 597 nm using an extinction coefficient of 58 mM^{-1} for 597–645 nm. The cytochrome c in these fractions was determined from the absorbance at 550 nm, after subtracting 0.45 times the absorbance at 597 nm, to correct for the contribution of the reaction center bacteriopheophytin absorbance at this wavelength. The bound cytochrome was then taken as the total minus the free and the results were presented as Scatchard plots.

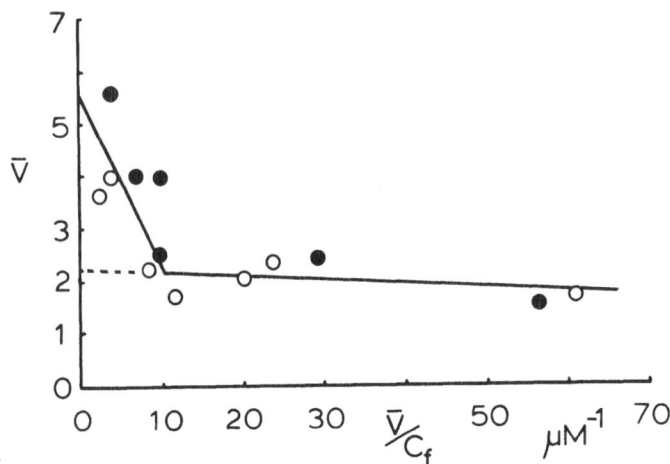

Figure 1. Binding of cytochrome c to reaction centers, determined by equilibrium exclusion chromatography. $\bar{V}$ is the number of cytochromes bound per reaction center; c_f is the free cytochrome c concentration. ○ – reaction centers solubilized in < 0.05% LDAO (RC 280/800 = 1.45). ● – RC's incorporated into egg *PC* vesicles (RC 280/800 = 1.2)

Kinetic measurements of cytochrome c oxidation were performed on a locally built spectrophotometer using the wavelength pairs 550–540 nm or 520–480 nm. The sample was deoxygenated and the pH and ambient redox potential were monitored in the cuvette and controlled. 20 μM naphtho-quinone (NQ) was added to mediate the redox potential and serve as a slow ($t_{1/2} > 1$ s) shuttle from the reaction center acceptor region to the oxidized cytochrome (see Figure 2c). Four minutes were allowed between actinic flashes which were provided by a rhodamine 6 G liquid dye laser with a 400 ns pulse width (Phase-R, Model DL 1100, Durham, N.H. 03855). The laser flash was more than 95% saturating. The faster kinetic measurements, with a time resolution of 30 ns, were done in single beam mode at 417 nm on the apparatus of D. DeVault, using a ruby laser with a 20 ns pulse width.

Results

Equilibrium binding studies

Binding determinations by equilibrium chromatography at low ionic strength showed that cytochrome c bound to reaction centers at multiple sites. Scatchard plots, shown in Figure 1, indicated 2 ± 1 cytochromes tightly bound ($K_D \leq 10^{-8}$ M) and 3 ± 2 cytochromes bound with $K_D \simeq 0.3 \times 10^{-6}$ M. Binding was similar for reaction centers solubilized in low levels of detergent (LDAO) or incorporated into egg phosphatidylcholine vesicles (Figure 1). The reaction centers used in the vesicle preparation were of high purity (see

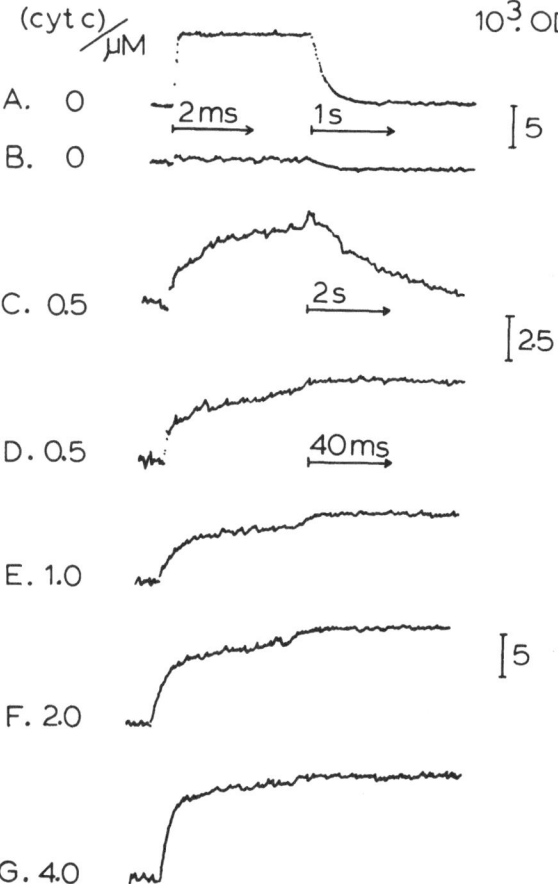

Figure 2. Kinetics of cytochrome c oxidation. A. P$^+$ absorbance change following a laser flash, in the absence of cytochrome c, single beam measurement at 540 nm; 1 μM reaction centers in 10 mM Tris, pH 8, 20 μM naphthoquinone, approx. 0.05% LDAO. Time constant, 20 μs. B, as for A, but dual wavelength measurement at 550–540 nm showing that the P$^+$ change is almost entirely cancelled. C–G. as for B, except for additions of cytochrome c (μM) as indicated. Note dual time base. The 2 ms marker is for the first half of all traces; the second half time base is 1 s for A and B, 2 s for C, and 40 ms for D through G. Note, also, the change in vertical sensitivity (2.5 × 10^{-3}) for C and D

legend to Figure 1) and it seems unlikely, therefore, that either of the binding sites is artifactual in nature.

Binding of cytochrome c was ionic strength dependent and could not be detected (*i.e.* $K_D \geq 10^{-5}$ M) in 100 mM NaCl. This is consistent with the binding at low ionic strength being due to an electrostatic association between the negative reaction center (pI $\sim$ 6.0) [12] and the positive cytochrome (pI $\sim$ 10.6) [6].

The characteristics described here for the binding of cytochrome c to

isolated reaction centers are at least partly in conflict with other reports
in the literature. Using reaction centers suspended in 0.1% triton X-100, at
low ionic strength, Rosen et al. [14] reported a single binding site with a
dissociation constant of 0.4×10^{-6} M. For reaction centers incorporated into
phosphatidyl choline vesicles, Pachence et al. [10] observed three cytochromes
bound per reaction center, with a single discernible dissociation constant,
also of 0.4×10^{-6} M. The binding affinity decreased as the ionic strength
was raised, becoming undetectable above 100 mM KCl. This latter study
agrees well with our observations on the weaker binding site, and there is
generally good agreement on an appropriate value for this dissociation
constant. However, neither of the previous studies detected the very tight
binding that we observed. Multiple binding sites for cytochrome c, including
a tight site with $K_D \leq 10^{-8}$ M, have been reported for cytochrome oxidase
[5]. We do not have any suggestions for this discrepancy except to point to
the rather high level of detergent used in ref. 14. Our data cannot claim
great precision in the number of cytochromes bound, but the existence of
two classes of binding site, one with very high affinity, was very clear and is
supported by the kinetics of oxidation of cytochrome c.

Kinetic measurements

The kinetics of oxidation of horse heart cytochrome c by reaction centers,
following a saturating laser flash at low ionic strength, are shown in Figure 2.
At low concentrations the reaction is second order, as shown earlier [12]. At
higher concentrations, however, a fast phase appears. The magnitude of the
fast phase titrates in with an apparent dissociation constant of about 10^{-6} M.
This phase is properly resolved in Figure 3 where the time base covers three
orders of magnitude following the excitation pulse. At low ionic strength
the apparent halftime was about $30 \mu s$, in good agreement with the value of
$25 \mu s$ first reported by Ke et al. [6]. The halftime was independent of
concentration, indicating a collision-independent process.

With addition of salt, the $30 \mu s$ phase was retarded, revealing an even
faster process (Figure 3E). When first clearly resolved, at 20 mM NaCl, this
phase exhibited a halftime of about $2 \mu s$. It slowed slightly and diminished
in magnitude as the salt concentration was increased and was no longer
observable at 60 mM NaCl.

In addition to the appearance of the $30 \mu s$ fast phase of cytochrome
oxidation, the slow phase of oxidation accelerated as the cytochrome
concentration was increased (Figure 2). This effect was more readily apparent
at moderate ionic strength due to the better separation of the kinetic phases.
In Figure 4, the inverse halftime of the slow phase in 100 mM NaCl is shown
as a function of cytochrome concentration, using reaction centers incorporated into phosphatidyl choline vesicles. We previously reported identical
behavior for cytochrome c_2 [8]. As pointed out by Bashford et al. [1]
this particular behavior, with a minimum at the equimolar point, is very

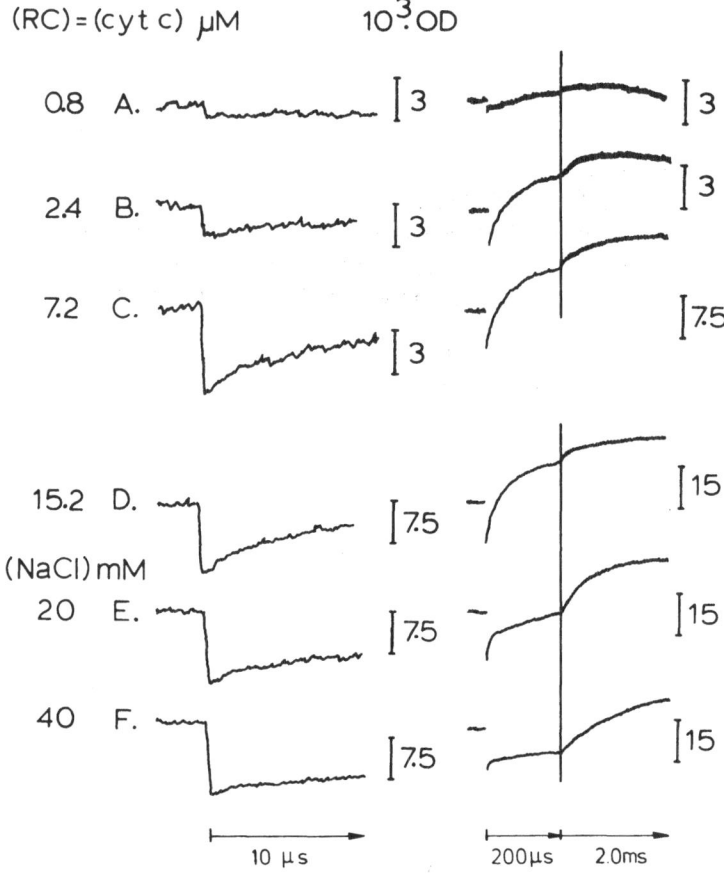

Figure 3. Concentration dependence of the rapid kinetics of oxidation of cytochrome c by reaction centers at equimolar concentration. Single beam measurement at 417 nm, low ionic strength (10 mM Tris, pH 8.0). The time course follows three orders of magnitude: the traces on the left represent the first 10 μsec; those on the right show the first 200 μsec and the next 2 msec for the same flash. A–D, the concentrations of both cytochrome and reaction centers were increased together as indicated (μM). Note the changes in vertical sensitivity. D–F, the effect of addition of salt: a faster phase (2 μs) is revealed (E), which was hidden under the 30 μsec phase in trace D. 20 μM NQ, 2 μM TMPD, and 1 mM ascorbate were used to re-reduce the flash oxidized cytochrom c on a slow time scale

characteristic of a second order process. For the slow phase of cytochrome oxidation by isolated reaction centers in solution, under identical conditions to Figure 4, we confirmed that the reaction is first order in both cytochrome c and reaction center (unpublished data) and, thus, second order overall. The bimolecular rate constant was determined to be $2 \cdot 10^7 \, M^{-1} \cdot s^{-1}$. In the absence of salt the rate of cytochrome oxidation was much faster but was also multiphasic, rendering the kinetic analysis more complicated. At low

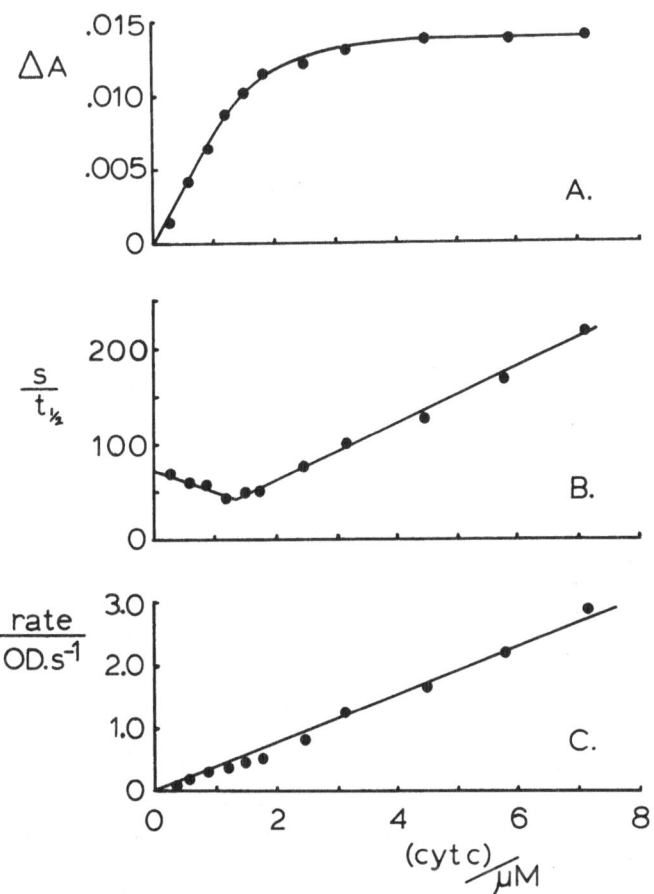

Figure 4. Concentration dependence of cytochrome oxidation by reaction centers in egg phosphatidylcholine vesicles. RCs ($RC_{out} = 1.3 \mu M$) were titrated with cytochrome c in 100 mM NaCl, 10 mM Tris, pH 8.0. Cytochrome oxidation was measured at 520–480 nm, which completely cancels the P^+ absorbance changes. A, the extent of cytochrome c oxidation after a single flash. B, inverse half time of the monotonic kinetics. C, the product of the extent (A) and half time (B); for first order or pseudo first order kinetics this is a measure of the initial rate

concentrations the reaction was second order, with a value for the second order rate constant at low ionic strength ($k \simeq 6 \cdot 10^9 \, M^{-1} \cdot s^{-1}$; see also ref. 8) close to the von Smoluchowski limit predicted from the size of the protein reactants. This is considerably higher than the $4 \cdot 10^8 \, M^{-1} \cdot s^{-1}$ reported by Prince et al. [11]. This discrepancy is almost certainly due to the very high concentration of LDAO (2%) used in the earlier study. We have found the reaction to be severely inhibited by concentrations of LDAO in excess of 0.2%, and Rickle and Cusanovich [13] have reported that cytochrome c is

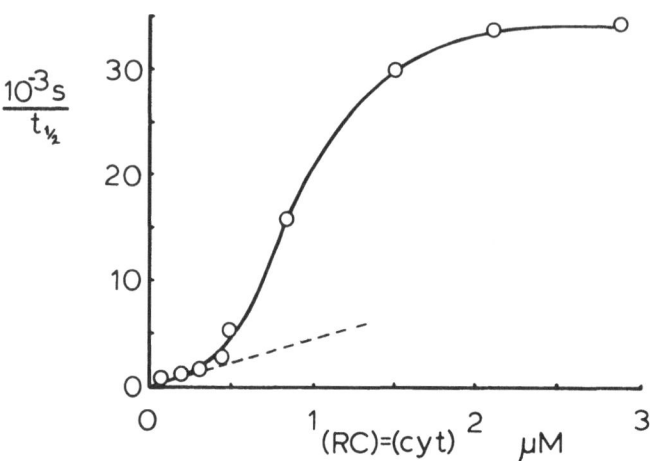

Figure 5. Concentration dependence of cytochrome c oxidation by reaction centers in solution. Reaction centers and cytochrome c were increased at equimolar concentrations in 10 mM Tris, pH 8.0, 20 μM NQ and $\leqq$ 0.05% LDAO. The net inverse half time of the multiphasic process is plotted. At low concentrations, the linear slope is indicative of a second order process with $k_{12} = 3 \times 10^9 \, M^{-1} \, s^{-1}$. At high concentrations the process becomes collision-independent. Cytochrome oxidation was measured at 550–540 nm

denatured by such levels of LDAO, causing a massive drop in the oxidation-reduction midpoint potential.

The multiphasic nature of cytochrome oxidation at low ionic strength was clearly revealed by the dependence of the overall halftime on concentration (Figure 5). Reaction centers and cytochrome c were kept at equimolar concentrations and increased together, as in Figure 3, and the net halftime of oxidation was determined. At low concentrations, even though binding was indicated by the equilibrium determinations ($K_D \leqq 10^{-8}$ M), the inverse net halftime increased linearly with concentration. The second order rate constant, determined from the slope of this line, was $3 \cdot 10^9 \, M^{-1} \cdot s^{-1}$ in good agreement with that given above. At intermediate concentrations, the net halftime of oxidation decreased due to contributions to the overall time course from the faster kinetics of cytochromes bound with $K_D \sim 10^{-6}$ M. These dominated the reaction at high concentrations and the oxidation rate became limited in a pseudo-first order process with an apparent rate constant of $2.3 \times 10^4 \cdot s^{-1}$.

Light-saturation of flash-induced cytochrome oxidation

The mobility of the cytochrome c can be examined by the dependence on flash intensity of cytochrome oxidation by reaction centers present in excess [19]. In Figure 6, the concentrations of reaction centers and cytochrome c were 1 μM and 0.5 μM, respectively. The fraction of reaction centers activated

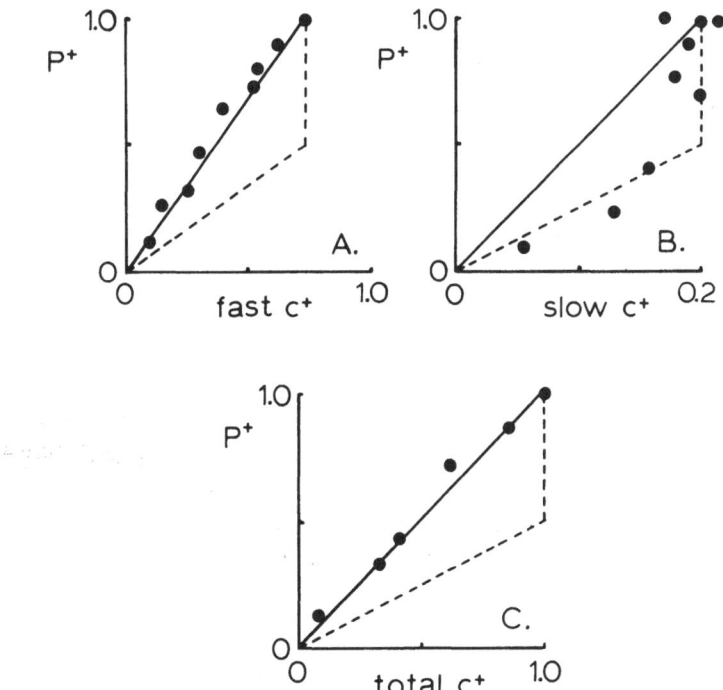

Figure 6. Light saturation behaviour of cytochrome c oxidation by reaction centers in solution. The intensity of the laser flash was attenuated with neutral density filters and the extents of reaction center activation (P^+) (at 540 nm) and cytochrome oxidation (c^+) (at 550–540 nm) were determined relative to the maximum amounts seen with saturating flashes: reaction centers (1.0 μM), cytochrome c (0.5 μM), 20 μM naptho-quinone. The amplitudes of fast (A) and slow (B) phases of cytochrome oxidation at low ionic strength (10 mM Tris, pH 8.0) were determined in dual time base. Note the different scales on the abscissa. C, the extent of the single, slow phase of cytochrome c oxidation observed in the presence of 100 mM NaCl. The dashed lines represent the expected result if the cytochrome were ideally free to diffuse; the solid lines represent the expectation for a long-lived complex

was varied by attenuation of the laser flash and the extent of cytochrome oxidation was recorded. At low ionic strength (Figure 6a) the extent of the fast cytochrome oxidation phase ($t_{1/2} \sim 30 \mu$s) increased linearly with the proportion of reaction centers activated. Since activation of all the reaction centers was necessary to ensure oxidation of all the rapidly reacting cytochrome we may conclude that this cytochrome complement was not free to diffuse between reaction centers on the time scale of the charge recombination reaction (> 0.1 s).

The slow phase oxidation, on the other hand, exhibited some degree of mobility as it became maximal at less than saturating flash intensities (Figure 6b). However, the equilibrium binding studies showed dissociation constants of 10^{-8} M and 10^{-6} M at low ionic strength and, therefore, at the concentrations used here essentially all the cytochrome was bound. Since the reaction kinetics of the slow phase indicated a collisional process (Figure 5) we must conclude that the tightly bound cytochrome is able to donate an electron to a different reaction center than the one with which it is associated and that the diffusion involved in this concentration range is that of the reaction center-cytochrome c complex. It should be noted that the relative amplitudes of fast and slow donation phases are not in simple agreement with competition by two independent binding sites for the low level of cytochrome. The rather large amplitude of fast oxidation, for which $K_D \sim 10^{-6}$ M, suggests that a cytochrome c bound to a tight site $(K_D \leq 10^{-8}$ M) can also bind to the weaker site. The possibility that cytochrome c_2 can simultaneously bind two reaction centers, inducing dimer formation, has recently been suggested for chromatophores by Snozzi and Crofts [16].

In 100 mM NaCl, the equilibrium chromatography could detect no binding $(K_D > 10^{-5}$ M) and kinetic analysis and concentration dependences (Figure 4) showed the reaction to be second order. However, the extent of cytochrome oxidation in subsaturating flashes fell off linearly with the fraction of activated reaction centers (Figure 6c), indicating a long-lived complex even at moderate ionic strength.

Discussion

We have attempted to fit the kinetic results to the following simple scheme for cytochrome c oxidation, which we previously used to describe the oxidation kinetics of cytochrome c_2 [9] :

$$RC + c \underset{k'_{21}}{\overset{k'_{12}}{\rightleftharpoons}} [RC \cdot \cdot c] \underset{k'_{32}}{\overset{k'_{23}}{\rightleftharpoons}} [RC \cdot c]$$

$$h\nu \Big\updownarrow k_b \qquad h\nu \Big\updownarrow k_b \qquad h\nu \Big\updownarrow k_b$$

$$RC^+ + c \underset{k_{21}}{\overset{k_{12}}{\rightleftharpoons}} [RC^+ \cdot \cdot c] \overset{k_{23}}{\longrightarrow} [RC^+ \cdot c] \overset{k_{34}}{\longrightarrow} [RC \cdot c^+]$$

where

$k_{12} \sim 3 \times 10^9 \, M^{-1} \, s^{-1}$

$k_{21} \sim 10^3 \, s^{-1}$

$k_{23} \sim 2 \times 10^4 \, s^{-1}$

$k_{34} \sim 3 \times 10^5 \, s^{-1}$

$k'_{12} \simeq 10^6 \, M^{-1} \, s^{-1}$

$k'_{21} \leq 10 \, s^{-1}$

$k'_{32} \simeq 5 \cdot k'_{23}$

k_b represents the backreaction, or charge recombination, between the oxidized donor, P^+, and reduced acceptor, Q^-. The scheme shows two bound states for the cytochrome, an 'inner complex' (indicated by one dot), from which electron transfer occurs rapidly and directly, and an 'outer complex' (indicated by two dots), which reacts only via the 'inner complex'. It should be noted that these two states both relate to the weak binding $(K_D \approx 10^{-6}$ M) detected by the equilibrium chromatography.

The scheme is incomplete in the sense that the unbinding of products ($[RC \cdot c^+] \rightleftharpoons RC + c^+$) is not shown explicitly. It, therefore, does not appropriately represent the overall energetics of the electron transfer equilibrium, but this does not affect the kinetic description. The rate constants given include values taken directly from the photoactivated reaction kinetics and from the light saturation behavior. Because of the rapidity of the first order processes, the magnitudes of the different kinetic phases was largely determined by the distribution of states established prior to flash activation. The rate constants given are for low ionic strength: $k_{34} \sim 0.3 \times 10^6 \, s^{-1}$ was obtained from the very fast ($2 \, \mu sec$) phase of Figure 3E and $k_{23} \sim 2.3 \times 10^4 \, s^{-1}$ from the $30 \, \mu s$ phase; $k_{12} \sim 3 \times 10^9 \, M^{-1} \, s^{-1}$ from the second order kinetic region at low concentration (e.g. Figure 5). The relative magnitudes of k'_{23} and k'_{32} were estimated from the apparent equilibrium between $[RC \cdot \cdot c]$ and $[RC \cdot c]$ which seems to favor the former by a factor of about five (see e.g. Figures 3D, E). The binding studies indicate that $[RC \cdot \cdot c]$ could be a multiple binding site in which case k_{23} may not be a true rate constant. The dissociation processes may be approached through the light-saturation experiments which show that the cytochrome is not free to diffuse on the time scale of the lifetime of P^+ ($> 0.1 \, s; k_b = 1-10 \, s^{-1}$). This technique actually analyses the ability of the cytochrome to reach additional reaction centers after first encountering an unactivated one and it is, therefore, a measure of k'_{21}, the dissociation rate constant in the dark adapted state. The upper limit for k'_{21} is about $10 \, s^{-1}$.

We have modelled this scheme with the IBM Continuous Systems Modelling Program (CSMP-3) and found it to simulate the observed kinetics rather well. However, the scheme encounters some problems from the equilibrium binding determinations. The principle of detailed balance requires that the dissociation constant be at least as large as the ratio of the association and dissociation rate constants. The measured high affinity binding with $K_D \leq 10^{-8}$ is apparently consistent with the diffusionally limited value for k_{12} ($3 \cdot 10^{-9} \, M^{-1} \, s^{-1}$) and a low value for k_{21}, similar to that suggested for k'_{21} ($\leq 10 \, s^{-1}$). However, the functional binding, associated with the first order kinetics, was found to have a K_D of $0.3 \times 10^{-6} \, M$. The discrepancy also exists at moderate ionic strength (100 mM NaCl) where, although k_{12} is reduced to $2 \times 10^7 \, M^{-1} \, s^{-1}$, the flash saturation experiment indicates that k'_{21} is still $\leq 10 \, s^{-1}$. An assumption of a similar value for k_{21} would indicate binding with $K_D \leq 10^{-6} \, M$ but estimations, both from the equilibrium chromatography and from the lack of a fast oxidation phase, show $K \geq 10^{-5} \, M$.

In an attempt to accommodate both 'weak' binding and slow dissociation we have considered a series of intermediate complexed states between the free reactants and the reactive state $[RC^+ \cdot c]$. However, although slower backward rate constants may be introduced by distributing the overall binding in this way, the reflection coefficients from these states become very high. For example, if $k_{21} \gg k_{23}$ the conversion from $[RC^+ \cdot \cdot c]$ to

$[RC^+ \cdot c]$ becomes very inefficient i.e. reflection from $[RC^+ \cdot \cdot c]$ back to free reactants becomes high. This leads to a severe reduction in the rate of oxidation of the free cytochrome and is quite incompatible with the very high second order rate constant observed for this process.

Clearly we must relinquish any attempt to compare the observed dark binding constants (whether determined by equilibrium methods or from the amplitudes of the kinetic phases) with the rate constants for the light activated processes. The very slow dissociation rate constant $(k'_{21} \leqslant 10\,s^{-1})$ indicated by the flash saturation experiment, however, is a dark value for dissociation of cytochrome from *unactivated* reaction centers following a subsaturating flash, and it is appropriate to compare k'_{21} to the dark equilibrium binding constant. Thus, if the anomaly is to be resolved in this fashion it is necessary to suppose that k'_{12} in the dark is much lower than the von Smoluchowski limit observed in the light (k_{12}) and that the oxidation state of the reaction center affects the forward rate constant by at least two orders of magnitude. A similar conclusion was necessitated by the kinetics of oxidation of cytochrome c_2 [11]. It could arise from a change in the electrostatic interaction between the reaction center and the cytochrome, although it is rather larger than is commonly encountered. Electrostatic influences generally affect on and off rates roughly equally [4], in which case k_{21} may be roughly $10^3\,s^{-1}$. This is similar to the value for cytochrome c_2 from *Rhodospirillum rubrum* determined by an entirely different method [13].

It is apparent from the data presented here that restricted freedom of mobility of cytochrome c is not associated with tight binding and is *not* incompatible with a highly efficient second order reaction mechanism. It is noteworthy that the only oxidation phase that exhibited free mobility of the cytochrome was observed under conditions of very tight binding to a site which, we have suggested, is inactive for electron transfer to the associated reaction center. In this bound state, however, the cytochrome may react with another reaction center in a diffusion controlled process similar to that of free cytochrome. This unexpected behavior, which is not displayed by the native cytochrome c_2 [8, 9], is only observed at low ionic strength. It is presumably not relevant to any physiological situation.

Acknowledgements

This work was initiated while R.E.O. was supported by a National Institute of Health training grant (USPH GM 7283-05) and was funded by NSF grants PCM 77-22086 and PCM 83-16487 to C.A.W.

References

1. Bashford CL, Prince RC, Takamiya K and Dutton PL (1979) Biochim Biophys Acta 545, 223–235

2. Brunner J, Skabal P and Hauser H (1976) Biochim Biophys Acta 455, 322–331
3. Dutton PL, Petty KM, Bonner HS and Morse SD (1975) Biochim Biophys Acta 389, 536–556
4. Eigen M (1974) in Quantum Statistical Mechanisms in the Natural Sciences, Plenum Press, New York
5. Ferguson-Miller S, Brautigan DL and Margoliash E (1978) J Biol Chem 253, 149–159
6. Ke B, Chaney TH and Reed D (1970) Biochim Biophys Acta 216, 373–383
7. Matsuura K and Nishimura M (1985) Biochim Biophys Acta In press
8. Overfield RE and Wraight CA (1980) Biochemistry 19, 3322–3327
9. Overfield RE, Wraight CA and DeVault D (1979) FEBS Lett 105, 137–142
10. Pachence JM, Dutton PL and Blasie JK (1983) Biochim Biophys Acta 724, 6–19
11. Prince RC, Bashford CL, Takamiya K, van den Berg WH and Dutton PL (1978) J Biol Chem 253, 4137–4142
12. Prince RC, Cogdell RJ and Crofts AR (1974) Biochim Biophys Acta 347, 1–13
13. Rickle G and Cusanovich M (1979) Arch Biochem Biophys 197, 589–598
14. Rosen D, Okamura MY and Feher G (1980) Biochemistry 19, 5687–5692
15. Singleton WS, Gray MS, Brown ML and White JL (1965) J Amer Oil Chemists Soc 42, 53–56
16. Snozzi M and Crofts AR (1985) Biochim Biophys Acta In press
17. Van Grondelle R, Duysens LNM and van der Wal HN (1976) Biochim Biophys Acta 449, 169–187
18. Van Grondelle R, Duysens LNM, van der Wel JA and van der Wal HN (1977) Biochim Biophys Acta 461, 188–201
19. Van Grondelle R (1978) Ph.D. Thesis, State University, Leiden

Photosynthesis Research 9, 181–195 (1986)
© *1986 Martinus Nijhoff/Dr. W. Junk Publishers, Dordrecht.*

Isolation of cytochrome bc_1 complexes from the photosynthetic bacteria *Rhodopseudomonas viridis* and *Rhodospirillum rubrum*

R. MAX WYNN, DALE F. GAUL[1], WON-KI CHOI[2], ROBERT W. SHAW and DAVID B. KNAFF*

Department of Chemistry and Biochemistry, Texas Tech University, Lubbock, Texas 79409, USA

(*Received 28 September 1985*)

Key words: cytochrome bc_1 complex, Rieske iron-sulfur protein, photosynthetic bacteria

Abstract. Cytochrome bc_1 complexes have been isolated from wild type *Rhodopseudomonas viridis* and *Rhodospirillum rubrum* and purified by affinity chromatography on cytochrome *c*-Sepharose 4B. Both complexes are largely free of bacteriochlorophyll and carotenoids and contain cytochromes *b* and c_1 in a 2:1 molar ratio. For the *Rps. viridis* complex, evidence has been obtained for two spectrally distinct *b*-cytochromes. The *R. rubrum* complex contains a Rieske iron-sulfur protein (present in approximately 1:1 molar ratio to cytochrome c_1) and catalyzes an antimycin A- and myxothiazol-sensitive electron transfer from duroquinol to equine cytochrome *c* or *R. rubrum* cytochrome c_2. Although an attempt to prepare a cytochrome bc_1 complex from the gliding green bacterium *Chloroflexus aurantiacus* was not successful, membranes isolated from phototrophically grown *Cfl. aurantiacus* were shown to contain a Rieske iron-sulfur protein and protoheme (the prosthetic group of *b*-type cytochromes).

Introduction

It has recently become clear that the membrane-bound electron transfer chains of mitochondria, chloroplasts, cyanobacteria and bacteriochlorophyll *a*-containing photosynthetic bacteria contain similar multi-peptide complexes that catalyze electron flow from quinol to a soluble cytochrome *c* or its functional equivalent, the copper-containing protein plastocyanin [18, 10, 17, 9]. This complex contains cytochrome c_1, two *b* cytochromes and the Rieske iron-sulfur protein. Preliminary evidence supported the presence of similar complexes in a photosynthetic green sulfur bacterium [19], in a gliding green bacterium [43, 44] and in the bacteriochlorophyll *b*-containing, photosynthetic purple non-sulfur bacterium *Rhodopseudomonas viridis* [40].

[1] Current address: Department of Biochemistry, University of Wisconsin, Madison, Wisconsin 53706, USA
[2] Permanent address: Department of Chemistry, Chonnam National University, Kwangju, Chonnam 500, Republic of Korea
*Address for offprints and all correspondence: D.B. Knaff, Department of Chemistry and Biochemistry, Texas Tech University, Lubbock, Texas 79409 USA

Dedicated to Prof. L.N.M. Duysens on the occasion of his retirement.

We have now developed an isolation protocol, utilizing affinity chromatography on cytochrome c covalently linked to Sepharose 4B after detergent solubilization of the complex, for purifying cytochrome bc_1 complexes from *Rps. viridis* and the bacteriochlorophyll *a*-containing, photosynthetic purple non-sulfur bacterium *Rhodospirillum rubrum*. We have also obtained evidence for the possible existence of a similar complex in the gliding green bacterium *Chloroflexus aurantiacus*.

Materials and methods

Rps. viridis (strain NHTC 133) and *R. rubrum* (strain S1) were grown and membrane fragments ('chromatophores') prepared as described previously [36, 12]. Bacteriochlorophyll *a* and *b* (BChl *a* and *b*) were determined following extraction into 7:2 acetone:methanol [7, 21]. Whole membrane fragments (200 KP) from phototrophically grown *Cfl. aurantiacus*, prepared as described in [14], were generous gifts of Prof. R.E. Blankenship and R.C. Fuller. Solubilization of the cytochrome complexes was accomplished by treating the membrane fragments essentially as described in [15], except that dodecylmaltoside (Behring Diagnostics) was substituted, at the same concentration, for octylglucoside. Purification of the solubilized complex was accomplished as described in [15], through the sucrose density gradient centrifugation step. The complex was then dialyzed against 30 mM glycyl-glycine buffer (pH 7.4) containing 0.05% sodium cholate and subjected to affinity chromatography as described in Results.

Protein was determined according to the method of Bradford [3], using bovine serum albumin as a standard. Heme c and protoheme were determined as in [31]. Ubiquinone content was determined according to Takamiya and Dutton [33]. The content of non-heme iron was determined by the method of Massey [22] and that of acid-labile sulfide according to Brumby et al. [6]. Polyacrylamide gel electrophoresis (PAGE) was conducted in the presence of sodium dodecylsulfate (SDS) on 10–20% gradient gels (1.5 mm thickness) prepared according to O'Farrell [23]. Gels were stained for protein with Coomassie Brilliant Blue and for heme with 3,3′,5,5′-tetramethylbenzidine plus hydrogen peroxide [34].

Cytochrome c_2 from *R. rubrum* was purified according to Bartsch [1]. Equine cytochrome c (type VI) was obtained from Sigma. The equine cytochrome c was used without further purification for enzyme assays and preparation of the affinity chromatography matrix. Cyanogen bromide-activated Sepharose 4B was purchased from Pharmacia. Equine cytochrome c was coupled to the cyanogen bromide-activated Sepharose 4B as described previously [11], using 52 mg cytochrome c and 1.75 g cyanogen bromide-activated Sepharose 4B.

Optical spectra and activity assays were obtained using an Aminco DW-2a spectrophotometer. Electron paramagnetic resonance spectra were obtained

using a Varian E-109 spectrometer equipped with an Air Products Helitrans
flexible He transfer line and a Varian/Hewlett-Packard E-935 data acquisition
system. All spectra, activity measurements and chemical analyses were
performed using preparations that had been chromatographed on the cytochrome c affinity column.

Results

Figure 1 shows the absorbance spectra of the cytochrome bc_1 complexes
isolated from wild type *Rps. viridis* and *R. rubrum*. The low absorbances in
the wavelength regions near 480 and 600 nm, relative to the oxidized cytochrome Soret absorbance near 412 nm, indicates the successful separation
of most of the BChl and carotenoids during the purification of the complexes.
Table 1, which summarizes the complexes' chemical compositions, confirms
the low BChl content of both preparations. The BChl:cytochrome c_1 ratios
of 0.26 and 0.11 for *Rps. viridis* and *R. rubrum*, respectively, compare
favorably to those obtained in the preparation of similar complexes from
carotenoid mutants of the purple non-sulfur bacterium *Rps. sphaeroides*
[42, 32]. These BChl:cytochrome c_1 ratios are considerably lower than the
1:1 ratio obtained using a solubilization procedure similar to ours, but
omitting cytochrome c-affinity chromatography, with the GA mutant of
Rps. sphaeroides [15]. In the case of both the *R. rubrum* and *Rps. viridis*
complexes, although the BChl:cytochrome c_1 ratio after the sucrose density
gradient step varied somewhat between preparations, cytochrome c-affinity
chromatography always resulted in a considerable decrease in the BChl:
cytochrome c_1 ratio. In typical preparations, cytochrome c affinity
chromatography removed approximately 90% of the remaining BChl and
carotenoids. Like other cytochrome bc_1 complexes [18, 15, 42, 32] the
Rps. viridis and *R. rubrum* complexes contain ubiquinone in an approximately
1:1 molar ratio to cytochrome c_1 (see Table 1).

Table 1 shows that the cytochrome b:cytochrome c_1 ratios, as calculated
from pyridine hemochrome determinations of protoheme and heme c,
approached 2:1 values for both complexes (1.75:1 for *Rps. viridis* and
1.8:1 for *R. rubrum* respectively). However, such measurements reveal no
details concerning the number of different b- and c-type cytochromes that
may be present. To approach this problem, a number of reduced *minus*
oxidized difference spectra were obtained. Figure 2A and B show the hydroquinone reduced *minus* ferricyanide oxidized difference spectra for the two
complexes. As hydroquinone $(E'_m = + 290 \, mV)$ is a relatively weak reductant,
such difference spectra show only high potential $(E'_m \geqslant ca. + 200 \, mV)$
components. The difference spectra both reveal the presence of a component
with absorbance maxima (α-band = 553 nm and β-band = 523–524 nm)
consistent with the presence of cytochrome c_1 [18, 10, 17, 9, 15, 42, 32]
in both the *Rps. viridis* and *R. rubrum* complexes. There is no evidence in the

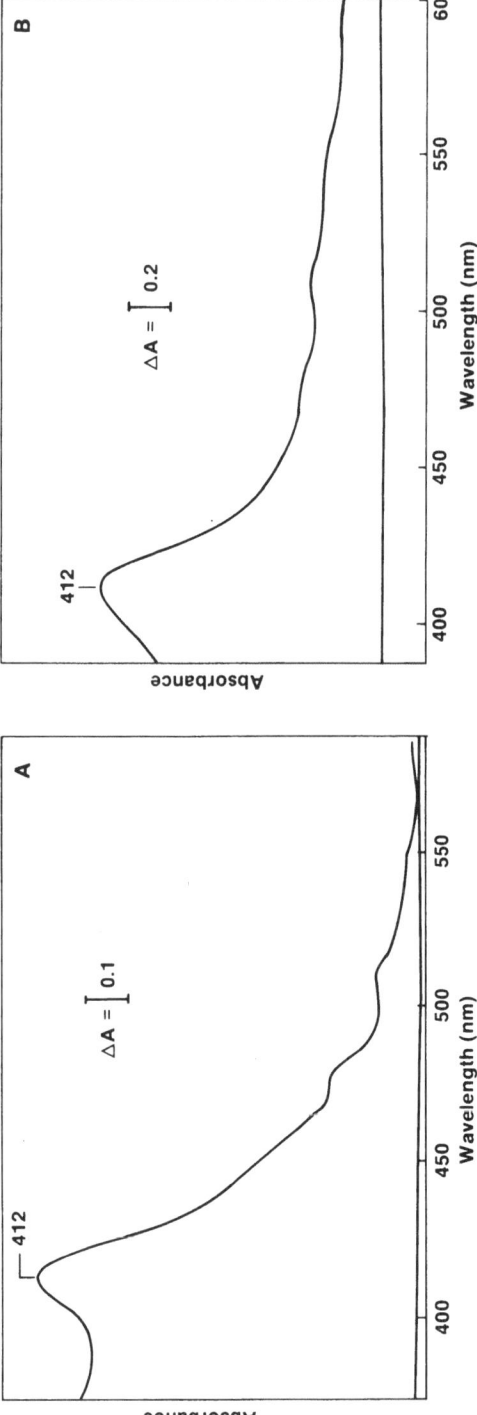

Figure 1. Absorbance spectra of the cytochrome bc_1 complexes from *Rps. viridis* (A) and *R. rubrum* (B). The spectra were obtained in 30 mM glycyl-glycine buffer (pH 7.4) containing 0.05% cholate. Spectral resolution = 1 nm. The bottom traces represent the zero absorbance (buffer *vs* buffer) instrumental baselines.

Table 1. Compositions of the purified cytochrome bc_1 complexes from *Rps. viridis* and *R. rubrum*

	Rps. viridis	*R. rubrum*
Cytochrome c_1 (heme c)	6.0	7.1
Cytochrome b (protoheme)	10.5	12.7
Acid-labile sulfide	0.83	15.9
Non-heme iron	2.4	49.4
Ubiquionene	9.4	7.4
Bacteriochlorophyll (BChl)	1.6	0.77

All values are expressed as nmol/mg protein. BChl is BChl b in the case of *Rps. viridis* and BChl a in the case of *R. rubrum*

Rps. viridis preparation for a shoulder at 558 nm, indicating the absence of the high potential ($E_m' = + 300$ mV) cytochrome c_{558} associated with the *Rps. viridis* reaction center [8, 29].

Figure 2A shows the [ascorbate *plus* phenazine methosulfate (PMS)] *minus* hydroquinone difference spectrum of the *Rps. viridis* complex. Such difference spectra would be expected to reveal components with E_m' values between approximately + 50 and + 200 mV. The absorbance features at 531 and 559 nm are consistent with the presence of a *b*-type cytochrome with 559 nm α-band and 531 nm β-band maxima in the *Rps. viridis* complex. The (dithionite) *minus* (ascorbate *plus* PMS) difference spectrum of the *Rps. viridis* complex (Figure 2a) exhibits, in the α-band region, an absorbance maximum at 559 nm and a shoulder at 565 nm. Similar features had previously been observed in (dithionite) *minus* (Ascorbate *plus* PMS) difference spectrum of unfractionated *Rps. viridis* membranes [40]. As dithionite is a considerably stronger reductant than ascorbate *plus* PMS and ascorbate *plus* PMS does not produce any absorbance feature at 565 nm, these results indicate the likely presence of a second lower potential *b*-type cytochrome in the *Rps. viridis* complex. The absence of any absorbance feature at 552 nm indicates that the *Rps. viridis* complex is free of the low potential ($E_m' = 0.0$ V) cytochrome c_{552} associated with the *Rps. viridis* reaction center [8, 29]. Addition of ascorbate *plus* PMS to the *R. rubrum* complex (Figure 2B) results in the appearance of absorbance features (absent in the hydroquinone-reduced complex) at 532 nm and 561 nm, indicating the presence of at least one *b*-type cytochrome with $E_m' < 200$ mV. Subsequent addition of dithionite results in a considerable increase in the magnitude of the 532 and 561 nm absorbance features but produces no shift in the position of the maxima. Thus, the data of Figure 2B do not allow one to conclude that more than one *b*-type cytochrome is present in the *R. rubrum* complex. However, it is possible that the *R. rubrum* complex contains more than one species of *b*-cytochrome since *R. rubrum* chromatophore membranes are known to contain at least two different *b*-type cytochromes active in photosynthetic electron transport [37, 13].

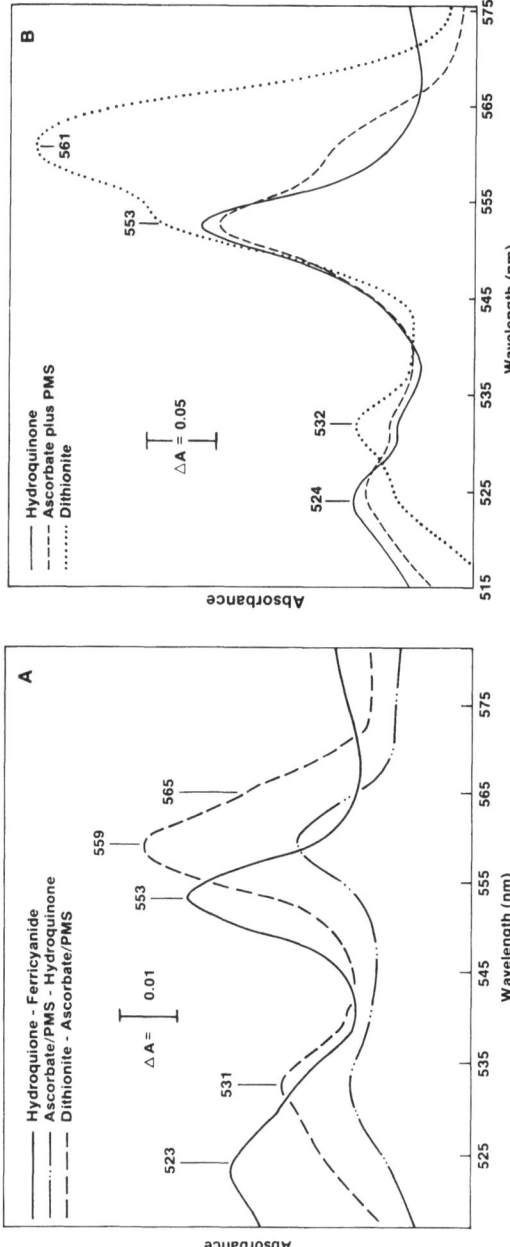

Figure 2. α- and β-band difference spectra of *Rps. viridis* (A) and *R. rubrum* (B) cytochrome bc_1 complexes. The difference spectra of the complexes in 30 mM glycylglycine buffer (pH 7.4) containing 0.05% cholate were recorded after addition of the indicated reductant. Hydroquinone, potassium ferricyanide, sodium ascorbate and sodium dithionite were added as small aliquots until no further absorbance changes were detected. PMS was present as a final concentration of 5 μM. Spectral resolution = 1 nm. In B, potassium ferricyanide was present in the reference cuvette for all three spectra, while the sample cuvette contained the indicated reductant.

As cytochrome bc_1 complexes isolated from a variety of sources contain the Rieske iron-sulfur protein [18, 10, 17, 9], the acid-labile sulfide and non-heme iron contents of the Rps. viridis and R. rubrum complexes were examined to determine whether these components of the prosthetic group of the Rieske protein [18] were present. Table 1 shows that the R. rubrum complexes were examined to determine whether these components of the prosthetic group of the Rieske protein [18] were present. Table 1 shows that the R. rubrum complex contained sulfide in a 2.2:1 molar ratio compared to cytochrome c_1. As the Rieske protein is known to contain equimolar acid-labile sulfide and non-heme iron [18], the considerably larger amount of non-heme iron, compared to sulfide, suggests the presence of significant amounts of adventitious iron in the preparation. Previously characterized cytochrome bc_1 complexes generally contain equimolar amounts of the Fe_2S_2-containing Rieske protein and cytochrome c_1 [18]. Thus the S^{-2}: cytochrome c_1 ratio near 2 observed for the R. rubrum preparation suggests a similar 1:1 stoichiometry for the Rieske protein and cytochrome c_1 in the R. rubrum complex. Direct evidence for the presence of a Rieske iron-sulfur protein in the R. rubrum complex is shown in the EPR spectrum of Figure 3. Reduction of the complex with hydroquinone produced EPR features at g = 2.03 and 1.89 characteristic of a reduced Rieske protein [18] and similar to those reported previously for the Rieske protein in unfractionated R. rubrum chromatophores [41]. (The EPR signal(s) at g = 2.0 probably arise from one or more unidentified free radicals.) No increase in the magnitude of the g = 1.89 EPR signal was observed in samples of the R. rubrum complex that were reduced with the strong reductant, dithionite, compared to that observed in the hydroquinone-reduced sample. This suggests a E'_m value $\geqslant +260\,mV$ for the Rieske iron-sulfur protein in the R. rubrum complex. In contrast to the R. rubrum complex, the reduced Rps. viridis complex showed no features attributable to a Rieske iron-sulfur protein, although unfractionated Rps. viridis membranes are known to contain a Rieske protein [40]. Consistent with the absence of a Rieske-like EPR signal is the low acid-labile sulfide content (S^{-2}:cytochrome c_1 = 0.14) of the Rps. viridis complex (see Table 1). It thus appears that the Rps. viridis Rieske protein [40] is not solubilized by dodecylmaltoside or that either the Rieske peptide and/or its Fe_2S_2 cluster is lost during subsequent purification of the cytochrome bc_1 complex.

Cytochrome bc_1 complexes isolated from a number of sources catalyze electron transfer from a variety of quinols to cytochromes c. As can be seen in Table 2, the R. rubrum complex possesses duroquinol:cytochrome c oxidoreductase activity with either the native R. rubrum cytochrome c_2 or the closely related [27] equine cytochrome c serving as an electron acceptor. These rates are similar to those reported for the cytochrome bc_1 complex isolated from Rps. sphaeroides GA. As shown in Table 2, electron transfer catalyzed by the R. rubrum complex was markedly inhibited by low

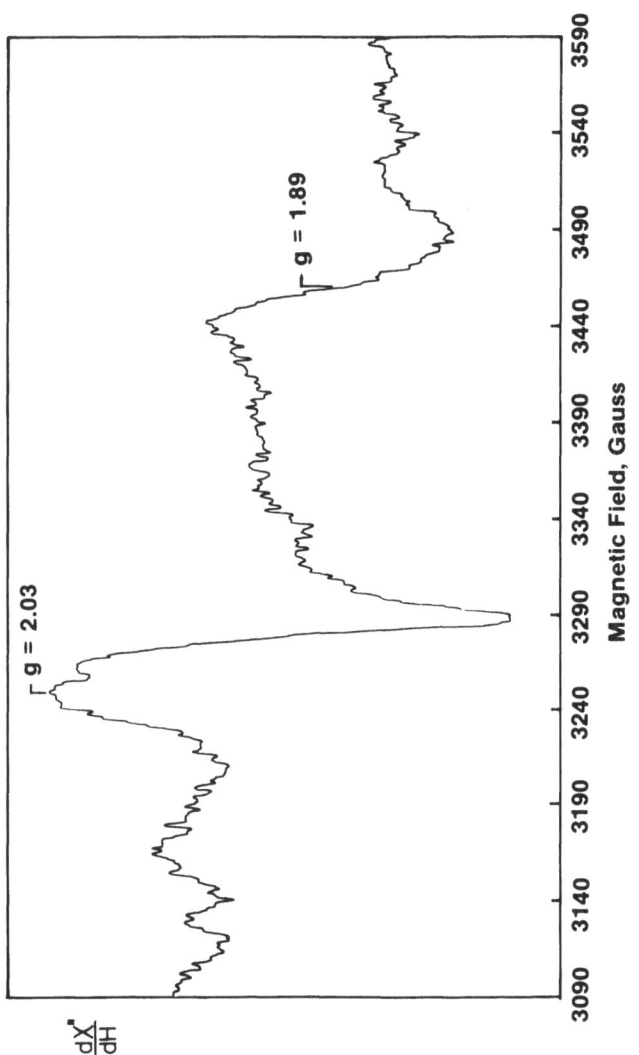

Figure 3. EPR spectrum of hydroquinone-reduced *R. rubrum* cytochrome bc_1 complex. The complex, equivalent to 2 μM cytochrome c_1, was present in the buffer of Figure 1, with hydroquinone added. The conditions of EPR spectroscopy were: Microwave frequency, 9.150 GHz; Microwave power, 3 mW; Modulation frequency, 100 kHz and Temperature, 12 K. The spectrum is an average of 32 scans.

Table 2. The duroquinol:cytochrome $c(c_2)$ cytochrome oxidoreductase activity of the R. rubrum cytochrome bc_1 complex

Additions	Activity (μmol $\times$ nmol Cyt $c_1^{-1} \times$ h^{-1})
Equine cytochrome c	5.2
R. rubrum c_2	6.0
Equine cytochrome c + 1 μM Antimycin A	1.0
Equine cytochrome c + 0.1 μM Myxothiazol	0.9
R. rubrum c_2 + 1 μM Antimycin A	1.2
R. rubrum c_2 + 0.1 μM Myxothiazol	1.3

Duroquinol (durohydroquinone), in an ethanol stock solution, was added to the reaction mixture to a final concentration of 90 μM. Cytochrome c or c_2 were added to the reaction mixture to give a final concentration of 25 μM. The assays were conducted in 50 mM Tricine buffer at pH 6.0. Cytochrome c reduction was monitored at 550–540 nm.

concentrations of antimycin A and myxothiazol, two specific inhibitors of electron transport through the cytochrome bc_1 region of a number of organisms [18, 10, 9], including R. rubrum [37]. The Rps. viridis complex did not exhibit any detectable duroquinol: cytochrome c-oxidoreductase activity, probably due to the fact that it is lacking the Rieske iron-sulfur protein.

Since equine cytochrome c is an effective electron acceptor for the R. rubrum complex and affinity chromatography using immobilized cytochrome c has proven effective in purifying other cytochrome bc_1 complexes [32, 2], it seemed possible that such a step would be useful in purifying either the Rps. viridis or R. rubrum complexes. As discussed above, chromatography of the complexes on cytochrome c covalently attached to a Sepharose 4B matrix is quite effective in removing contaminating carotenoid and BChl from the complexes. Figures 4A and 4B show the elution profiles from the cytochrome c-Sepharose affinity column for the Rps. viridis and R. rubrum complexes, respectively. Typical recovery of the complexes (based on cytochrome c_1) was > 95%, indicating that cytochrome c-affinity chromatography can be an extremely high yield step. The fact that the cytochrome bc_1 complexes could be eluted from the cytochrome c-Sepharose 4B column by increasing the ionic strength, suggests that electrostatic forces are involved in the binding of the complexes to cytochrome c.

Figure 5A shows the peptide composition of the Rps. viridis and R. rubrum complexes, as determined by staining for protein after SDS–PAGE. Each complex contains three major peptides with apparent molecular weights of 43, 30 and 18 kDa in the case of Rps. viridis and 36, 30 and 17 kDa in the case of R. rubrum. (Additional minor bands at 27 and 24 kDa were observed in the R. rubrum complex.) Similar peptide compositions have previously been reported for cytochrome bc_1 complexes isolated from two different mutants of Rps. sphaeroides [15, 42], although there is evidence for an additional, lower molecular weight peptide (10–12 kDa) in the Rps. sphaeroides

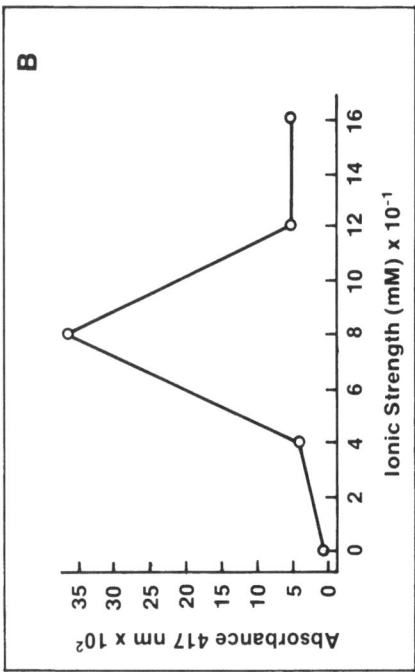

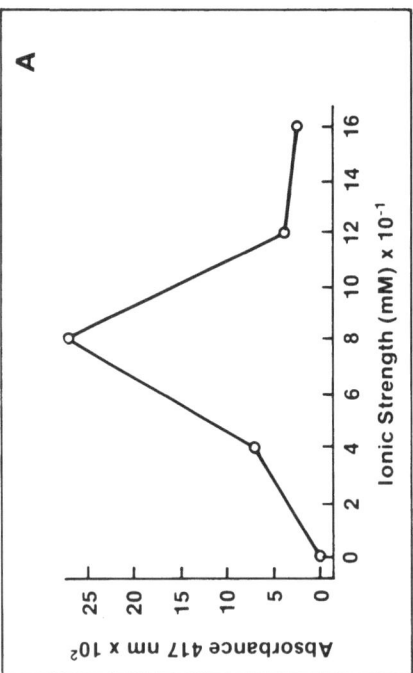

Figure 4. Binding of *Rps. viridis* (A) and *R. rubrum* (B) cytochrome bc_1 complexes to an equine cytochrome c-Sepharose 4B affinity column. 50 nmol of the complexes in 30 mM glycylglycine buffer (pH 7.4) containing 0.05% cholate were applied to the affinity column (bed volume 5.3 ml) which had been pre-equilibrated with the same buffer. The complexes were eluted with 5 ml aliquots of the indicated ionic strength buffers (NaCl was added to the equilibrating buffer to give the desired ionic strength.). 2 ml fractions were collected

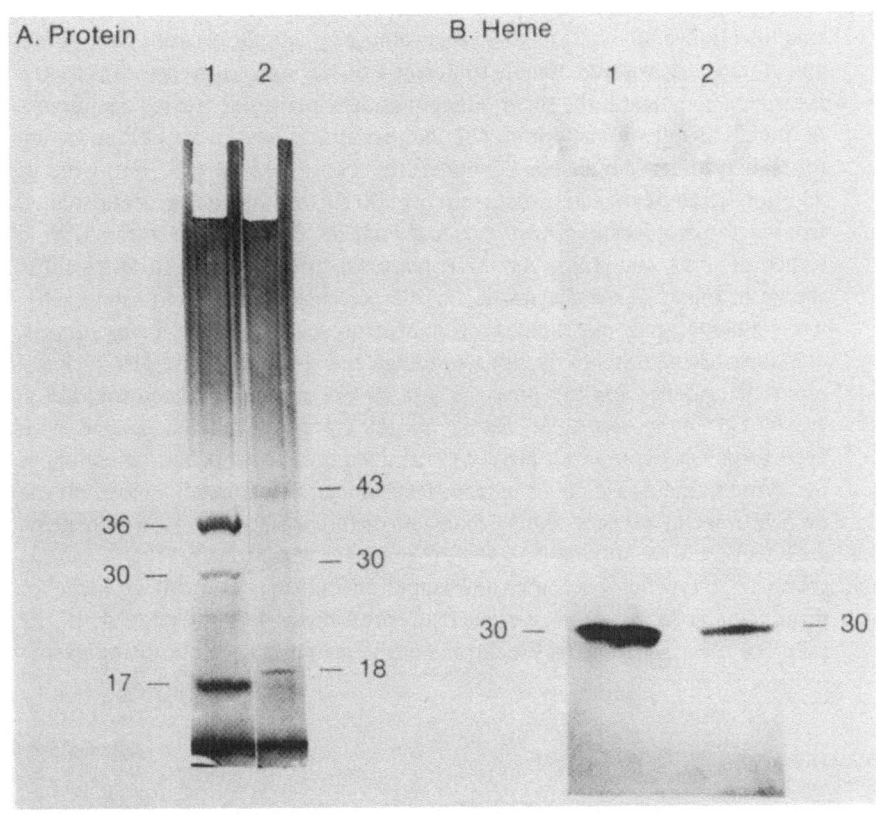

Figure 5. Peptide composition of cytochrome bc_1 complexes. A. Gels stained for protein with Coomassie Brilliant Blue: Lane 1, *R. rubrum* complex equivalent to 0.4 nmol cytochrome c_1; Lane 2, *Rps. viridis* complex equivalent to 0.2 nmol cytochrome c_1. B. Gels stained for heme: Lane 1, *R. rubrum* complex equivalent to 0.8 nmol cytochrome c_1; Lane 2, *Rps. viridis* complex equivalent to 0.2 nmol cytochrome c_1. Molecular weight standards used were: Bovine serum albumin (M_r = 66 kDa); Ovalbumin (M_r = 45 kDa); Carbonic anhydrase (M_r = 31 kDa); Soybean trypsin inhibitor (M_r = 21.5 kDa) and Lysozyme (M_r = 14.4 kDa)

complex [18, 42]. As shown in Figure 58, staining the gels for covalently-bound heme [34] to identify cytochrome c_1 [18], revealed only a single peptide ($M_r = 30$ kDa) in either the *Rps. viridis* or *R. rubrum* complex. The 30 kDa apparent molecular weight for cytochrome c_1 in the two complexes is similar to M_r values obtained for the cytochrome in several other cytochrome bc_1 complexes [18, 17, 15, 42, 32]. The absence of $M_r = 24$ and 35 kDa peptides and of a 38 kDa heme-staining peptide in the *Rps. viridis* complex indicates that the complex is not appreciably contaminated by any reaction center constituents [8, 35].

Despite the fact that the procedure described above was successful for the solubilization and isolation of cytochrome bc_1 complexes from *Rps. viridis* and *R. rubrum*, we were unable to detect any of the electron carriers expected for such a complex in the supernatant of dodecylmaltoside-treated membranes of the gliding green bacterium, *Cfl. aurantiacus*. However, an EPR spectrum (obtained under conditions identical to those of Figure 3, but without addition of any exogenous reductant) of 200 KP membrane fragments isolated from phototrophically-grown *Cfl. aurantiacus* contained features with g values of 2.03 and 1.90. As these features, essentially identical to those shown in Figure 3, are diagnostic for the presence of a Rieske protein [18], we conclude that membranes of phototrophically-grown *Cfl. aurantiacus* contain such a protein. Zannoni and Ingledew [43, 44] have also recently reported evidence for the presence of a Rieske protein in phototrophically-grown *Cfl. aurantiacus*. As Rieske proteins are generally associated with cytochrome bc_1 complexes [18], we examined the *Cfl. aurantiacus* membranes for cytochrome b. We could detect no evidence for features typical of b-type cytochromes in reduced *minus* oxidized difference spectra. However, heme analysis of the membranes showed the presence of protoheme, the prosthetic group of b-type cytochromes (310 nmol protoheme/mg protein, heme c: protoheme $= 28:1$). The presence of protoheme is consistent with the presence of at least one b-cytochrome in the membranes of phototrophically-grown *Cfl. aurantiacus*.

Discussion

Cytochrome bc_1 complexes with the characteristic $2:1$ cytochrome $b:c_1$ stoichiometry found in several other cytochrome bc_1 complexes [18] have been isolated from the photosynthetic purple non-sulfur bacteria *Rps. viridis* and *R. rubrum*. This represents the first such isolations from these species and also the first ones from wild type strains of any photosynthetic purple bacteria. Previous preparations from *Rps. sphaeroides* had utilized a carotenoid-free mutant [42] or a strain with an altered carotenoid complement [15, 32]. Although the *Rps. viridis* complex described herein lacks quinol:cytochrome oxidoreductase activity, probably due to the loss of the Rieske iron-sulfur protein [18, 10], the isolation of such a complex confirms the presence of cytochromes b and c_1 in a bacterium that, until recently, was thought to contain neither component [40]. In particular, the relatively low cytochrome bc_1 complex content (relative to reaction center) of unfractionated *Rps. viridis* membranes [40], resulted in absorbance contributions from reaction center-associated cytochromes [8, 29, 35] partially obscuring those from cytochromes b and c_1. Similarly, an unequivocal identification of the heme-staining 30 kDa peptide (see Figure 5B) as cytochrome c_1 is considerably less difficult in the *Rps. viridis* cytochrome bc_1 complex than in the unfractionated membranes because of the presence of such large amounts of the 38 kDa cytochrome c_{552}/c_{558} peptide in unfractionated *Rps. viridis* membranes [40, 8, 35]. Finally, the absence of interfering absorbance changes attributable to reaction center components has made possible a demonstration (see Figure 2A) that *Rps. viridis* appears to contain two spectrally distinct b-type cytochromes, with the lower potential cytochrome b exhibiting a maximum (or a shoulder) at 565 nm not found with the higher potential cytochrome b. Similar spectral differences between two b-type cytochromes have been reported in a number of other cytochrome bc_1 complexes, including that from mitochondria [18].

While *R. rubrum* was known to contain multiple b-cytochromes [13], at least two of which participate in light-dependent electron flow [37], and a Rieske protein [41], no evidence had been presented for the third component of a cytochrome bc_1 complex, cytochrome c_1, in this bacterium [37]. The spectral evidence of Figure 2B and the presence of a 30 kDa, heme c-containing peptide in *R. rubrum* (Figure 5B) membranes provide unequivocal evidence for the presence of cytochrome c_1 in *R. rubrum*.

As the Fe_2S_2 cluster of the Rieske protein and the non-covalently bound protoheme of cytochrome b are lost under the denaturing conditions of SDS–PAGE, it has not yet been possible for us to unambiguously identify the two major peptides (other than the 30 kDa cytochrome c_1) in each complex. Evidence that the cytochrome b peptide in the *Rps. sphaeroides* complex has a molecular weight between 40–48 kDa [42, 16 but see 20] suggests that the 43 kDa peptide in *Rps. viridis* and the 36 kDa peptide in *R. rubrum* may be cytochrome b peptides (Even though cytochrome bc_1 complexes contain two spectrally distinct hemes with different E_m values, both protohemes appear to be bound to the same peptide [18, 16, 20, 39, 28]). If these assignments are correct, the remaining major peptide in the *R. rubrum* complex ($M_r = 17$ kDa) can tentatively be identified as the Rieske iron-sulfur protein. The Rieske iron-sulfur protein of the *Rps. sphaeroides* GA cytochrome bc_1 complex has been reported [18] to have a somewhat higher apparent molecular weight (22–25 kDa).

The observation that the isolated *R. rubrum* complex exhibits quinol: cytochrome c-oxidoreductase activity that is sensitive to the two highly specific inhibitors antimycin A and myxothiazol [37] (see Table 2) suggests that the basic characteristics of electron flow through the native complex have survived the isolation procedure. The fact that the isolated $R.$ $rubrum$ cytochrome bc_1 complex apparently can form an electrostatic complex (see Figure 4B) with immobilized equine cytochrome c (a cytochrome with many structural similarities to the endogenous cytochrome c_2 [27]) suggests the presence of a cytochrome c/c_2 binding site in the $R.$ $rubrum$ complex. A possibly similar binding site, involving cytochrome c_1, has been characterized in the mitochondrial cytochrome bc_1 complex [30, 25, 4]. The $R.$ $rubrum$ cytochrome bc_1 complex may thus prove to be a useful tool for investigating questions that have recently been raised concerning the mode of binding of $R.$ $rubrum$ cytochrome c_2 to its membrane-bound reaction partners [26].

Our inability, and that of others [5], to see cytochrome b in difference spectra of $Cfl.$ $aurantiacus$ membranes could be explained by low amounts of cytochrome bc_1 complex present. (In contrast, Zannoni and Ingledew have reported that membranes from $Cfl.$ $aurantiacus$ grown at high light intensity contain several b-type cytochromes [43, 44].) It should be pointed out that our demonstration of protoheme in $Cfl.$ $aurantiacus$ does not unequivocally establish the presence of b-cytochrome(s), as protoheme is also the prosthetic group of o-type oxidases [24]. However, should cytochrome b indeed be present in $Cfl.$ $aurantiacus,$ it would seem likely that this gliding green bacterium (like representative species from all other known families of photosynthetic bacteria [18, 10, 17, 9, 19, 40]) does contain a cytochrome bc_1 complex. A more detailed description of the results obtained with the $Cfl.$ $aurantiacus$ membranes is forthcoming (R.M. Wynn, J.M. Foster, R.C. Fuller, R.E. Blankenship, R.W. Shaw and D.B. Knaff — Manuscript in preparation).

Acknowledgements

The authors would like to acknowledge the debt all workers interested in electron flow in photosynthetic bacteria owe to the pioneering work of Prof. L.N.M. Duysens. Helpful discussions on electron flow patterns in $R.$ $rubrum$ with Prof. Duysens, Prof. J. Amesz, Dr. R. van Grondelle and H.N. van der Wal are gratefully acknowledged. The authors would like to thank Prof. J.P. Thornber for providing the initial culture of $Rps.$ $viridis,$ Prof. R.E. Blankenship and Prof. R.C. Fuller for providing the $Cfl.$ $aurantiacus$ membranes and Dr. W. Trowitzsch for the generous gift of myxothiazol. We would also like to thank Dr. D. Zannoni for his generous access to manuscripts prior to publication. This research was supported by grants from the National Science Foundation (PCM8408564 to D.B.K.) and the Robert A. Welch Foundation (D-909 to R.W.S.).

References

1. Bartsch RG (1971) Methods Enzymol 23:344–363
2. Bill K, Broger C and Azzi A (1982) Biochim Biophys Acta 679:28–34

3. Bradford MM (1976) Anal Biochem 72:248–254
4. Broger C, Salardi S and Azzi A (1983) Eur J Biochem 131:349–352
5. Bruce BD, Fuller RC and Blankenship RE (1982) Proc Natl Acad Sci USA 79: 6532–6536
6. Brumby PE, Miller RW and Massey VJ (1965) J Biol Chem 240:2222–2228
7. Clayton RK (1963) In Bacterial Photosynthesis (Gest H. et al., eds.) p. 498. Yellow Springs, Ohio: Antioch Press
8. Clayton RK and Clayton BJ (1978) Biochim Biophys Acta 501:478–487
9. Coremans JMCC, van der Wal HN, van Grondelle R, Amesz J and Knaff DB (1984) Biochim Biophys Acta 807:134–142
10. Crofts AR and Wraight CA (1983) Biochim Biophys Acta 726:149–185
11. Davidson MW, Gray GO and Knaff DB (1985) FEBS Lett 187:155–159
12. Davidson VL and Knaff DB (1981) Biochim Biophys Acta 637:53–60
13. Dutton PL and Jackson JB (1972) Eur J Biochem 30:495–510
14. Feick RG, Fitzpatrick M and Fuller RC (1982) J Bacteriol 150:905–915
15. Gabellini N, Bowyer JR, Hurt E, Melandri A and Hauska G (1982) Eur J Biochem 126:105–111
16. Gabellini N and Hauska G (1983) FEBS Lett 154:171–174
17. Gaul DF and Knaff DB (1983) FEBS Lett 162:69-75
18. Hauska G, Hurt E, Gabellini N and Lockau W (1983) Biochim Biophys Acta 726: 97–133
19. Hurt EC and Hauska G (1984) FEBS Lett 168:149–154
20. Iba K, Takamiya K and Arata H (1985) FEBS Lett 183:151–154
21. Jones OTG and Saunders VA (1972) Biochim Biophys Acta 275:417–436
22. Massey VJ (1957) J Biol Chem 229:763–775
23. O'Farrell PH (1975) J Biol Chem 250:4007–4021
24. Pool RK (1983) Biochim Biophys Acta 726:205–243
25. Rieder R and Bosshard HR (1980) J Biol Chem 255:4732–4739
26. Rieder R, Wiemken V, Bachofen R and Bosshard HR (1985) Biochem Biophys Res Comm 128:120–126
27. Salemme FR (1977) Ann Rev Biochem 46:299–329
28. Saraste M (1984) FEBS Lett 166:367–372
29. Shill DA and Wood PM (1984) Biochim Biophys Acta 764:1–7
30. Smith HT, Ahmed AJ and Millett F (1981) J Biol Chem 256:4984–4990
31. Takaichi S and Morita S (1981) J Biochem 89:1513–1519
32. Takamiya K, Doi M and Okimatsu H (1982) Plant and Cell Physiol 23:987–997
33. Takamiya K and Dutton PL (1979) Biochim Biophys Acta 546:1–16
34. Thomas PE, Ryan D and Levin W (1976) Anal Biochem 75:168–176
35. Thornber JP, Cogdell RJ, Sefter REB and Webster GD (1980) Biochim Biophys Acta 593:60–75
36. Trosper TL, Benson DL and Thornber JP (1977) Biochim Biophys Acta 460: 318–330
37. van der Wal HN and van Grondelle R (1983) Biochim Biophys Acta 725:94–103
38. Weiss H and Kolb HJ (1979) Eur J Biochem 99:139–149
39. Widger WR, Cramer WA, Herrmann RG and Trebst A (1984) Proc Natl Acad Sci USA 81:674–678
40. Wynn RM, Gaul DF, Shaw RW and Knaff DB (1985) Arch Biochem Biophys 238: 373–377
41. Yoch DC, Carrithers RP and Arnon DI (1977) J Biol Chem 252:7453–7460
42. Yu L, Mei Q-C and Yu C-Y (1984) J Biol Chem 259:5752–5760
43. Zannoni D and Ingledew JW (1985) Abstr V Int Symp on Photosyn Prokaryotes Grindelwald, Sept 22–28, 229
44. Zannoni D and Ingledew JW, FEBS Lett, in press

Photosynthesis Research 9, 197–210 (1986)
© *1986 Martinus Nijhoff/Dr. W. Junk Publishers, Dordrecht.*

On the action of hydroxylamine, hydrazine and their derivatives on the water-oxidizing complex

VERENA FÖRSTER and WOLFGANG JUNGE

Abteilung Biophysik, Fachbereich Biologie/Chemie, Universität Osnabrück, Postfach 4469, D-4500 Osnabrück, Germany, FRG

(*Received 1 October 1985*)

Key words: hydrazine, hydroxylamine, photosynthesis, photosystem II, water oxidation, protons

Abstract. Photosynthetic water oxidation proceeds by a four-step sequence of one-electron oxidations which is formally described by the transitions $S_0 \to S_1$, $S_1 \to S_2$, $S_2 \to S_3$, $S_3 \to (S_4) \to S_0$. State S_1 is most stable in the dark. Oxygen is released during $S_3 \to (S_4) \to S_0$. Hydroxylamine and hydrazine interact with S_1. They cause a two-digit shift in the oxidation sequence as observed from the dark equilibrium, i.e. from $S_1 \to S_2$: $S_2 \to S_3$: $S_3 \to (S_4) \to S_0$: $S_0 \to S_1$: ... in the absence of the agents, to $S_1^* \to S_0$: $S_0 \to S_1$: $S_1 \to S_2$: $S_2 \to S_3$: ... in the presence of hydroxylamine or hydrazine.

We measured the concentration dependence of this two-digit shift via the pattern of proton release which is associated with water oxidation. At saturating concentrations hydroxylamine and hydrazine shift the proton-release pattern from $OH^+(S_1 \to S_2)$: $1H^+(S_2 \to S_3)$: $2H(S_3 \to S_0)$: $1H^+(S_0 \to S_1)$: ... to $2H^+(S_1^* \to S_0)$: $1H^+(S_0 \to S_1)$: $OH^+(S_1 \to S_2)$: $1H^+(S_2 \to S_3)$: $2H^+(S_3 \to S_0)$: ... The $2H^+$ were released upon the first excitation with a half-rise time of 3.1 ms, both with hydroxylamine and with hydrazine. The concentration dependence of the shift was rather steep with an apparent Hill coefficient at half saturation of 2.43 with hydroxylamien (Förster and Junge (1985) FEBS Lett. 186, 53–57) and 1.48 with hydrazine. The concentration dependence could be explained by cooperative binding of $n \geqslant 3$ molecules of hydroxylamine and of $n \geqslant 2$ molecules of hydrazine, respectively. Tentatively, we explain the interaction of hydroxylamine and hydrazine with the water-oxidizing complex (WOC) as follows: Two bridging ligands, possible Cl^- or OH^-, which normally connect two Mn nuclei, can be substituted by either 4 molecules of hydroxylamine or 2 molecules of hydrazine when the WOC resides in state S_1.

Abbreviations

DNP–INT, dinitrophenylether of iodonitrothymol; FWHM, full width at half maximum; NR, neutral red (3-amino-7-dimethylamino-2-methyl-phenazine·HCl); PS II, photosystem II; WOC or (in formulas:) W, water-oxidizing complex.

Dedicated to Prof. L.N.M. Duysens on the occasion of his retirement.

Introduction

When dark-adapted thylakoids are excited by a series of single-turnover flashes of light, water is oxidized in a four-step sequential process which is formally described by the transitions $S_1 \rightarrow S_2$ (S_1 stable in the dark), $S_2 \rightarrow S_3$, $S_3 \rightarrow S_4 \rightarrow S_0$, $S_0 \rightarrow S_1, \ldots$, [13]. Dioxygen is liberated during $S_3 \rightarrow S_4 \rightarrow S_0$. Hydroxylamine ($NH_2OH$), hydrazine ($NH_2NH_2$) and certain derivatives of these molecules interact with state S_1 in the dark. We call the resultant modified state S_1^*. (This formal assignment does not infer that the redox state of the WOC is the same in S_1 and S_1^*.) Illumination then induces the oxidation sequence $S_1^* \rightarrow S_0$, $S_0 \rightarrow S_1$, $S_1 \rightarrow S_2$, $S_2 \rightarrow S_3$, $S_3 \rightarrow S_0, \ldots$ This two-digit shift was observed via the pattern of oxygen evolution [1, 16] and via the pattern of proton release [8] associated with the water-oxidation cycle. The stoichiometric pattern of proton release (into the thylakoid lumen) is shifted from $OH^+(S_1 \rightarrow S_2): 1H^+(S_2 \rightarrow S_3): 2H^+(S_3 \rightarrow S_4): 1H^+(S_0 \rightarrow S_1): \ldots$ (unmodified) to $2H^+$ (first flash): $1H^+(S_0 \rightarrow S_1): 0H^+(S_1 \rightarrow S_2): 1H^+(S_2 \rightarrow S_3) \ldots$ in the presence of NH_2OH or NH_2NH_2.

The mechanism of the action of these agents is not understood. The water-oxidizing complex is supposed to be a binuclear or tetranuclear manganese complex [3]. It has been speculated by Radmer and Ollinger [16, 17] that NH_2OH and NH_2NH_2 act as 'water analogues', i.e. that they substitute the oxygen precursor, $2H_2O$, at the active site of the complex. Previously, we studied the concentration dependence of the two-digit shift with the aim to determine the coordination number of hydroxylamine to the WOC. We found a cooperative action of at least 3 molecules of NH_2OH [9]. Here, we present new studies on the action of NH_2OH, NH_2NH and derivatives of these molecules. In the light of the present results we discuss the implications for structure and function of the WOC.

Materials and methods

Thylakoids were prepared from peas (Pisum sativum), frozen and stored under liquid nitrogen until use [5, 7].

pH changes in the thylakoid lumen were monitored via absorption changes of neutral red ($\pm 13 \mu M$ NR), the external phase being buffered by 1.3 g/l bovine serum albumin (for the method see [11, 12]). Photometric measurements were carried out as described in detail in [7].

Hydroxylamine hydrochloride (H-9876), hydrazine dihydrochloride (H-6628), O-methylhydroxylamine hydrochloride (M-1139) and O-sulfonylhydroxylamine (H-0134) were purchased from Sigma and dried under vacuum before use. N-methylhydroxylamine hydrochloride was purchased from Merck (No. 820802). In stock solutions of these salts hydrochloride was neutralized by KOH. They were used for up to 10 h at longest.

For measurements, thylakoids were thawed and diluted in the dark to a final concentration of $20\,\mu M$ chlorophyll. The medium contained $2\,mM$ potassium hexacyanoferrate(III), $25\,mM$ KCl, $3\,mM$ MgCl$_2$, $5\,\mu M$ 2,6-dimethyl-p-benzoquinone (pH 7.2); $4\,\mu M$ DNP–INT was present in all samples in order to abolish internal pH transients due to photosystem I [7]. The complete suspension was kept in the dark at room temperature for at least five minutes before measurement. A given dilute suspension was used for 10–20 min.

Results

Modification of the proton-release pattern associated with water oxidation by different 'water analogues'

Figure 1, trace a shows the typical pattern of proton release by PS II (DNP–INT present) into the thylakoid lumen which is observed upon excitation of dark-adapted thylakoids with a series of short flashes (compare [7]). The rapid pH jumps which are not time resolved in this particular measurement (half-rise time $< 2\,ms$) exhibit a damped periodical pattern as function of flash number (compare Figure 2a). In terms of S-state transitions this is interpreted by the release per PS II of OH$^+$($S_1 \rightarrow S_2$): 1H$^+$($S_2 \rightarrow S_3$): 2H$^+$($S_3 \rightarrow S_4 \rightarrow S_0$): 1H$^+$($S_0 \rightarrow S_1$) . . . [7, 10]. Since we have to assume that slower phases are not due to water oxidation [7] we shall use the term 'proton yield' only for the fraction of absorption changes with half-rise times of less than $2\,ms$.

For the sake of reproducibility aliquots of large stocks of homogeneous frozen thylakoids (2–4 mM Chl) were used for all experiments presented in previous papers [6–9] as well as in this study. It should be noted that the kinetics of the internal pH changes (repetitive excitation [5]) as well as the pattern of proton release (Förster, unpublished) are the same independently of whether the starting material is freshly prepared or frozen, provided that 30% ethylene glycol has been present in the freezing medium.

Figure 1, traces b–f show the proton-release pattern measured in the presence of 'water analogues'. We limited the measurements to concentrations at which the compounds did not destroy more than 10% of the water-oxidizing centers irreversibly (tested via oxygen evolution rates at continuous illumination after washing out these agents by a single centrifugation step). $14\,\mu M$ NH$_2$OH, $60\,\mu M$ NH$_2$OSO$_3$H and $120\,\mu M$ NH$_2$NH$_2$ (Figure 1, traces b, c, f and Figure 2b, c) caused a two-digit shift in the proton-release pattern, which then was 2H$^+$ (first flash): 1H$^+$($S_0 \rightarrow S_1$): OH$^+$($S_1 \rightarrow S_2$): 1H$^+$($S_2 \rightarrow S_3$): 2H$^+$($S_3 \rightarrow S_4 \rightarrow S_0$) . . . (compare Figure 2a, b, for hydroxylamine see also [9]). $500\,\mu M$ CH$_3$–NHOH (Figure 1, trace e) had only a moderate effect, and $250\,\mu M$ NH$_2$–O–CH$_3$ (Figure 1, trace d) had no effect on the proton-release pattern.

With NH$_2$OH, NH$_2$OSO$_3$H and NH$_2$NH$_2$ present at saturating concentrations proton release upon the first flash was time resolved (Figure 3). With

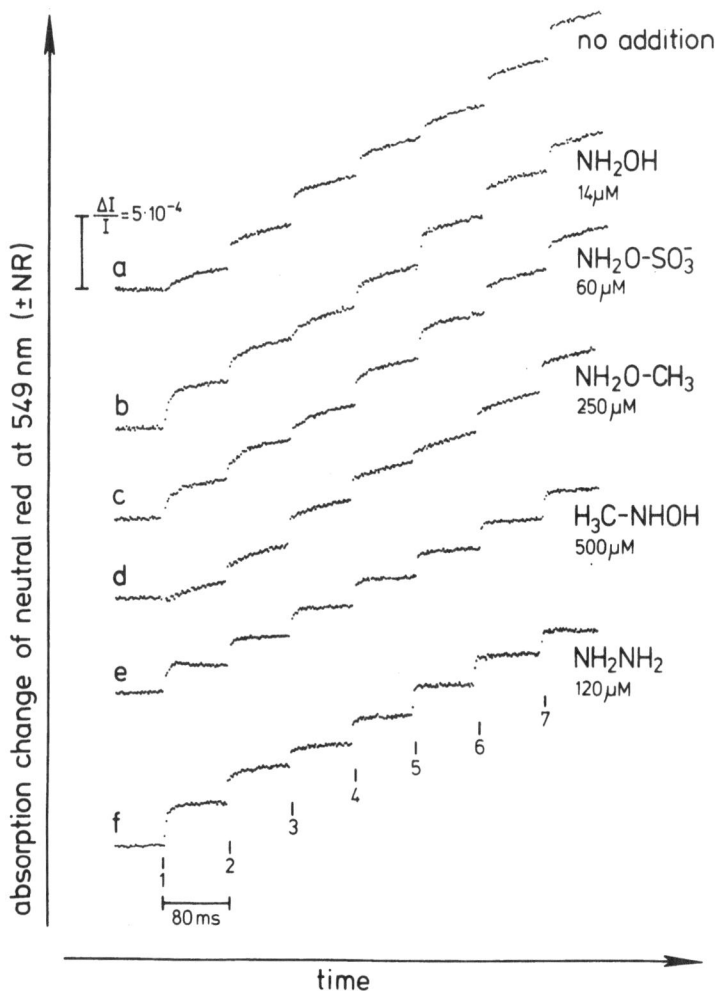

Figure 1. Absorption changes of neutral red at 549 nm (± NR) in response to a series of saturating flashes, measured in the absence and in the presence of NH_2OH, NH_2OSO_3H, NH_2OCH_3, CH_3NHOH and NH_2NH_2 at concentrations as indicated. Flashes were provided by a PRA 610B Xenon flash lamp (2 μs FWHM), the measuring light intensity was 90 μW at 548 nm.

each of these agents the same proton yield at the same half rise time was found: $2H^+$ at 3.1 (± 0.5) ms. We will discuss this phenomenon below. (The difference between this half-rise time and that published previous for NH_2OH (1.7 ms in [8]) is explained by the following: This kinetic component is followed by a slow drift which is supposedly not due to water oxidation [7, 8]. With increasing concentration of amines the slow drift vanished. In Figure 3 it is virtually absent, which allowed a more accurate determination of the half rise time from these traces than formerly, where the rapid phase was apparently faster due to superposition of the slow drift.)

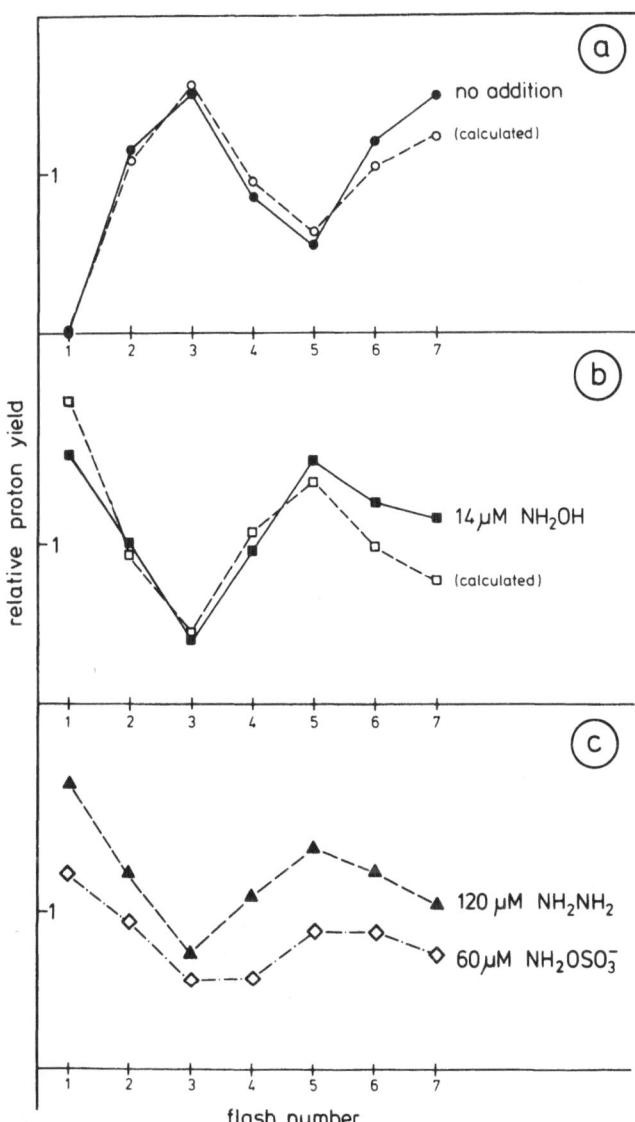

Figure 2. Amplitudes of internal pH changes of $\tau < 2$ ms half-rise time as function of flash number, as obtained by kinetic analysis of the traces in Fig. 1; additives as indicated. Open symbols in (a) and (b): patterns calculated under the assumption of 100% S_1 (a) and 100% S_1^* (b) in the dark, 10% misses and 10% double hits.

Concentration dependence of the two-digit shift of the proton-release pattern by water analogues

Previously, we studied the concentration dependence of the two-digit shift of the proton-release pattern as effected by NH_2OH [8, 9]. We find that the

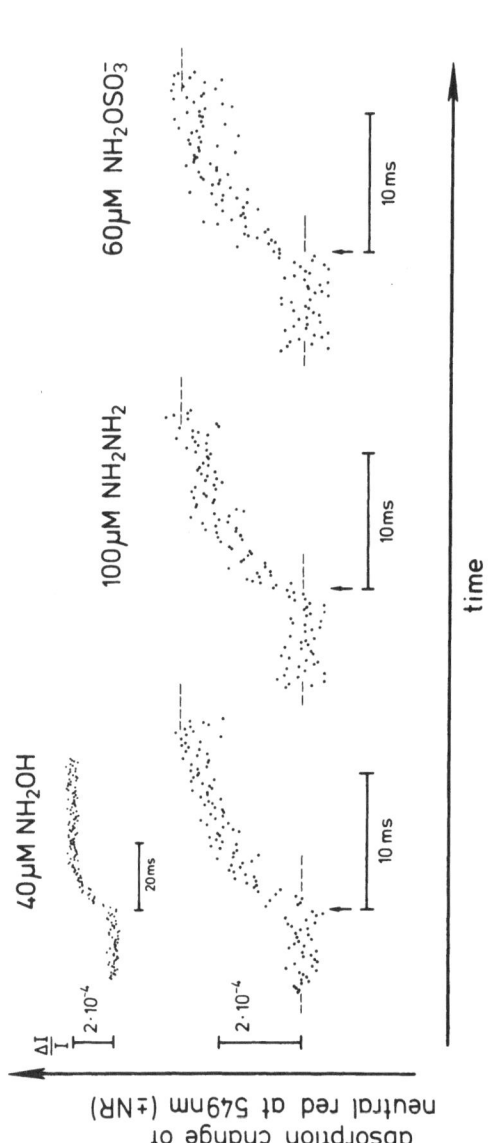

Figure 3. Time-resolved absorption changes of neutral red at 549 nm (± NR) upon the first flash given to thylakoids which had been incubated in the dark for at least five minutes with 40 μM NH_2OH, 100 μM NH_2NH_2, and 60 μM NH_2OSO_3H, respectively.

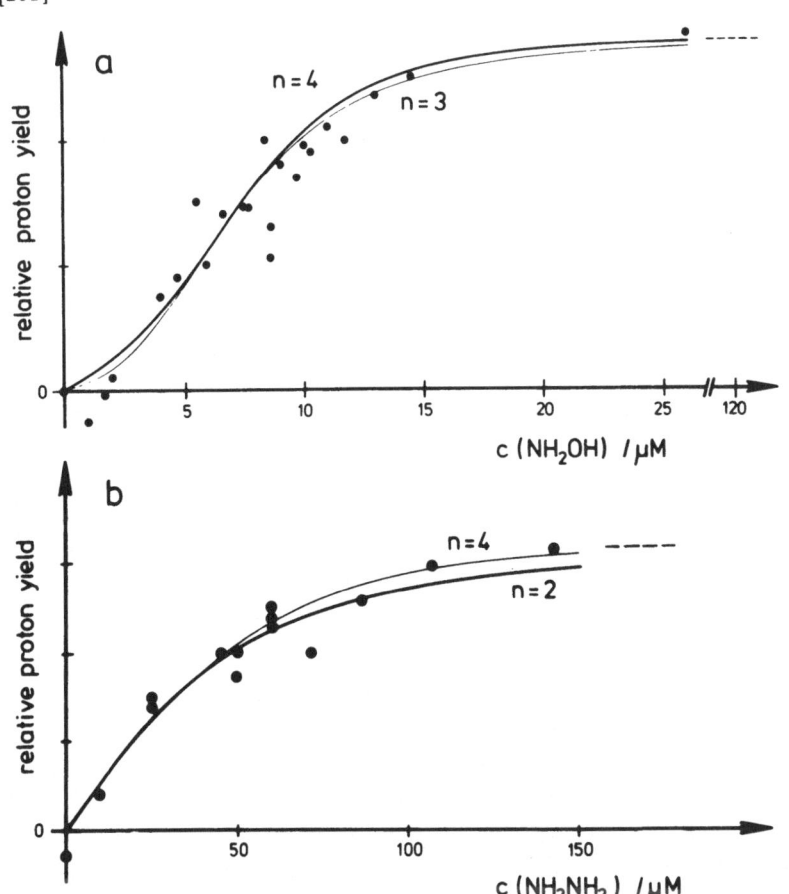

Figure 4. Proton yields upon the first flash as function of the concentration of hydroxylamine (a) and hydrazine (b), (Figure 4a taken from [9]). The solid curves were calculated under assumption of cooperative binding of 3 and 4 molecules of hydroxylamine, and of 2 and 4 molecules of hydrazine (see text).

concentration dependence (for comparison shown in Figure 4a) is rather sharp with an apparent Hill coefficient of 2.43 at half saturation. A quantitative fit (see lines in Figure 4a) is obtained by the model of sequential interaction of allosteric enzymes under the following assumptions: There are four allosterically interacting coordination sites for hydroxylamine at the WOC. The dissociation constant is $K_d = 100\,\mu M$ and the allosteric interaction factor from one state to the state of next higher occupation is $a = 1/5$. The proton-release pattern is one and the same no matter whether the WOC is loaded with 1, 2, 3 or 4 NH_2OH molecules and it is independent of their particular distribution of over the four sites [9].

   In the present study, we investigated the concentration dependence of two-digit shift of the proton-release pattern as caused by NH_2NH_2. As in the case of hydroxylamine, the pattern could be considered as the sum of the $(0:1:2:1:)_n: \ldots$ and the $2:1:(0:1:2:1:)_n: \ldots$ pattern at different percentages, depending on the concentration (data not shown). Figure 4b shows the rise of the first-flash yield as function of NH_2NH_2 concentration. The proton yields at zero concentration upon the first flash, $Y_1(c = 0)$, as well as the respective asymptotic yields at high concentration, $Y_1(\infty)$ were marked by dashed lines. The percentage, s, of water-oxidizing complexes which gives rise to the 'shifted' pattern of proton liberation, $2:1:(0:1:2:1:) \ldots$, is

$$s(c) = \frac{Y_1(c)}{Y_1(\infty)}$$

From a Hill plot, $\log(s/1-s)$ versus $\log c$, we obtained an apparent Hill coefficient at half saturation of 1.48 for NH_2NH_2, (compare 2.43 for NH_2OH [9]). The concentration at half saturation with NH_2NH_2 was $34\,\mu M$, higher than with NH_2OH ($7\,\mu M$); (for the solid lines in Figure 4 see Discussion). A relative decrease of the yield at the third flash paralleled the relative increase of the yield at the first flash with the same concentration dependence (for NH_2OH see [9], for NH_2NH_2 not shown).

We also measured the proton-release pattern at various concentrations of NH_2OSO_3H. It turned out that the effect was always saturated at $60\,\mu M$ NH_2OSO_3H. At intermediate concentrations, however, results varied considerably. In aqueous solution NH_2OSO_3H is known to hydrolyse in the hour time range, in the presence of nucleophilic agents even faster [14]. Thus, it is conceivable that in a thylakoid suspension NH_2OSO_3H hydrolyses within some minutes. Therefore, we suspect that the active species in this case was not hydroxylamine-O-sulfonic acid but its hydrolysis product NH_2OH.

Decomposition of the hydroxylamine-loaded complex and rebinding of hydroxylamine

When NH_2OH is added to dark-adapted thylakoids, it primarily reacts with state S_1 of the WOC. When, after $5-10\,min$ incubation, thylakoids are submitted to a series of flashes they undergo the transitions $S_1^* \cdot nNH_2OH \rightarrow S_0$, $S_0 \rightarrow S_1$, $S_1 \rightarrow S_2, \ldots$ [1, 8]. We studied the turnover time of the reaction $S_1^* \cdot nNH_2OH \rightarrow S_0$ by measuring the proton-release pattern at diminished dark times between the first and second flash. While the interval between the first and second flash was varied, the intervals between the following flashes was kept constant at $80\,ms$. With and interval of $30\,ms$ between the first and second flash we found the centers only partially relaxed to the supposed S_0 state. (We could not determine this accurately since the proton-release pattern at the $30\,ms$ interval was not just the sum of the '$80\,ms$ pattern' and the undisturbed pattern). In consequence, proton release upon the first flash ($\tau_{1/2} = 3.1$ ms) resulted from an early product of $[S_1^* \cdot NH_2OH]^+$ while the native state S_0 was reached only at dark times greater than 30 ms.

We studied the recovery of the NH_2OH-modified proton-release pattern as function of the dark time between a group of two conditioning flashes (30 ns FWHM, flash distance 100 ms) and a subsequent series of probing flashes (spaced 80 ms apart). Except for a deviation upon the first probing flash, the probing flash series evoked a proton-release pattern which was similar to the pattern observed in NH_2OH-free thylakoids. This was independent of the dark time between the conditioning flashes and the probing flashes in the range from 80 ms to 10 s. No rebinding seemed to occur in this time range. This is compatible with observations made on a longer time scale by other investigators [19]. The deviation consisted in an additional proton yield ($\tau_{1/2} < 1$ ms) upon the first probing flash.

Discussion

Classical biochemical methods in the investigation of enzymatic mechanisms take advantage of the variation of substrate concentration or of substituents of the substrate. Since in photosynthetic oxygen evolution H_2O is solvent as well as substrate these methods fail. It has been hoped to gain insight into the mechanism of water oxidation via application of 'water analogous' artificial electron donors. It has been expected that these artificial donors bind in competition with water to the WOC, which is assumed to be a manganese complex with a $2(+ 2)$ arrangement of the metal nuclei.

H_2N-OH, H_2N-NH_2 and some derivatives shift the phase of the period-of-four oscillation of water-oxidation under a series of exciting flashes applied to the dark-adapted material from $S_1 \rightarrow S_2, S_2 \rightarrow S_3, S_3 \rightarrow (S_4) \rightarrow S_0, S_0 \rightarrow S_1, \ldots$ to $S_1^* \rightarrow S_0, S_0 \rightarrow S_1, S_1 \rightarrow S_2, S_2 \rightarrow S_3, \ldots$ [1, 7, 16]. We investigated this two-digit shift as reflected in the proton-release pattern (Figure 1, 2). In their respective concentration ranges of reversible interaction with the WOC, NH_2OH and NH_2NH_2 caused a complete two-digit shift. The hydroxylamine derivative $CH_3-NH-OH$ was only moderately effective. Among the O-substituted hydroxylamine derivatives, the hydrophobically substituted compound $H_2N-O-CH_3$ was entirely ineffective. The hydrophilically substituted compound $H_2N-O-SO_2H$, on the other hand, was effective (compare [18]). However, the effective species in the latter case may be the hydrolysis product NH_2OH (see above). Thus, it appeared that the interaction with the WOC was restricted to those molecules which did not or only little exceed the sterical dimensions of the NH_2OH molecule.

The similarity between NH_2OH, $NH_2NH_2, \ldots$ and two H_2O suggested to consider these molecules as competitive substrates at the active site of the WOC, as outlined by Radmer and Ollinger [1, 17]. We obtained the number of interacting molecules from the concentration dependence of the two-digit shift of the water-oxidation cycle effected by NH_2OH and NH_2NH_2.

From the Hill coefficients n at half saturation it was directly evident that $\geqslant 3$ NH_2OH molecules (n = 2.43) or $\geqslant 2$ NH_2NH_2 molecules (n = 1.48) interacted with the WOC in the dark. In [9] we fitted the effect of NH_2OH by a sequential interaction model for multi-site enzymes. It is worthwhile to summarize briefly this model before discussing the results in the light of the literature on water oxidation and coordination chemistry.

The sequential interaction model (see for instance section 4 in [20]) considers the following equilibria:

$$W \underset{K_d}{\overset{+D}{\rightleftharpoons}} WD_1 \underset{aK_d}{\overset{+D}{\rightleftharpoons}} WD_2 \underset{bK_d}{\overset{+D}{\rightleftharpoons}} WD_3 \underset{cK_d}{\overset{+D}{\rightleftharpoons}} WD_4 \rightleftharpoons \ldots$$

W = water-oxidizing complex;
K_d = dissociation constant;
a, b, c = sequential interaction factors;

The water-oxidizing complex had n equivalent binding places. The degeneracy of the states WD_i followed the pattern:

$(1, 2, 1)$ for n = 2

$(1, 3, 3, 1)$ for n = 3

$(1, 4, 6, 4, 1)$ for n = 4

With these assumptions the following 'normalized reactivities' s_n, i.e. the percentages of water-oxidizing centers modified in proton liberation, for enzymes with 2, 3 and 4 sites were:

$$s_2 = (2X + X^2/a)/(1 + \text{nominator})$$

$$s_3 = (3X + 3X^2/a + X^3/a^2b)/(1 + \text{nominator})$$

$$s_4 = (4X + 6X^2/a + 4X^3/(a^2b) + X^4/(a^3b^2c))/(1 + \text{nominator})$$

$X = [D]/K_d$ = effector concentration, [3], as normalized to the dissociation constant, K_d;

It was assumed that each of the ligands was equally effective but that the effect was the same independent of the number of ligands coordinated to the WOC.

Coordination of 'water analogues' to the water-oxidizing complex

Assuming that the sequential interaction factors were equal (a = b = c), we obtained fits to the experimental points with several sets of parameters (listed in Table 1). In Figure 4 these fits are shown by solid lines. Although the fits obtained with coordination numbers of 3 and 4 for NH_2OH and of 2 and 3 for NH_2NH_2 were not distinguishable at the experimental resolution we considered binding of 4 NH_2OH or of 2 NH_2NH_2 as most probable, since they required only a moderate allosteric interaction factor (a = 1/5). This

Table 1. Fit parameters for the concentration dependence of the two-digit shift of the water-oxidation cycle by NH_2OH and NH_2NH_2

Compound	Number of bound molecules	K_d	Allosteric factor $a = b = c$
NH_2NH_2	2	117 μM	0.2
	3	232 μM	0.05
NH_2OH	3	152 μM	0.05
	4	97 μM	0.2

factor decreased the binding constant from 100 μM, valid for the first ligand, to 800 nM for the fourth ligand. If the allosteric factor was a = 1/20, as assumed for the alternative fit curves, the binding constant would have been reduced from 100 μM to 12.5 nM. For comparison it may be noted that the allosteric interaction factor of hemoglobin (4 binding sites, apparent Hill coefficient 2.8) is in the moderate range (a = 1/6.7). Despite of these arguments, and if one considers that manganese is dimeric or tetrameric in the WOC [3], uneven coordination numbers are unlikely. Therefore, we assume that the WOC can bind up to 4 NH_2OH or up to 2 NH_2NH_2, respectively.

Implications for the structure of the water-oxidizing complex

Coordination of 4 NH_2OH to the WOC would imply 4 or 8 binding sites on manganese, depending on whether it reacted as a mono- or as a bidentate ligand. While NH_2OH may act as chelating ligand NH_2NH_2 usually does not so, it rather bridges two metal centers [15]. Assuming that NH_2NH_2 occupied the same binding sites as NH_2OH we arrived at the following tentative structural model for a two-manganese complex (Figure 5): In the undisturbed complex two manganese are ligated by a bridging ligand (Figure 5A). NH_2NH_2 replaces the bridging ligand which leads to a distorted complex with enlarged Mn–Mn distance. Structural elements as in Figure 5B have been found in $[Mn(NH_2NH_2)_2Cl_2]_n$ crystals [4]. Similarly, NH_2OH might break the metal–metal bridges (Figure 5C). The bridging ligand replaced by NH_2OH and NH_2NH_2 may be Cl^- or OH^-. In the case of OH^- this may result in a displacement of the O_2 precursor. The two protons observed upon the first flash in the presence of NH_2OH or NH_2NH_2 as well as the two protons observed during $S_3 \rightarrow S_0$ may be due to the insertion of new bridging ligands after release of NH_2OH, NH_2NH_2 or O_2, respectively, according to $2H_2O \rightarrow 2OH_{bound}^- + 2H^+$.

Radmer and Ollinger [17] speculated on the size of the catalytic site of the WOC based on the assumption of a structural and functional analogy between NH_2OH and the O_2 precursor. Since they assumed that the reactive site was occupied by only one molecule of NH_2OH they postulated a minimum size of 2.5 × 4 Å. In the light of the coordination of 4 NH_2OH to the WOC, however, a functional analogy is questionable. Contrary to Radmer's and

A

B

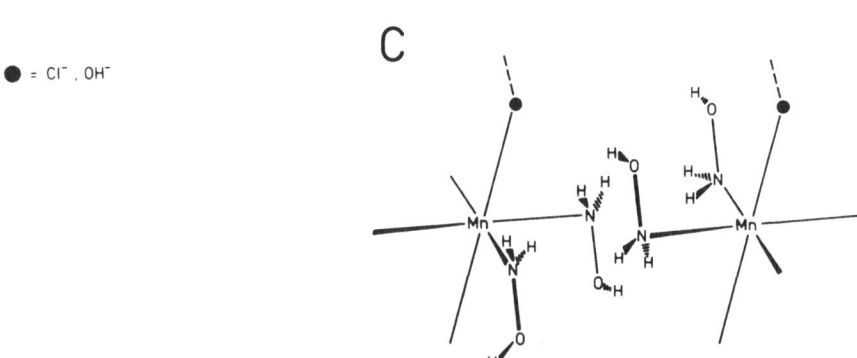

● = Cl⁻ , OH⁻

C

Figure 5. Hypothetical structural elements of the WOC. A: in native state S_1; B and C: after reaction with NH_2NH_2 and NH_2OH, respectively; (for explanation see text).

Ollinger's assumption of a rigid conformation, the binding of four molecules NH_2OH entails conformational changes in the protein. In the hypothetical complex shown in Figure 5 the Mn—Mn distance of the native chloride-bridged complex A is ≈ 3.8 Å. It has to be enlarged to 4.5 Å for the binding of 2 NH_2NH_2 (B) and to > 5.5 Å for the binding of NH_2OH (C). The relatively long time which is needed for relaxation to the S_0 state after the first flash (note, that 30 ms were insufficient) may be due to the displacement of NH_2OH and to the reversal of conformational changes.

Reactivity of the hydroxylamine-loaded complex

It is remarkable that with either compound, NH_2OH and NH_2NH_2, we found a monophasic release of two protons per PS II reaction center with a half rise time of 3.1 ms. This half-rise time is longer than those observed during regular water oxidation. Since neither the amount nor the rate of proton release depended on the chemical nature of the 'water analogue' it is reasonable to assume that the first flash causes the oxidation of the respective bound 'water analogue', but that the proton release resulted from a secondary

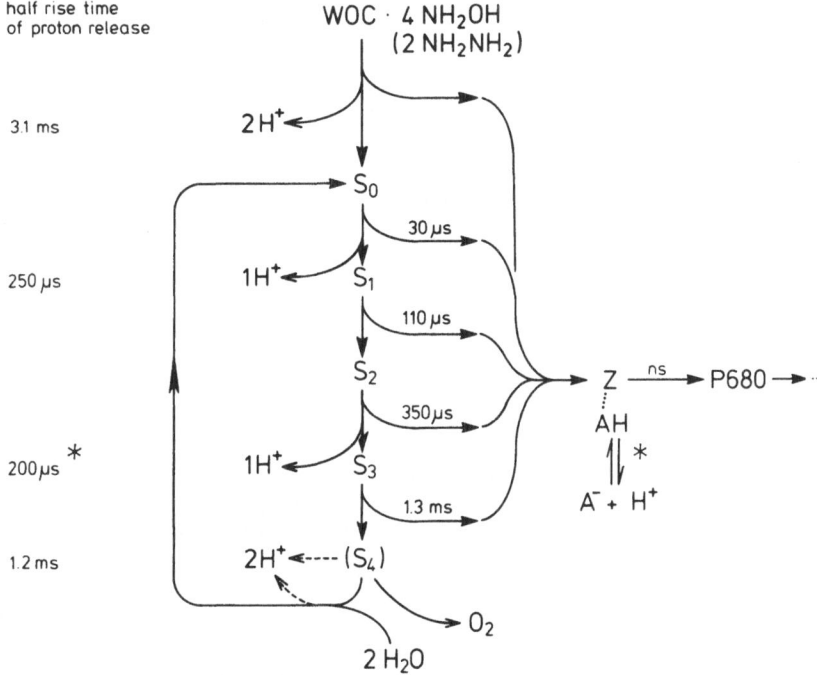

half rise time
of proton release

Figure 6. Scheme summarizing electron transfer and proton release at the donor side of PS II, including modification by hydroxylamine and hydrazine.

reaction of the WOC. This secondary reaction might be the reinsertion of water after ejection of the NH_2OH-oxidation products according to $2H_2O_{free} \rightarrow 2H^+ + 2OH^-_{bound}$ in a state which is precursor to S_0.

Synopsis

The scheme of Figure 6 may help to summarize our present understanding of photosynthetic water oxidation. By successive one-electron reactions via an intermediate electron carrier, Z, the water-oxidizing complex is oxidized from state S_0 to the higher oxidation states S_1, S_2, S_3 and S_4 at reaction half-rise times between $30\,\mu s$ to $1.3\,ms$, increasing with the oxidation state [2] (Fig. 6). State S_4 spontaneously relaxes to S_0, thereby releasing O_2. Proton release is associated with the transitions $S_0 \rightarrow S_1$ ($1H^+$, $250\,\mu s$), $S_2 \rightarrow S_3$ ($1H^+$, $200\,\mu s$) and $S_3 \rightarrow (S_4) \rightarrow S_0$ ($2H^+$, $1.2\,ms$) [7]. The chemical nature of the protolytic reactions is unknown. However, differentiations can be made by comparison of the half-rise time of proton release with the kinetics of electron abstraction from the water-oxidizing complex. While proton release during $S_0 \rightarrow S_1$ and during $S_3 \rightarrow S_4 - S_0$ appears to occur on the level of the water-oxidizing complex, an additional proton push and pull coupled to oxidoreduction of Z has to be assumed in order to explain the

$200\,\mu s$ component of proton release during $S_2 \to S_3$ (marked by an asterisk in Figure 6) [7].

In the dark, 1–4 molecules of NH_2OH and 1 or 2 molecules of NH_2NH_2 cooperatively bind to the WOC. Abstraction of only one electron apparently leads to state S_0 [8]. This is accompanied by release of $2H^+$ with $\tau_{1/2} = 3.1$ ms, independently of the compound. Complete relaxation to S_0 obviously takes considerably longer (30 ms were insufficient). We suggest that NH_2OH and NH_2NH_2 act on state S_1 by displacing bridging ligands between two manganese centers under enlargement of the Mn–Mn distance and under conformational changes in the protein, which both should be measurable in further work. ⋅

Acknowledgements

We thank Norbert Spreckelmeier for expert electronic help, Dr. Hans van Gorkom for critical comments on the manuscript and Hella Kenneweg for the photographs. DNP–INT was a gift by Prof. A. Trebst. Financial aid has been provided by the Deutsche Forschungsgemeinschaft (Sonderforschungsbereich 171-84, Projekt A2).

References

1. Bouges B (1971) Biochim Biophys Acta 234, 103–112
2. Dekker JP, Plijter JJ, Ouwehand L and van Gorkom HJ (1984) Biochim Biophys Acta 767, 176–179
3. Dismukes GC (1985) Photochem Photobiol, in press
4. Ferrari A, Braibanti A, Bigliardi A, Dallavalle F (1963) Z Krist 119, 284–***
5. Förster V (1984), Thesis, Universität Osnabrück (FRG)
6. Förster V, Hong Y-Q and Junge W (1981) Biochim Biophys Acta 638, 141–152
7. Förster V and Junge W (1985) Photochem Photobiol 41, 183–190
8. Förster V and Junge W (1985) Photochem Photobiol 41, 191–194
9. Förster V and Junge W (1985) FEBS Lett 186, 53–57
10. Fowler CF (1977) Biochim Biophys Acta 462, 414–421
11. Hong YQ and Junge W (1983) Biochim Biophys Acta 722, 197–208
12. Junge W, Ausländer W, McGeer A and Runge Th (1979) Biochim Biophys Acta 546, 121–141
13. Kok B, Forbush B and McGloin M (1970) Photochem Photobiol 11, 457–475
14. Matsugama HJ and Audrieth LF (1959) J Inorg Nucl Chem 12, 186–192
15. Purcell KF and Kotz JC (1977) Inorganic Chemistry, WB Saunders, Philadelphia
16. Radmer R and Ollinger O (1982) FEBS Lett 144, 162–166
17. Radmer R and Ollinger O (1983) FEBS Lett 152, 39–43
18. Radmer R and Ollinger O (1984) in: Advances in Photosynthesis Research, (C Sybesma, ed), Martinus Nijhoff/Dr. W. Junk Publ., The Hague; Vol I, 269–272
19. Renger G and Hanssum B Biochim Biophys Acta, in press
20. Segel IH (1975) Enzyme Kinetics, John Wiley and Sons Inc., New York

Photosynthesis Research 9, 211–227 (1986)
© 1986 Martinus Nijhoff/Dr. W. Junk Publishers, Dordrecht.

Photosynthetic free energy transduction related to the electric potential changes across the thylakoid membrane

OLAF VAN KOOTEN, JAN F.H. SNEL and WIM J. VREDENBERG

Laboratory of Plant Physiological Research, Agricultural University, Wageningen, Gen. Foulkesweg 72, 6703 BW Wageningen, The Netherlands

(Received 5 August 1985)

Key words: ATPase flux, $\Delta\tilde{\mu}_{H^+}$, electron transport, model simulation, microelectrode measurements, thylakoid membrane

Abstract. A model based on our present knowledge of photosynthetic energy transduction is presented. Calculated electric potential profiles are compared with microelectrode recordings of the thylakoid electric potential during and after actinic illumination periods of intermediate duration. The information content of the measured electric response is disclosed by a comparison of experimental results with calculations. The proton flux through the ATP synthase complex is seen to markedly influence the electric response. Also the imbalance in maximum turnover rate between the two photosystems, common to obligate shade plants like *Peperomia metallica* used in the microelectrode experiments, is clearly reflected in the electric potential profile.

Introduction

Ever since the advent of the chemiosmotic theory [33], many techniques have been used to measure the light-induced electrochemical potential across the thylakoid membrane in higher plant chloroplasts [59, 51, 26, 60, 28]. The common technique for measuring the electric potential changes, due to actinic light flashes of relatively short duration ($\leqslant 1$ ms), is by monitoring the electrochromic absorbance change at 515 nm [26, 60, 28, 53]. This absorbance change was first observed by Duysens [15]. However, for actinic illuminations of longer duration ($\geqslant 10$ ms), the absorbance change is obscured by scattering changes and this technique is not applicable [38, 12]. The only other kinetic technique available to measure the electric potential changes with sufficient time resolution, during actinic illumination periods of intermediate duration, is the open ended glass capillary microelectrode technique [6–11, 49–52]. Recent results, comparing flash-induced potential changes measured with both of the above mentioned techniques in the same specimen, further enhance our trust in the reliability of the microelectrode measurements [47].

The time-course of the electric potential during actinic illumination periods of several seconds is complex [11]. This complexity suggests that a closer

---

Dedicated to Prof. L.N.M. Duysens on the occasion of his retirement.

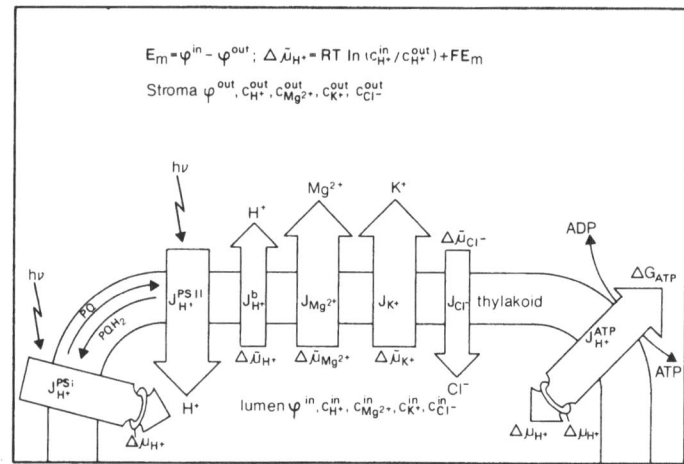

Figure 1. Schematic representation of the forces and fluxes involved in our model. The two photosystems are described as two photon-induced proton pumps interconnected by a mobile hydrogen carrying redox pool and attenuated by the $\Delta\tilde{\mu}_{H^+}$. Active dissipation of $\Delta\tilde{\mu}_{H^+}$ is through the ATPase, controlled by the chemical free energy of phosphorylation and by $\Delta\tilde{\mu}_{H^+}$. The activation state of the ATPase is also regulated by $\Delta\tilde{\mu}_{H^+}$. The passive dissipation of any potential build-up is caused by the ions and protons present. The main formulae are given and the sign convention is taken from [29].

analysis of the electric potential time-course should provide us with a wealth of information on the precise mechanism of energy transduction in thylakoids. To perform this analysis it is necessary to know the interactions between the forces and fluxes across the thylakoid membrane related to the electric potential at all times during the illumination. We have developed a model describing these forces and fluxes, enabling us to calculate the time-resolved changes in potentials and concentrations on both sides of the thylakoid membrane during and after actinic illumination. The forces and fluxes on which the model is based are schematically depicted in Figure 1. Two light driven electron transport complexes, photosystem I (PSI) and photosystem II (PSII), interconnected by a mobile pool of hydrogen carrying redox mediators, cause the transport of protons from the outside (stroma) to the inside (lumen) of the thylakoid during illumination. The magnitude of this active proton flux $(J_{H^+}^{PSI} + J_{H^+}^{PSII})$ is attenuated by the rise in the internal electrochemical proton potential $(\tilde{\mu}_{H^+})$ [24, 44]. On the basis of our knowledge of electron transport in thylakoid membranes [60, 13] we have developed an algorithm describing the light-induced proton flux as a function of the redox state of intermediary redox pool and the backpressure of $\Delta\tilde{\mu}_{H^+}$. The passive fluxes of protons and other ions present across the thylakoid membrane are capable of dissipating the $\Delta\tilde{\mu}_{H^+}$. In our model we restrict ourselves to fluxes of H^+, Mg^{2+}, K^+ and Cl^-. These fluxes are adequately described by the Goldmann-Hodgkin-Katz equation [17]. $\Delta\tilde{\mu}_{H^+}$ can also be dissipated by the ATP synthase protein complex (CF_0–CF_1). The proton flux through this complex in either direction is dependent on the free energy of ATP in the stroma (ΔG_{ATP}) and $\Delta \tilde{\mu}_{H^+}$. The activation of CF_1 is determined by $\Delta \tilde{\mu}_{H^+}$. We have slightly modified an algorithm derived by Gräber et al. [18–21, 25, 41], which enables us to operationalize it within our model.

When all the fluxes can be calculated as a result of a known set of electrochemical potential differences then the concomitant changes in concentrations and electric potential can be derived as follows:

$$\frac{dE_m}{dt} = \frac{F}{C_m} (J_{H^+}^{PSI} + J_{H^+}^{PSII} + J_{H^+}^{b} + J_{H^+}^{ATP}$$
$$+ 2J_{Mg^{2+}} + J_{K^+} - J_{Cl^-}) \tag{1}$$

$$\frac{dc_{H^+}^{in}}{dt} = \frac{A}{V} \beta (J_{H^+}^{PSI} + J_{H^+}^{PSII} + J_{H^+}^{b} + J_{H^+}^{ATP}) \tag{2}$$

$$\frac{dc_i^{in}}{dt} = \frac{A}{V} J_i \tag{3}$$

In eq. 1 the sum of all fluxes, multiplied by the valency of their constituent ions, is multiplied by the ratio of the Faraday constant (F) and the membrane capacitance (C_m). This results in a time-dependent change of the electric potential E_m across the thylakoid membrane (sign convention is taken from [29]). In eq. 2 the sum of all proton fluxes is multiplied by the surface to volume ratio (A/V) of the thylakoid and by a buffering capacity parameter (β) for the lumen. This results in a free proton concentration change in the lumen. The concentrations and potentials of the stroma are kept constant for reasons which will be explained further on. Finally the changes in concentration of ion species i are simply calculated with eq. 3. When all the potentials, concentrations and fluxes are known or can be calculated at a certain moment t, then we can assume that eqs. 1–3 are constant during an incremental period Δt, provided Δt is small enough. As a result a new set of potentials and fluxes can be calculated at the moment t + Δt. With the aid of a computer the time resolved potential changes during and after illumination can thus be calculated.

The basic assumptions of the model are the chemiosmotic theory as proposed by Mitchell [33, 34], which implies that both stroma and lumen are isopotentials at all times. All potential differences fall linearly and exclusively across the thylakoid membrane, and all processes can be described by semi-continuous functions. As a result of these assumptions the model is unfit to describe so-called "semi-localized" chemiosmotic theories [56, 58] and short term phenomena, such as single-turnover actinic flashes, for which discontinuous algorithms based on a more mechanistic model should be derived.

However on the ms to s time scale the model performs a synthesis of our knowledge of photosynthetic energy transduction gathered so far.

Materials and methods

Microelectrodes were made from borosilicate glass (Clark, GC 200F-15) and pulled on a BB-CH type puller [4]. The electrodes were filled with 1 M KCl and impaled in giant chloroplasts *in situ* of leaf cuts of the shade plant *Peperomia metallica*. The leaf cuts were immersed in 15 mM sorbitol + 20 mM 4-(2-hydroxyethyl)-1-piperazine ethane-sulphonic acid/KOH pH 7.5. After impalement at ambient room temperature (i.e. about 25 °C), both the resistance and the risetime of the measuring circuit were determined. These were about 50 MΩ and less than 2 ms (0–90% of a square voltage pulse), respectively. Potential changes between the electrode tip and a reference, consisting of an agar-1 M KCl bridge situated near the leaf cut, were measured by a high-impedance (10^{13} Ω) unity gain amplifier as described elsewhere [40]. The signals were recorded on a transient recorder (Nicolet 2090–111 A) and relayed to a microcomputer (HP-86) for further processing.

Peperomia metallica plants were grown in a laboratory built climate chamber under TL-33 fluorescent lamps (Philips) and received 12 hours of 1.2 W/m² light daily. Transmission electron microscopy pictures of these chloroplasts stained with osmiumtetroxide revealed densely stacked thylakoid membranes with large grana common to shade plant chloroplasts [32].

Partial destruction of thylakoid membranes by impalement was checked by comparing the half-time of the decaying electric field after a short flash with electrochromic P515 measurements in intact preilluminated leaves [35]. If the halftime of decay measured with the micro-electrode was an order of magnitude faster than that measured with the P515 technique (about 60 ms at 25 °C) the measurements were discarded.

Model calculations were performed on a PDP11/23 minicomputer (Digital) under Fortran programming. Parameters subjected to continuous addition, such as membrane potential E_m and the concentrations (see eqs. 1–3) were kept in double precision (i.e. 64 bits per real variable) in order to minimize systematic errors. The incremental time step Δt varied between 0.5 μs and 50 ms, depending on the magnitude of the fluxes.

Results and interpretation

We will present our model and its algorithms in 4 parts, the light driven proton pump ($J_{H^+}^{PSI} + J_{H^+}^{PSII}$), the passive dissipation of $\Delta \tilde{\mu}_{H^+}$, the buffering capacity of the lumen and the active ATPase-dependent proton flux ($J_{H^+}^{ATP}$).

1. The light driven proton pumps

Our algorithm for light-induced proton pumping in thylakoid membranes is based on the Z-scheme of electron transport [60]. Two photosystems, i.e. PSI and PSII, each with its own apparent absorption cross section, are interconnected by a pool of mobile redox mediators. The rate limiting step for electron transport is assumed to be in the intermediary redox pool, i.e. at the site of oxidation of plastohydroquinone (PQH_2) by the cytochrome b_6–f complex (cyt b–f) [44, 24, 22]. The turnover of a photosystem results in the translocation of a proton from the stroma to the lumen. The maximum turnover rate (k_2) of PSII, when provided with a proper electron acceptor, is about $1000\,s^{-1}$ [60]. The maximum turnover rate (k_1) of PSI is dependent on the plant species. Extremely high values $k_1 = 200\,s^{-1}$ have been found in spinach, but low values were found in the shade plant *Asarum europaeum* $k_1 = 20\,s^{-1}$ [14]. The number of photosystems per unit membrane area, expressed as Chl/P700 for PSI and Chl/Q_A for PSII, can also vary considerably between sun and obligate shade plants [32, 5, 31, 57]. All these differences can be accounted for by expressing the number of electron transport chains per unit area (c_{elch}) as the number of chlorophyll molecules per cytochrome f molecule, i.e. c_{elch} = 400 Chl/cyt f in sun leaves and c_{elch} = 1000 Chl/cyt f in shade leaves [5].

The two turnover rates can then incorporate the differences in concentration ($Q_A/P700$) and the differences in absorption cross section. Since we will compare our model with experiments on chloroplasts of the obligate shade plant *Peperomia metallica* we will use the following values: c_{elch} = 1000 Chl/cyt f = $1.14\ 10^{-13}$ mol P700 cm^{-2} (based on 2 P700/cyt f [5] and on $1.75\,m^2/mg$ Chl [3]), $k_1 = 100\,s^{-1}$, $k_2 = 1000\,s^{-1}$, n = 32/P700 (ref. 5). The last parameter n represents the approximate number of electron equivalents the intermediary redox chain can contain. The redox state of the intermediary electron transport chain will attenuate the proton fluxes caused by PSI and PSII as follows:

$$J_{H^+}^{PSI} = \frac{n_{rd}}{n} k_1 c_{elch} \tag{4}$$

$$J_{H^+}^{PSII} = \frac{n_{ox}}{n} k_2 c_{elch} \tag{5}$$

It follows from eqs. 4 and 5 that the sum of the fractions reduced (n_{rd}/n) and oxidized (n_{ox}/n) intermediary redox equivalents must always equal 1. The change in time of the concentration in oxidized redox equivalents is enhanced by PSI turnovers and diminished by PSII turnovers.

$$\frac{d}{dt}(n_{ox}c_{elch}) = J_{H^+}^{PSI} - J_{H^+}^{PSII} \tag{6}$$

combining eqs. 4–6 yields

$$\frac{dn_{ox}}{dt} = \frac{n_{rd}}{n} k_1 - \frac{n_{ox}}{n} k_2 \tag{7}$$

This differential equation can be solved if we assume n_{ox} = X at t = 0.

$$\frac{n_{ox}}{n} = \frac{k_1}{k_1 + k_2} + \left(X - \frac{k_1}{k_1 + k_2}\right) e^{-(k_1 + k_2)t/n} \tag{8}$$

If we define $t = 0$ as the moment saturating actinic illumination starts and $X = 1$, i.e. a fully oxidized intermediary redox pool, the two fluxes can be calculated as shown in Figure 2A.

For convenience the concomitant rise in proton motive force (p.m.f. expressed in mV) is also shown. The calculation of the p.m.f. will be explained later. It is clear from Figure 2A that the two fluxes reach a steady-state within 0.2 s. However it is also clear that this kind of proton pumping would lead to a breakdown of the thylakoid membrane within several seconds of illumination, due to a prodigious rise in p.m.f. This is prevented in the thylakoid by attenuation of the turnover rate k_1, i.e. decreasing the internal pH leads to a decreased turnover rate of PQH_2 oxidation [24, 44]. The so-called "backpressure" effect has been shown to be approximately linear with the pH [24], which implies that it can be incorporated in k_1 as follows:

$$k_1' = \frac{k_1}{\left(1 + \alpha \frac{\Delta\tilde{\mu}_{H^+}}{RT}\right)} \tag{9}$$

The basic assumption in eq. 9 is that E_m has an equivalent effect with regard to the backpressure as ΔpH. By dividing $\Delta\tilde{\mu}_{H^+}$ with the gas constant (R) and the absolute temperature (T) this quotient becomes dimensionless. The parameter α expresses the coefficient of coupling. When α equals 1, eq. 9 implies an attenuation of k_1 by a factor of 10 when $\Delta\tilde{\mu}_{H^+}$ reaches a value equivalent to $\Delta pH = -4$, a value reached after about 0.5 s of illumination in Figure 2A. This can be considered as a reasonable value since the ratio of uncoupled to coupled electron transport under saturating continuous illumination in freshly prepared thylakoids is usually around 7 to 8 (personal observation). In Figure 2B the same calculation as in Figure 2A is shown except that eq. 9 is now included with $\alpha = 1$. The two fluxes become equal in magnitude after about 0.3 s of illumination. But the total magnitude $J_{H^+}^{PSI} + J_{H^+}^{PSII}$ at $t = 0.3$ s is much lower than in Figure 2A and it continues to decline as the p.m.f. rises. If the calculation in Figure 2B were to be continued the proton flux would reach steady-state within 10 s and the p.m.f. stays below 250 mV.

2. The passive dissipation of $\Delta\tilde{\mu}_{H^+}$

When the p.m.f. rises due to light induced proton pumping, it can be dissipated by passive ion movement across the membrane. These passive ion fluxes can be described by the Goldmann-Hodgkin-Katz equation [17].

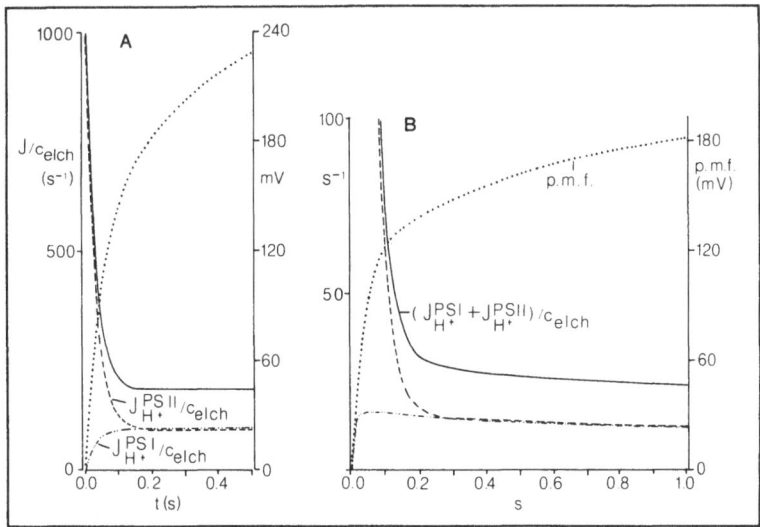

Figure 2. The photon-induced proton flux of both PSI and PSII calculated in a thylakoid fully permeable to all ions except protons ($P_{H^+} = 2.10^{-5}$ cm s^{-1}). The concomitant rise in p.m.f. is completely based on a change in pH in the lumen. A: in this calculation illumination is started at t = 0 and no "backpressure" of $\Delta\tilde{\mu}_{H^+}$ on $J_{H^+}^{PSI}$ is assumed. B: equivalent to A but now a backpressure on $J_{H^+}^{PSI}$ is assumed. As a consequence it takes much longer to reach a steady state photon-induced proton flux and the p.m.f. will stay within a reasonable range. Note the difference in left hand scaling between A and B.

$$J_i = \frac{z_i F E_m}{RT} \, P_i \, \frac{(c_i^{out} - c_i^{in} e^{(z_i F E_m/RT)})}{(1 - e^{(z_i F E_m/RT)})} \tag{10}$$

The subscript in eq. 10 denotes the ion species i. The permeability coefficients (P_i) were taken from the literature ($P_{Cl^-} = 1.8 \ 10^{-8}$ cm s^{-1} [1, 51] $P_{K^+} = 3.6$ 10^{-8} cm s^{-1} [1, 51], $P_{H^+} = 2.10^{-5}$ cm s^{-1} [51, 42] or $2.5 \ 10^{-4}$ cm s^{-1} [21]). The permeability coefficient of magnesium is unknown. We assume $P_{Mg^{2+}} = 0.5 \ 10^{-8}$ cm s^{-1}. The concentrations of all ions in the stroma is assumed to be constant. In reality transient concentration changes will occur, but for all ions except protons these changes will be marginal [51]. The pH of the stroma may rise by less than 1 pH unit during illumination [23]. This rise is related to a light-induced extrusion of protons across the chloroplast envelope [55]. Since the stromal pH rises from 7.03 in the dark to 7.91 in the light [38], with kinetics unknown to us, we keep pH^{out} = 7.5 and constant. For the other ion concentrations it is known that chloroplasts in leaves contain between 20 and 50 mM of K$^+$ and Mg^{2+} [2]. However the thylakoid membrane has a large negative surface charge density, $\sigma^{in} = -3.67 \, \mu C$ cm^{-2} and $\sigma^{out} = -2.06 \, \mu C$ cm^{-2} at an external pH 7.5 [30]. This leads to a surplus of positive ions in the water adjacent to the membrane surface.

Calculations of the space charge density result in a surplus positive charge concentration of at least 0.2 M in a layer of 10 nm thickness adjacent to the membrane [39]. When potassium and magnesium are present in equal concentration in the stroma, the surplus charge near the membrane will predominantly be formed by the divalent cation [3]. These considerations have led us to choose our concentrations as follows: $c_{Mg^{2+}}^{out} = 0.1$ M, $c_K^{out} = 20$ mM and $c_{Cl}^{out} = 20$ mM. The concentrations were so chosen that E_m, according to the calculation of a passive discharge of the membrane after a flash, will decay with a halftime of 60 ms (see materials and methods).

When eq. 10 is used it must be replaced by an approximation whenever the exponentials approach one. The following approximation was used whenever $|E_m| \leqslant 0.5$ mV [43].

$$J_i = P_i(c_i^{out}e^{-z_iFE_m/RT} - c_i^{in}e^{z_iFE_m/RT}) \qquad (11)$$

By using the equations given one can calculate the time resolved potential and flux changes in a system consisting of a light driven proton pump and passive dissipation by protons and three other ions. The results of such a calculation are given in Figure 3. The time resolved changes in E_m, -2.3 $(RT/F)\Delta pH$ and p.m.f. are shown. The values of parameters not mentioned so far are $T = 293.16$ K, $C_m = 1 \mu F cm^{-2}$ [50] and $A/V = 2.9 \ 10^6 cm^{-1}$ (equivalent to $5 \mu l/mg$ Chl). The buffering parameter β used in eq. 2 was taken as 1, which explains the steep rise in p.m.f. The state of the intermediate redox pool at $t = 0$ was not taken as fully oxidized, since cytochrome f, plastocyanin and P700 normally stay reduced in the dark ($X = 0.867$). The value of the turnover rate of PSII was halved compared to the calculations shown in Figure 2. This was done to prevent E_m from reaching a prodigious peak value. In Figure 3B all the time courses of the changes in fluxes are given on an enlarged time scale. It is clear that the main passive fluxes are those of the cations and the anion in the first 200 ms. After that the passive proton flux becomes predominant. Before the light is turned off the system has reached a true steady-state in that the passive proton flux equals the sum of the two light-induced fluxes. After the light is turned off the passive proton flux remains the main dissipative flux for many seconds. The time course of E_m in Figure 3A resembles experimental results from microelectrode measurements in *P. metallica* chloroplasts *in situ* immersed in a medium with a high concentration of dicyclohexylcarbodiimide (DCCD), i.e. more than 1 mM (see inset Figure 3A). DCCD is known to inhibit the ATP-ase complex at low concentrations ($50 \mu M$). At these high concentrations its effects are quite unpredictable. However one effect must be the covalent binding to carboxyl residues (COO^-) or proteins. These residues form the main buffering constituent of the lumen [54]. Since our calculation in Figure 3 does not contain any buffering effect, i.e. $\beta = 1$, this might explain the apparent similarity between calculation and experiment.

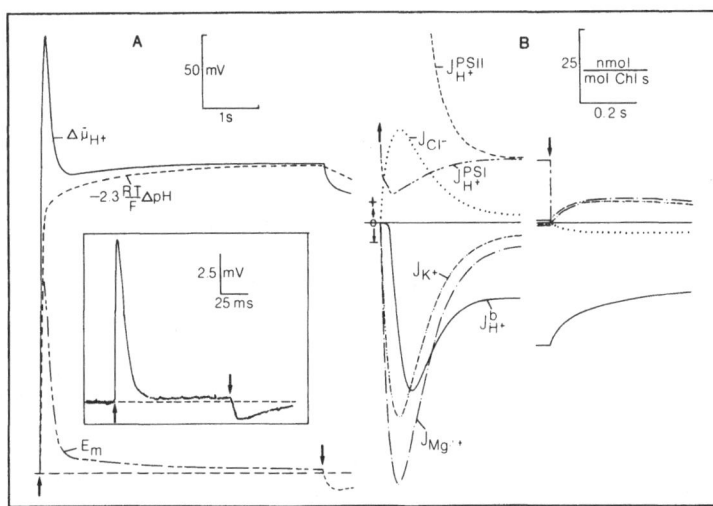

Figure 3. A: calculation of the three potentials influencing the proton flux in a thylakoid. The upward arrows indicate the start and the downward arrows the end of illumination. The inset shows an E_m profile measured in a *P. metallica* chloroplast immersed in a suspension of DCCD ($\simeq 1$ mM). The apparent similarity between the measured E_m profile and the calculated one is probably explained by the lack of buffer in the lumen. B: the fluxes of all ions and protons are given on an enlarged time scale. Of the passive fluxes the non-protonic are the major fluxes when E_m is large. After 0.3 s the passive proton flux $J_{H^+}^b$ is the major flux and this remains so even after the illumination period.

3. The buffering capacity of the lumen

In order to describe the buffering capacity in the lumen, it is necessary to derive a proper algorithm for β in eq. 2. From literature data [23, 27, 54] it is possible to deduce such parameters as the concentration of certain buffer groups in the lumen B_i and their dissociation constants K_i. These parameters are related to a buffer reaction as follows

$$H^+ + A_i^- \rightleftharpoons HA_i$$

$$K_i = [A_i^-] [H^+]/[HA_i]$$

$$B_i \equiv [HA_i] + [A_i^-]$$

When the above equations are combined an expression can be derived for the total concentration of protonated buffer groups as a function of free proton concentration.

$$[HA_i] = \frac{B_i[H_i]}{K_i + [H_i]} = \frac{B_i c_{H^+}^{in}}{K_i + c_{H^+}^{in}} \tag{12}$$

If the change in both bound and free proton concentrations can be considered to be infinitesimal, both sides in eq. 12 can be differentiated with respect to the free proton concentration

$$d[HA_i] = \frac{B_i K_i}{(K_i + c_{H^+}^{in})^2} dc_{H^+}^{in} \tag{13}$$

The sum of bound and free proton concentration changes (dh^+) is related to the net proton flux ΣJ_{H^+} and the incremental time period Δt as follows:

$$dh^+ = \frac{A}{V} \Sigma J_{H^+} \Delta t = d[HA_i] + dc_{H^+}^{in} \tag{14}$$

When more than one buffer type is present eq. 14 can easily be expanded (to incorporate all buffers present)

$$dh^+ = \sum_i d[HA_i] + dc_{H^+}^{in} \tag{15}$$

By combining eqs. 13 and 15 the expression for β in eq. 2 results

$$\beta = \left[1 + \sum_i \frac{B_i K_i}{(K_i + c_{H^+}^{in})^2} \right]^{-1} \tag{16}$$

This expression for β is not the proper buffer capacity as defined in [27], but can be transformed into it. However eq. 16 is useful once we know the values of B_i and K_i in the lumen of the thylakoid. When we try to extract the concentrations B_i from literature data we must keep in mind that different authors have determined the lumenal buffer capacity under different conditions, i.e. in vivo [23], in isotonic suspension [54], or in hypotonic suspension [27]. Since almost all buffers are either protein residues or lipid headgroups, a change in surface to volume ratio (A/V) will result in an equivalent change in B_i. The maximum surface to volume ratio reported on up till now is $A/V = 5.3 \ 10^6 \text{ cm}^{-1}$ (equivalent to $3.3 \ \mu l/mg$ Chl [23]). We will estimate the concentrations B_i for this surface to volume ratio and multiply them by the ratio of A/V actually used and this maximum value. In our case this ratio is 3.3/5. In Table 1 the concentrations B_i and dissociation constants of the different buffer types in the lumen are given. The sum of the concentrations $\Sigma B_i = 0.3$ M, which is based on the value of 1 mole buffer per mole chlorophyll found in [54] and $3.3 \ \mu l/mg$ Chl. In the pK range from 8 to 6.5 the buffering is constant [27]. Below pH 6.5 there is a steep rise in buffering capacity peaking at pH 5.5 [23, 54]. Below pH 4.5 the buffering capacity rises again very steeply [54], probably due to the fact that most phospholipids and sulpholipids are oriented to the lumenal side of the membrane [46]. The values given in Table 1 must be considered as a crude approximation of reality. However the values give rise to a satisfactory fit of the experimental curve presented in [54].

Calculations of the time-course of the potential changes incorporating the buffers of Table 1 are shown in Figure 4. The effect of the buffering on the potential changes is clear when Figure 4 is compared with Figure 3. The rise

Table 1. Buffer concentrations B_i and their concomitant dissociation constants pK_i in the lumen estimated from literature data [23, 27, 54] and based on a volume to chlorophyll ratio of $3.3\ \mu l/mg$ Chl and $\Sigma_i B_i = 300\ mM$

B_i	10	10	10	10	15	40	40	45	50	70	mM
pK_i	8.0	7.5	7.0	6.5	6.0	5.5	5.0	4.5	4.0	3.5	

in chemical proton potential is slower and more in accordance with the associated light scattering changes found in [12]. It takes about 50 s of saturating actinic illumination before steady-state is attained (Figure 4). The electric membrane potential (E_m) in the steady-state is higher due to the fact that the non-protonic ion fluxes have created a Nernst-potential which is non-negligible. After 50 s of illumination (Figure 4A) the calculation is continued to simulate the processes during 4 minutes dark adaptation. The reoxidation of the intermediary redox pool was assumed to be a pseudo first order process with a rate constant of $0.038\ s^{-1}$ [37].

In Figure 4B a simulation of the potential changes in the light after the 4 minutes dark adaptation of Figure 4A is shown. These calculations were done to simulate the experimental procedure used in microelectrode measurements. Before a microelectrode recording, as shown in Figure 4C, can be

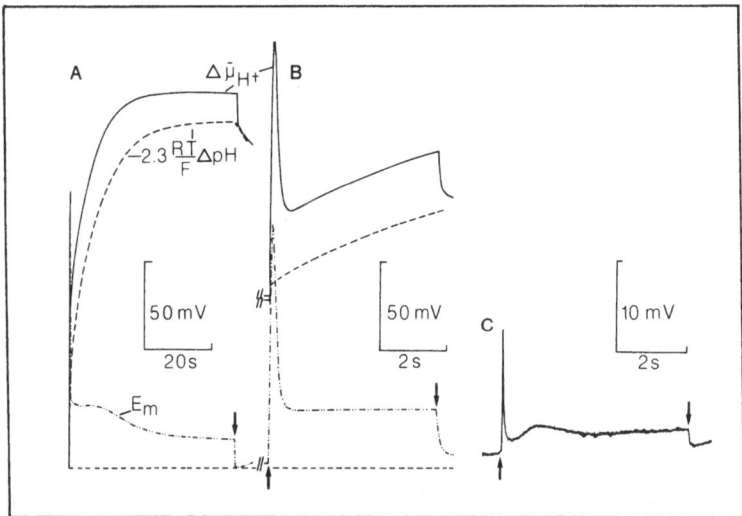

Figure 4. A: calculated time-courses of the potentials in a fully buffered thylakoid (Table 1). The upward arrows indicate the start and the downward arrows the end of illumination (illumination period = 50 s). After the light is turned off the calculation is continued simulating 4 min dark adaptation. B: calculated time-courses of the potentials after 4 min dark adaptation. This potential time course should be compared with the experimental findings. C: measured time-course of E_m in *P. metallica*. The complexity of this time-course is not reflected in the calculated profile in B. Note the difference in time scale between A and B.

performed the chloroplast is preilluminated for at least a few minutes, due to the requirement of light for visualisation of the impalement. Successive dark adaptation periods are usually about 4 minutes [11]. When Figures 4B and 4C are compared it is clear that the complexity of the experimental time-course of E_m is not simulated by our model as yet. This is mainly due to the lack of active dissipation of $\Delta\tilde{\mu}_{H^+}$ by a simulated proton conducting ATPase complex, as will be shown below.

4. The ATPase dependent proton flux

In order to make our model a better reflection of reality it is necessary to incorporate the ATPase dependent proton flux. A full algorithm for this flux, containing both activation and chemical reaction rate, has been presented in ref. 18. The validity of this algorithm was corroborated in ref. 21. The description of the ratio of activated to total (E_a/E_t) ATPases, could be retained without modification in our calculations.

$$\frac{E_a}{E_t} = \frac{K_E^0 e^{-b\Delta\tilde{\mu}_{H^+}/RT}}{1 + K_E^0 e^{-b\Delta\tilde{\mu}_{H^+}/RT}} \tag{17}$$

The values $pK_E^0 = 5.9$ and $b = 1.7$ of ref. 21 have been used here. However the algorithm for the relative rate of ATP synthesis (v/v_{max}) could not be retained, because of the fact that the values of some parameters therein need to be estimated and moreover will change with pH. For our calculations we approach the relative rate of ATP-synthesis with a simple first order discontinuous function.

$$\frac{v}{v_{max}} = \begin{cases} 1 & \text{for } \Delta\tilde{\mu}_{H^+} > \dfrac{\Delta G_{ATP}}{n} + \dfrac{RT}{A} \\[2ex] \dfrac{A}{RT}\left(\Delta\tilde{\mu}_{H^+} - \dfrac{\Delta G_{ATP}}{n}\right) & \\[2ex] -\tfrac{1}{3} & \text{for } \Delta\tilde{\mu}_{H^+} < \dfrac{\Delta G_{ATP}}{n} - \dfrac{RT}{3A} \end{cases} \tag{18}$$

In eq. 18 n equals the stoichiometry number (H^+/ATP) for which we used the well accepted value of 3 [43]. The value of the slope $A = 0.53$ was taken from ref. 25 (i.e. ΔpH dependent ATP synthesis rate in preactivated thylakoids). The value of $v_{max} = 380$ mmol ATP (mole Chl)$^{-1}$ s^{-1} [21] converted to our coordinate system was $- 2.2 \; 10^{-11}$ mole ATP cm^{-2} s^{-1}. A combination of eqs. 17 and 18 leads us to the ATPase dependent proton flux

$$J_{H^+}^{ATP} = n\frac{E_a}{E_t}\frac{v}{v_{max}}v_{max} \tag{19}$$

In principle our model allows a calculation of ΔG_{ATP} as a function of $J_{H^+}^{ATP}$ and changes in P_i, ADP and ATP concentrations. But for a proper simulation incorporating the changes in ΔG_{ATP} it is necessary to incorporate the functioning Calvin cycle and the adenylate kinases as well. Since it has been found that ΔG_{ATP} is not subject to large fluctuations (e.g. between 42 in the dark and 46 kJ mole^{-1} in the light [16]) in intact chloroplasts we have left it constant for the calculations shown in Figure 5, i.e. $\Delta G_{ATP} = 46$ kJ mole^{-1} When we compare the experiment (Figure 5D) and the calculations in Figure 5B the qualitative equivalence is evident. The amplitudes of E_m however differ by about a factor 10. The amplitude of the peak value of E_m differs strongly from impalement to impalement (our record to date is a maximum of 118 mV). In the experiment shown in Figure 5 the total resistance of the microelectrode rose by more than 10 MΩ after impalement, while a rise of 0.5 MΩ can be expected for a chloroplast of 30 μm diameter and a membrane resistance of $R_m = 10^4 \Omega$ cm^2 [51]. This rise in impedance could indicate a clogging up of the electrode tip and might explain a certain attenuation of E_m.

The time course of E_m (Figure 5A, B, D) can now be explained by close inspection of the flux changes (Figure 5C). The transient initial peak of E_m after turning on the light is caused by the excessive imbalance in turnover rate of PSII and PSI common to obligate shade plants. The second peak in E_m is mainly caused by a change from ATP hydrolysis to ATP synthesis (Figure 5C) when $\Delta\tilde{\mu}_{H^+}$ surpasses $\Delta G_{ATP}/n$. This is corroborated by the fact that the second peak is absent when the chloroplast is preenergized or when DCCD is present (e.g. Figure 3 and [11]).

Conclusion and perspectives

The model presented in this paper is a rough approximation of photosynthetic free energy transduction in thylakoids. The description of the light-induced proton pump rates (eqs. 4, 5 and 8) meets the determined boundary conditions. The time-dependent change of the redox state of the intermediary PQ pool is too simple to describe the actual process. To improve this formulation it will be necessary to incorporate a set of differential equations describing charge separation and charge movement at the reducing side of PSII, as was done by Renger and Schulze [37]. Due to the fact that we compare saturating actinic illumination experiments with our simulations, eq. 8 gives a reasonable fit. Whenever the model will be used to simulate the fluxes at low light intensities, a more precise description of $J_{H^+}^{PSI}$ and $J_{H^+}^{PSII}$ will probably be necessary.

The use of $\Delta\tilde{\mu}_{H^+}$ instead of ΔpH [24] in eq. 9 is not experimentally sustained. When $\Delta\tilde{\mu}_{H^+}/RT$ was substituted by $-2.3\,\Delta pH$ the initial electric membrane potential E_m reached an 'unphysiological' transient level after the

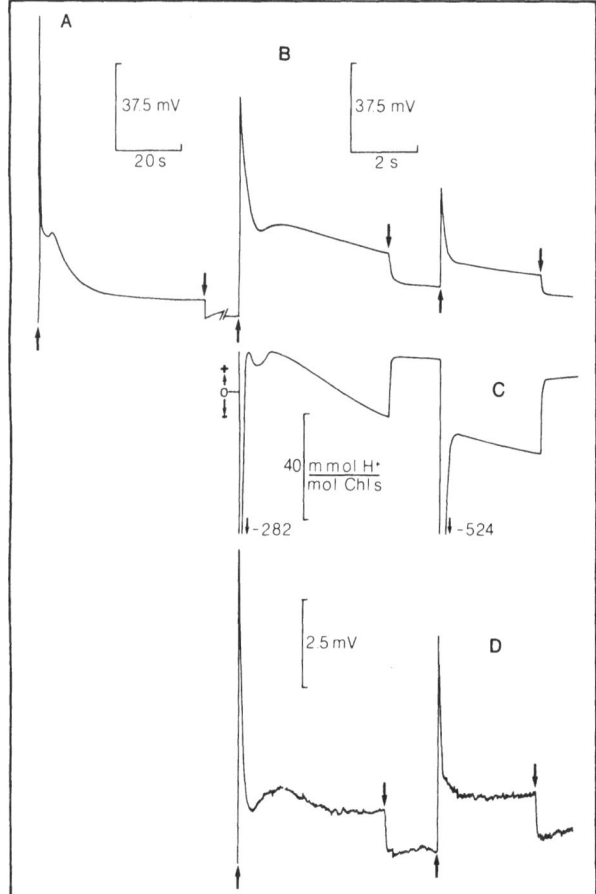

Figure 5. A: calculated E_m time-course equivalent to Figure 4A except that now an active dissipation of $\Delta\tilde{\mu}_{H^+}$ is possible by a simulated ATPase. After a 50 s illumination period the light is shut off and dark adaptation is simulated for 4 min. B: two subsequent illumination periods simulated after 4 min. dark adaptation. Upward arrows indicate the start and downward arrows the end of illumination. The complex time-course of the two illuminations is a result of the steep slope in eq. 18. C: the calculated ATPase dependent proton flux from the same simulation as in B. Positive flux is equivalent to ATP hydrolysis and a negative flux is equivalent to ATP synthesis. The flux should be divided by the stoichiometry number $n(=3)$ in order to obtain the rate of ATP synthesis or hydrolysis. D: measured E_m time-course of two subsequent illumination periods after 5 minutes dark adaptation in a *P. metallica* chloroplast. The complex kinetics of this time-course is clearly reflected in the calculations (B). Note that the time axis is the same in B, C and D, but different in A.

light was turned on. A transformation of E_m into ΔpH is assumed to occur at the PQH_2 oxidation site. A similar mechanism has been suggested to occur within the ATPase complex [18].

The presence of different buffer groups in the model is a point of consideration. Although Table 1 is derived from the available experimental data, these data are insufficient for our purpose. The buffering capacity has a profound influence on the time-resolved E_m (e.g. compare Figures 3 and 4). After the initial peak in the first 100 ms of illumination, E_m becomes a continuous descending function in the light (i.e. $dE_m/dt \leqslant 0$). This decay rate will be slower in the presence of lumenal buffers. However, this will never result in a transient rise. As shown in Figures 4 and 5 a second transient rise in E_m during illumination occurs. To explain this rise an (extra) active influx of positive charges is required. This influx is caused by the hydrolysis of ATP (Figure 5C). The relative magnitude of the second rise in E_m during illumination will, amongst other things, be strongly dependent on ΔG_{ATP} (see eqs. 17–19). Microelectrode recordings in chloroplasts of the liverwort *Anthoceros sp.* reveal a much higher secondary potential rise [11]. This can be explained with our model either by assuming a higher ΔG_{ATP} in the stroma of these chloroplasts, or by a shift of the activation ratio curve (eq. 17) to lower values of $\Delta\tilde{\mu}_{H^+}$.

In this paper we have presented the basic version of our model. The model can also accomodate the effects of a Q-cycle, a Donnan potential, membrane modifying reagents and phosphorylation of the light-harvesting complex. Results of these will be shown elsewhere. However, the model needs further verification with regard to other readily measurable phenomena, such as proton uptake, oxygen evolution and electron transport. Since it can already calculate ΔpH and the redox state of the PQ pool, the model seems promising for simulating time dependent changes in fluorescence yield. If used in this manner the model will be more prone to experimental verification. Furthermore it may be useful to discriminate between quantitative aspects of different hypotheses on energy coupling. The model as yet is unable to discriminate between localized and delocalized chemiosmosis [48, 58]. It is possible to calculate lag times for photophosphorylation (Figure 5C). The results up till now are not in contradiction to the chemiosmotic theory as proposed by Mitchell [33, 34]. Future refinements are certainly required to give an answer on this matter.

Acknowledgements

We would like to acknowledge the assistance of W.J.M. Tonk, without whom we would not have been able to measure the electric potential changes at the precision required. We are indebted to dr. W. Junge whose fundamental criticism of an earlier version of our model has spurred us on and resulted in the model presented here. This research was partly supported by the Netherlands Foundation of Biophysics financed by the Netherlands Organization for the Advancement of Pure Research (Z.W.O.) and partly by a special grant from the Agricultural University at Wageningen.

References

1. Barber J (1972) Biochim Biophys Acta 275, 105–116
2. Barber J (1976) in: The Intact Chloroplast (Barber J, ed), Vol 1, pp. 89–134, Elsevier, North-Holland
3. Barber J (1980) Biochim Biophys Acta 594, 255–308
4. Bertrand D, C and P, Henauer R and Bader CR (1983) J Neurosci Meth 7, 171–183
5. Björkman O (1981) in: Encyclopedia of Plant Physiology vol. 16A (Lanse OL, Nobel PS, Osmond CB and Ziesler H, eds), pp. 57–107, Springer Verlag, Berlin
6. Bulychev AA, Andrianov VK, Kurella GA and Litvin FF (1971) Soviet Plant Physiol 18, 204–210
7. Bulychev AA, Andrianov, VK, Kurella GA and Litvin FF (1972) Nature 236, 175–177
8. Bulychev AA, Andrianov VK, Kurella GA and Litvin FF (1976) Biochim Biophys Acta 420, 336–351
9. Bulychev AA and Vredenberg WJ (1976) Biochim Biophys Acta 449, 48–58
10. Bulychev AA, Andrianov VK and Kurella GA (1980) Biochim Biophys Acta 590, 300–308
11. Bulychev AA (1984) Biochim Biophys Acta 766, 647–652
12. Coughlan SJ and Schreiber U (1984) Z Naturforsch 39c, 1120–1127
13. Cramer WA and Crofts AR (1982) in: Photosynthesis (Govindjee, ed), Vol I, pp. 387–467, Academic Press
14. Dietz KJ, Neimanis S and Heber U (1984) Biochim Biophys Acta 767, 444–450
15. Duysens LNM (1954) Science 120, 353–354
16. Giersch C, Heber U, Kobayashi Y, Inoue Y, Shibata K and Heldt HW (1980) Biochim Biophys Acta 590, 59–73
17. Goldmann DE (1943) J Gen Physiol 27, 37–60
18. Graeber P and Schlodder E (1981) in: Photosynthesis (Akoyunoglou G, ed), Vol II, pp. 867–879, Balaban Int., Philadelphia
19. Graeber P (1982) in: Current Topics in Membranes and Transport. Vol 16, pp. 215–245, Academic Press
20. Graeber P (1984) in: Charge and Field Effects in Biosystems (Allen MJ and Usherwood PNR, eds), pp. 227–242, Abacus Press
21. Graeber P, Junesch U and Schatz GH (1984) Ber Bunsenges Phys Chem 88, 599–608
22. Haehnel W (1984) Ann Rev Plant Physiol 35, 659–693
23. Heldt HW, Werden K, Milovancev M and Geller G (1973) Biochim Biophys Acta 314, 224–241
24. Huber HL and Rumerg B (1981) in: Photosynthesis I (Akoyunoglou G, ed), pp. 419–429, Balaban Int., Philadelphia
25. Junesch U and Graeber P (1984) in: Photosynthesis Research (Sybesma C, ed), Vol II, pp. 431–436, Dr. W. Junk Pub., The Hague
26. Junge W (1977) Ann Rev Plant Physiol 28, 503–536
27. Junge W, Ausländer W, McGeer A and Runge T (1979) Biochim Biophys Acta 546, 121–141
28. Junge W and Jackson JB (1982) in: Photosynthesis (Govindjee, ed), Vol I, pp. 589–646, Academic Press
29. Lowe AG and Jones MN (1984) Trends Biol Sci 9(1), 11–12
30. Mansfield RW, Nakatani HY, Barber J, Mauro S and Lannoye R (1982) FEBS Lett 137, 133–136
31. McCauly SW, Taylor SE, Dennenberg RJ and Melis A (1984) Biochim Biophys Acta 765, 186–195
32. Melis A (1984) J Cell Biol 24, 271–285
33. Mitchell P (1961) Nature 191, 144–148
34. Mitchell P (1966) Biol Rev 41, 445–502
35. Peters RLA, Bossen M, Van Kooten O and Vredenberg WJ (1983) J Bioenerg Biomembr 15, 335–346

36. Remish D, Bulychev AA and Kurella GA (1981) J Exp Bot 32, 979–987
37. Renger G and Schulze A (1985) Photobiochem Photobiophys 9, 79–87
38. Robinson SP (1985) Biochim Biophys Acta 806, 187–194
39. Rubin BT and Barber J (1980) Biochim Biophys Acta 592, 87–102
40. Schapendonk AHCM (1980) Doctoral Thesis, Agricultural University, Wageningen, the Netherlands
41. Schlodder E, Graeber P and Witt HT (1982) in: Electron Transport and Photophosphorylation (Barber J, ed), pp. 105–175, Elsevier
42. Schönfeld M and Schickler H (1984) FEBS Lett 167, 231–234
43. Schultz SG (1980) in: Basic Principles of Membrane Transport, p. 29, Cambridge University Press, Cambridge
44. Siggel U (1976) Bioelectrochem Bioenerget 3, 302–318
45. Strotman H and Schumann J (1983) Physiol Plant 57, 375–382
46. Sundby C and Larsson C (1985) Biochim Biophys Acta 813, 61–67
47. Van Kooten O, Leermaker FAM, Peters RLA and Vredenberg WJ (1984) in: Photosynthesis Research (Sybesma C, ed), Vol II, pp. 265–268, Dr. W. Junk, The Hague
48. Van Kooten O (1984) TIBS 9(5), 221–222
49. Vredenberg WJ, Homann PH and Tonk WJM (1973) Biochim Biophys Acta 314, 261–265
50. Vredenberg WJ and Tonk WJM (1975) Biochim Biophys Acta 387, 580–587
51. Vredenberg WJ (1976) in: The Intact Chloroplast (Barber J, ed), Vol I, pp. 53–88, Elsevier, North-Holland
52. Vredenberg WJ and Bulychev AA (1976) Plant Sci Lett 7, 101–107
53. Vredenberg WJ (1981) Physiol Plant 53, 498–502
54. Walz D, Goldstein L and Avron M (1974) Eur J Biochem 47, 403–407
55. Werdan K, Heldt HW and Milovancev M (1975) Biochim Biophys Acta 396, 276–292
56. Westerhoff HV, Helgerson SL, Theg SM, Van Kooten O, Wikström M, Skulachev VP and Danscsházy Zs (1983) Acta Biochim Biophys Acad Sci Hung 18, 125–149
57. Whitmarsh J and Ort DR (1984) Arch Biochem Biophys 231, 378–389
58. Williams RJP (1985) in: The Enzymes of Biological Membranes (Martonosi AN, ed), Vol 4, pp. 71–110, Plenum Press, New York
59. Witt HT (1971) Quart Rev Biophys 4, 365–477
60. Witt HT (1979) Biochim Biophys Acta 505, 355–427

Photosynthesis Research 9, 229–238 (1986)
© *1986 Martinus Nijhoff/Dr. W. Junk Publishers, Dordrecht.*
[227]

Energization and ultrastructural pattern of thylakoids formed under periodic illumination followed by continuous light

ÁGNES FALUDI-DÁNIEL[1], L.A. MUSTÁRDY[1], I. VASS[2] and J.G. KISS[2]

[1] Institute of Plant Physiology, Biological Research Center, Hungarian Academy of Sciences H-6701, Szeged, P.O.B. 521, [2] Department of Theoretical Physics, József Attila University H-6722, Szeged and [3] Clinic of ENT, Medical University H-6701, Szeged, Hungary

(*Received 13 August 1985*)

Key words: granum formation, light harvesting pigment-protein complex, thermoluminescence, membrane shrinkage, energy conservation

Abstract. Bean leaves grown under periodic illumination (56 cycles of 2 min light and 98 min darkness) were subsequently exposed to continuous illumination, and in connection with granum formation and accumulation of the light-harvesting pigment-protein complex thermoluminescence and light-induced shrinkage of thylakoid membranes were studied. Juvenile chloroplasts with large double sheets of thylakoids obtained under periodic light exhibited low temperature spectra of polarized fluorescence yielding fluorescence polarization (FP) values < 1 at 695 nm, characteristic for pheophytin emission. In the course of maturation under continuous light when normal grana appeared and the chlorophyll a/b light-harvesting photosystem II complex was incorporated into the membrane, at 695 nm the relative intensity of fluorescence dropped and FP changed to a value of > 1, suggesting an overlap between the emission of pheophytin and that of the chlorophyll a/b light-harvesting photosystem II complex. Thermoluminescence glow curves recorded with juvenile thylakoids displayed a relatively high proportion of emission at low temperatures (around −10°C) while with mature chloroplasts, more thermoluminescence originated from energetically deeper traps (discharged around 28°C). This means that during thylakoid development the capacity of the membrane to stabilize the separated charges increases, which might be favourable for the ultimate conservation of energy. The more extensive energization of mature thylakoids was also indicated by a light-induced decrease in the thickness of the membranes upon illumination; a change which could not be detected in juvenile thylakoids.

Abbreviations

EDTA, ethylenediamine tetraacetic acid; Hepes, 4-(2-hydroxy ethyl)-1-piperazine ethane sulfonic acid

Introduction

In the chloroplasts of green plants a characteristic of advanced stages in development is that thylakoids stack to form grana. Stacking is more than mere membrane adhesion as it involves a rearrangement of internal membrane

Address for offprints and all correspondence: Á. Faludi-Dániel, see address above.

Dedicated to Prof. L.N.M. Duysens on the occasion of his retirement.

components [13] which results in a lateral heterogeneity within the thylakoids [2]. In a regime where several minutes of illumination alternate with long dark periods photosynthetically competent membranes are formed, but proper stacking does not occur and the chlorophyll a/b light-harvesting photosystem II complex is not incorporated into the membranes. Such juvenile chloroplasts obtained under periodic light have particular induction curves of fluorescence with a high variable fluorescence but a much slower rise than normally developed mature chloroplasts [1]. Fluorescence induction kinetics have been shown by Duysens and Sweers [6] to reflect the status of Photosystem II reaction centers, so one might expect energization of juvenile thylakoids to have particular features.

In this work we studied the thermoluminescence and light-induced changes in the membrane dimensions together with the granum formation and accumulation of the light-harvesting pigment-protein complex, in an effort to understand the functional changes which occur during the development of protochloroplasts into chloroplasts.

Materials and methods

Plant material

Phaseolus vulgaris (cv. Red Kidney) was grown for five days under complete darkness at $24 \pm 1\,°C$. The seedlings were then illuminated by periodic white light of $5\,W m^{-2}$ obtained by $75\,W$ tungsten lamps in cycles of 2 min light and 98 min darkness [3]. After exposure to 56 light-dark cycles the plants were further exposed to continuous light of $5\,W m^{-2}$ for various periods of time.

Chloroplast isolation

Leaves were infiltrated in vacuum with 0.2% bovine serum albumin and homogenized gently in a medium containing $0.4\,M$ sorbitol, $10\,mM$ NaCl, $1\,mM$ $MnCl_2$, $5\,mM$ $MgCl_2$, $2\,mM$ EDTA and $50\,mM$ Hepes, pH 7.5 [12]. Chloroplasts were collected by a 10 min centrifugation at $3000 \times g$, and the pellet suspended in the isolation medium. The suspension was adjusted to 100 nm chlorophyll per ml (for thermoluminescence measurements) or to 10 nm per ml (for polarized fluorescence measurements).

Electron microscopy

Small pieces of leaves were fixed in a Karnovsky solution [8] postfixed is osmic acid [9], dehydrated and embedded in Araldite. Ultrathin sections from the sample were stained with uranylacetate and lead citrate. Electron micrographs obtained with a JEOL 100/B electron microscope were analysed for granum formation and the light-induced decrease of thylakoid thickness. Granum formation was characterized by the number of thylakoids in stacks and expressed as the percentage of the total number of thylakoids found in a unit area of chloroplast section.

Long-term energization of thylakoids, manifested by a light-induced decrease in the thickness of the membranes, was studied in leaf pieces exposed to intense light ($60 \, W \, m^{-2}$, 30 min) by comparing them with leaves kept for 2 hours in darkness prior to fixation. The thickness of the membranes was recorded by tracing the electron micrograph negatives on a microdensitometer coupled with an XY recorder. The resolution of the measurement was satisfactory for detection of differences of less than 1 nm, corresponding to about 10 per cent of the total thickness of thylakoid membranes [10].

Measurement of polarized fluorescence spectra

This was carried out according to Garab et al. [7] using juvenile chloroplasts (or mature chloroplasts) oriented in a magnetic field and trapped by freezing at 77 K. Fluorescence was excited with naturally polarized light focussed on the front of the cuvette with the blue spectral bands of an HBO 200 high-pressure mercury arc lamp. Observation was from the front, applying polaroid sheets in positions transmitting vertically and horizontally polarized light. For the oriented samples, the fluorescence polarization ratio (FP) corresponded to the intensity of fluorescence emitted in a plane parallel, $I_{\parallel}$, relative to the intensity emitted in a plane perpendicular, $I_{\perp}$, to the idealized membrane plane – thus FP denotes $I_{\parallel}/I_{\perp}$.

The measurement of thermoluminescence

Glow curves of chloroplast suspensions were measured at $-80\,°C$ to $+80\,°C$ using an apparatus similar to that described by Tatake et al. [14]. Illumination was by various intensities of white light applied during cooling from $+20$ to $-80\,°C$ at a rate of $20\,°C \, min^{-1}$. Subsequent heating was set to a rate of $10\,°C \, min^{-1}$ which was optimal for peak resolution [14]. Each sample was characterized by the difference between two measurements: a "dark" and a "light" measurement. In the former the sample is not illuminated during cooling and reveals the luminescence from the energization remaining in the sample. This emission, which may correspond to some long-lived delayed luminescence component, can be eliminated from the sample only by long incubation at room temperature. As this incubation may have some chaotropic effect, it is better avoided. The light measurement displays the sum of charges remaining from previous excitations together with the recently stabilized ones. The difference between the glow curves originating from light and dark measurements reveals the trapping capacity of the membrane during actual energization.

By mathematical curve resolution 5 components were distinguished with peak positions between $-40\,°C$ and $+40\,°C$ [15], and the capacity

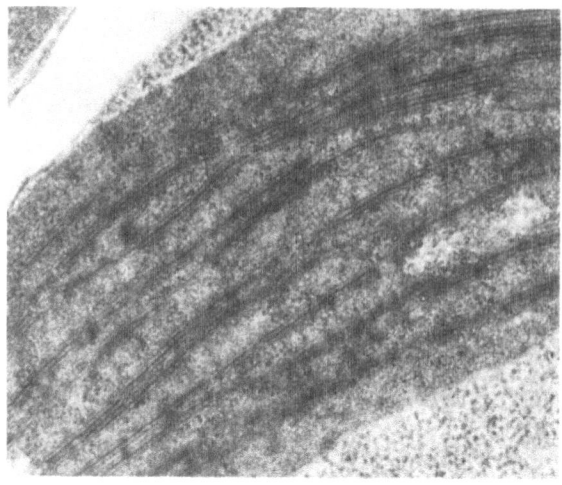

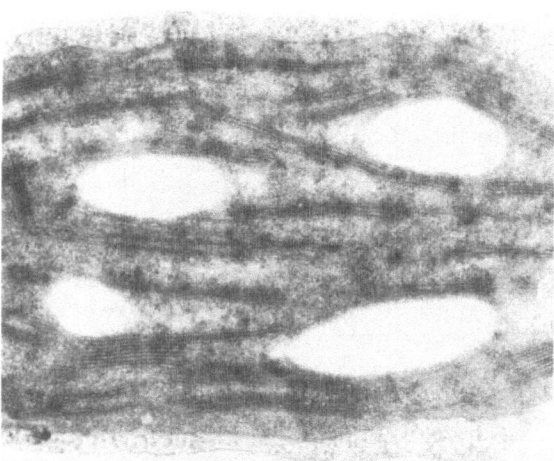

Figure 1. Ultrastructural changes during the juvenile chloroplast to mature chloro-plast transformation. A = after 56 cycles of periodic illumination. B = after a further 24 hours of maturation under continuous light.

of the membranes to stabilize the separated charges is characterized by the relative area of the band with a peak position at 28 °C.

Results

When seedlings grown in intermittent light were exposed to continuous light, the thylakoid pattern in the chloroplasts was rapidly rearranged. From

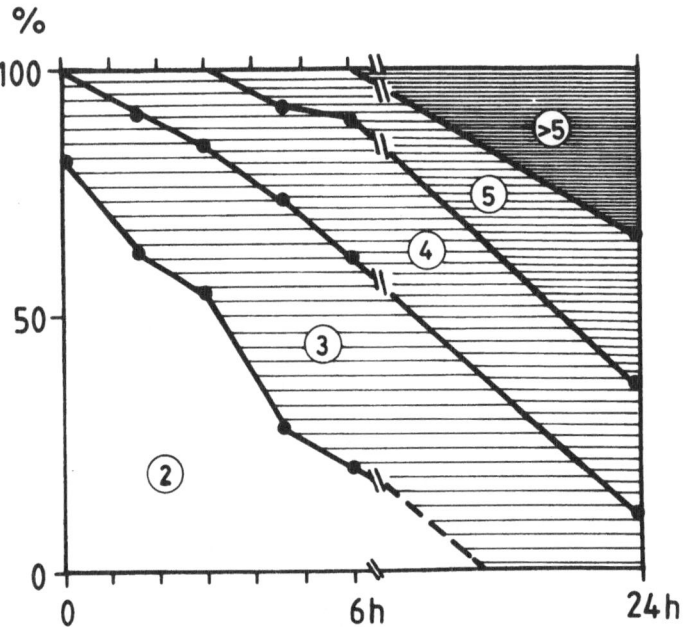

Figure 2. The number of thylakoids per developing granum at various stages of maturation under continuous light. The frequency distribution has calculated from 5 plastids at each point inspected.

large double sheets several stacks of membranes of smaller diameter were formed leading to a thylakoid pattern similar to that of mature chloroplasts (Figure 1). The total amount of thylakoid membrane did not change greatly ($6.3 \mu/\mu^2$ in the sections of juvenile chloroplasts and $7.5 \mu/\mu^2$ after 6 hours of continuous illumination), while during this period the mean number of thylakoids in stacks increased from 2.2 to 3.9. Later the double sheets disappeared completely and the number of thylakoids per stack exceeded an average value of 4.7 (Figure 2).

Ultrastructural rearrangement of the thylakoids was accompanied by an extensive accumulation of the chlorophyll a/b light-harvesting photosystem II complex and its incorporation into the membranes in a definite order of orientation. This was demonstrated by a decrease of the chlorophyll a to chlorophyll b ratio from 9.7 to 3.6, by the polarized fluorescence spectra and changes in FP, the fluorescence polarization ratio (Figure 3). The most characteristic difference between the spectra of juvenile chloroplasts and mature chloroplasts was in the emission band at 695 nm, where juvenile thylakoids displayed an intense fluorescence component emitted perpendicular to the membrane plane resulting in a very low (< 1) FP. With mature chloroplasts the emission at 695 nm became weaker and changed its anisotropy, so that the FP increased to a value of > 1. The negatively

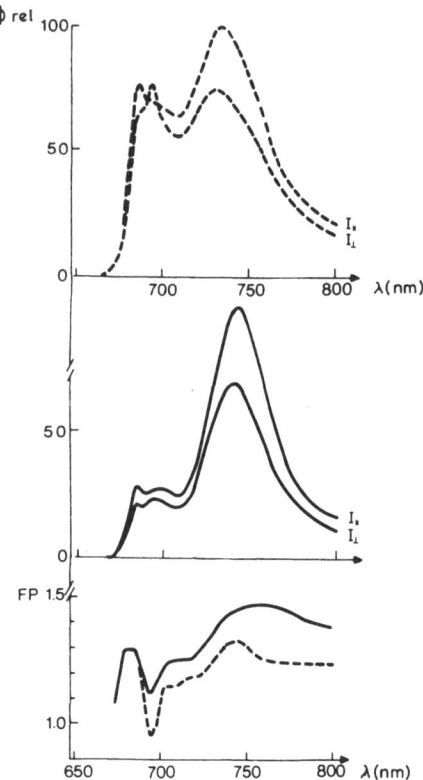

Figure 3. Polarized fluorescence and fluorescence polarization ratio (FP) spectra of juvenile chloroplasts obtained with periodic illumination (– – –) and chloroplasts developed after 24 hours of subsequent continuous illumination (——). Orientation of plastids was at magnetic field strength: 1.2 T, temperature 77 K, and the absorbance of the suspensions at 678 nm was 0.12. Polarized fluorescence spectra are normalized to the long wavelength peak of the fluorescence emitted parallel with the membrane plane. Leaves with juvenile chloroplasts contained 430 nm chlorophyll/g fresh weight of leaf; chlorophyll a/chlorophyll b = 9.7. Leaves with mature chloroplasts contained 910 nm chlorophyll/g fresh weight of leaf and chlorophyll a/chlorophyll b = 3.6.

polarized fluorescence has been attributed to emission from the reaction center pheophytin [5]. Our results indicate that in mature chloroplasts this emission is partly masked by fluorescence of light-harvesting chlorophyll.

Thermoluminescence glow curves showed band patterns depending on both the developmental stage of thylakoids and the exciting light intensities (Figures 4 and 5). Computer analysis could fit each curve with four components, corresponding with the bands positioned near −20, −10, +15 and +28 °C, but with various relative intensities (Table 1).

The juvenile thylakoids emitted relatively more light at subzero temperatures than the chloroplasts, in which most of the emission was generated from deeper traps. The difference due to thylakoid development was especially clear in the case of the band at 28 °C. The emission at 15 °C

Table 1. Computer resolution of the thermoluminescence glow curves of chloroplasts developed in intermittent light (I) and of mature (M) chloroplasts (normalized to the same chlorophyll content)

Band area in % of the total area under the glow curve*

Nominal band position in °C	Calculated band position in °C	Exciting light (mW m^{-2}) 10³			10⁴			10⁵		
		I	M	M/I	I	M	M/I	I	M	M/I
−20	−22 ± 2	3	4	1.3	4	3	0.8	2	4	2.0
−10	−7 ± 1	42	30	0.7	27	22	0.8	22	11	0.5
+15	+13 ± 1	23	34	1.5	46	43	0.9	43	48	1.3
+28	+25 ± 1	22	31	1.4	16	26	1.6	19	23	1.2

* The contribution of the respective band to the total glow between −40 and +40 °C.

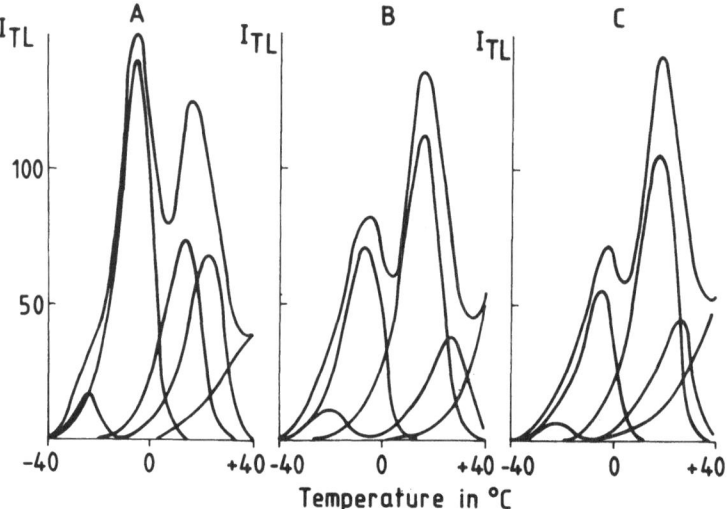

Figure 4. Mathematical curve resolution of the thermoluminescence glow curves ("light" minus "dark" measurements, see text) of juvenile chloroplasts isolated from bean leaves grown under periodic light. Actinic light, A: $10^3 \, mW \, m^{-2}$, B: $10^4 \, mW \, m^{-2}$, C: $10^5 \, mW \, m^{-2}$.

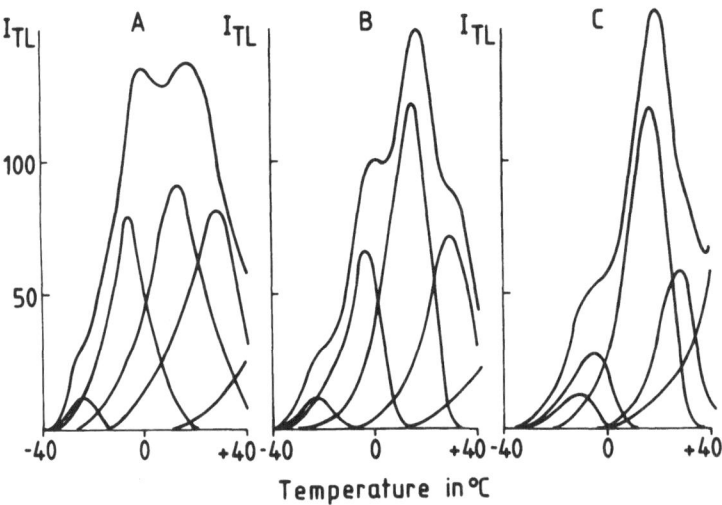

Figure 5. Mathematical curve resolution of the thermoluminescence glow curves ("light" minus "dark" measurements, see text) of mature chloroplasts isolated from bean leaves grown under periodic light and subsequently exposed to continuous light for 24 h. Actinic light, A: $10^3 \, mW \, m^{-2}$, B: $10^4 \, mW \, m^{-2}$, C: $10^5 \, mW \, m^{-2}$.

Table 2. The thickness of the photosynthetic membranes in juvenile and mature chloro-plasts of bean leaves kept in darkness and illuminated by $6 \cdot 10^5$ mW m^{-2} white light. N = the number of thylakoids measured. SE = standard error

	Juvenile chloroplast nm ± SE		Mature chloroplast nm ± SE
In the dark N = 36	13.5 ± 0.2	N = 32	14.0 ± 0.2
In the light N = 42	14.4 ± 0.2	N = 40	12.6 ± 0.3
Difference*	+0.9 ± 0.2		−1.4 ± 0.3

* Significant at P > 99%.

was influenced more by the intensity of excitation: with high actinic light the relative intensity of this band was enhanced both in the juvenile and mature thylakoids.

Architectural development of thylakoids was shown also by light-induced changes of the membrane dimensions (Table 2). In juvenile thylakoids strong illumination caused a slight swelling rather than shrinkage, while chloro-plasts developed under continuous illumination from the former displayed a clear tendency towards a decreased width of thylakoid membranes upon illumination.

Discussion

The developmental process which rearranges large sheets to granum stacks is largely unknown. Our knowledge of granum formation, membrane folding and appression originates from material in which thylakoid growth and membrane stacking proceeded simultaneously, so that these phenomena could not be studied separately [4]. Our present work, however, reveals that lateral ultrastructural and functional heterogeneity of thylakoids can develop also without an extensive growth of the membranes.

The structural rearrangement of thylakoids proceeds with a parallel accumulation and incorporation of the chlorophyll a/b light-harvesting photosystem II complex into the membranes. This might result in an in-creased efficiency for photosynthesis by reduction of energy losses by fluor-escence from photosystem II, the excess quanta being transferred to photo-system I [16].

A further result of the maturation of thylakoids is demonstrated by thermoluminescence showing an increase in the activation energy of charge recombination. This means a decrease in the probability of the back-reactions of energy conversion. In this process the light-harvesting pigment-protein complex might play an important role as indicated by the preferential sub-zero emission from a chlorophyll b-less mutant of barley [11].

Efficient conservation of the energy in mature thylakoids might be connected with the intense energization which leads to conformational changes and loss of water that cause a decrease of the membrane width. In contrast to this, juvenile thylakoids display an increase in the width of the membrane that might reflect light-induced damage of the chloroplast architecture, similar to that found in light-sensitive mutant chloroplasts [10].

References

1. Akoyunoglou G (1977) Arch Biochem Biophys 183:571–580
2. Anderson JM and Andersson B (1982) Trends in Biol Sci 6:289–292
3. Argyroudi-Akoyunoglou JH and Akoyunoglou G (1973) Photochem Photobiol 18:219–223
4. Brangeon B and Mustárdy LA (1979) Biol Cell 36:71–80
5. Breton J (1982) FEBS Lett 147:16–20
6. Duysens LNM and Sweers HE (1973) In: Studies on Microalgae and Photosynthetic Bacteria. Japan. Soc of Plant Physiologists, Tokyo pp 353–358
7. Garab GyI, Kiss JG, Mustárdy LA and Michel-Villaz M (1981) Biophys J 34:423–437
8. Karnovsky MJ (1965) J Cell Biol 27:137A–138A
9. Millonig G (1961) J Appl Physics 32:1637
10. Mustárdy LA, Machowicz E and Faludi-Dániel Á (1976) Protoplasma 88:65–73
11. Mustárdy LA, Sz-Rózsa Zs and Faludi-Dániel Á (1981) In: Proceedings of the Fifth International Congress on Photosynthesis Research Vol I. (G Akoyunoglou ed.) Balaban Internatl. Science Services, Philadelphia PA pp 665–671
12. Reeves SG and Hall DO (1973) Biochim Biophys Acta 314:66–78
13. Stachelin LA (1976) J Cell Biol 71:136–148
14. Tatake WG, Desai TS and Battacharjee SK (1971) J Phys Eng Sci Instruments 4:755–777
15. Vass I, Horváth G, Herczeg T and Demeter S (1981) Biochim Biophys Acta 634:140–152
16. Szitó T, Zimányi L and Faludi-Dániel Á (1985) Biochim Biophys Acta 808:428–436

Photosynthesis Research 9, 239–249 (1986)
© *1986 Martinus Nijhoff/Dr. W. Junk Publishers, Dordrecht.*

What role does sulpholipid play within the thylakoid membrane?

JAMES BARBER and KLEONIKI GOUNARIS

AFRC Photosynthesis Research Group, Department of Pure and Applied Biology,
Imperial College of Science and Technology, London SW7 2BB, UK

(*Received 27 August 1985*)

Abstract. Sulphoquinovosyldiacylglycerol is a negatively charged lipid which exists in the thylakoid membrane. It is proposed that a large proportion of this acidic lipid does not form a part of the bulk lipid matrix but is closely associated with protein complexes where it is tightly bound and participates in either optimising catalytic activities, or maintaining the complexes in a functional conformation. Experimental evidence for this proposal is emerging from studies with isolated photosystem 2, and coupling factor complexes.

Introduction

Sensitive fluorescence and absorption spectroscopy are the major experimental approaches for investigating the primary processes of photosynthesis. Classical studies using these techniques by Professor Duysens and his colleagues on a wide range of prokaryotic and eukaryotic photosynthetic organisms established: (i) a firm basis for our understanding of energy transfer mechanisms between light harvesting pigments [23, 27], (ii) knowledge of the photochemical trapping and primary charge separation processes [25, 29] and (iii) evidence for electron transfer pathways [24, 26]. These outstanding contributions were made in the absence of a detailed understanding of the structure of the proteins which harbour the chromophores and redox centres involved, and indeed did not take into account the fact that the primary processes of photosynthesis take place within a complex membrane system. The relationship between primary photosynthetic processes and membrane structure did not emerge until the realization that the redox reactions of electron transport could act as a proton pump able to create an electrochemical potential gradient for driving ATP synthesis [50]. Even at this stage the main challenge was to arrange the various redox components vectorially across the membrane and there was no need to consider the nature of the proteins and lipids which constituted the reaction matrix [70]. Today we are rectifying this latter deficiency. We now know that the redox reactions take place in supermolecular multipeptide complexes and that the light harvesting pigments are specifically interacting

---

Dedicated to Prof. L.N.M. Duysens on the occasion of his retirement.

with proteins. Some of these protein complexes are strikingly comparable, both in structure and function, within a wide range of photosynthetic organisms while others are not. In the case of higher plants and green algae five different major complexes have been identified [4, 6]: photosystem 1 (PS1), photosystem 2 (PS2), light harvesting chlorophyll a/b complex (LHC), cytochrome b-f complex (cyt b-f) and coupling factor complex (CF_0-CF_1). It is generally agreed that these complexes are segregated into the two different domains of the thylakoid membrane [2, 6]; PS2 and LHC are preferentially localized in the appressed regions of the grana while PS1 and CF_0-CF_1 are restricted to the unappressed regions. The distribution of cyt b-f is not so certain but the complex seems to occur in both membrane regions [1]. The identification of different functional protein complexes and their asymmetric distribution in the membrane has a number of important consequences and has focussed attention on the roles of plastoquinone, plastocyanin and ferredoxin as long-range diffusible redox carriers able to communicate between various complexes so as to facilitate electron flow [37, 49, 72]. In the past few years many of the polypeptides which constitute each complex have been identified and are now being further characterised by the techniques of molecular genetics [17]. Crystallization of isolated complexes suitable for high resolution X-ray studies is also possible as demonstrated by the success with reaction centres isolated from *Phodopseudomonas viridis* [20].

Clearly, as the level of understanding of the structure of the photosynthetic apparatus grows, spectroscopic studies of the type pioneered by Duysens find a new dimension for interpretation. For example, Duysens and Talens [28] identified a process occurring in intact tissue which has been termed State transitions. This work, together with that of Bonaventura and Myers [13] and Murata [53] established that the State transitions are a regulatory mechanism which exists within oxygen evolving organisms to optimise the delivery of quanta to PS2 and PS1 under limiting light conditions. Because of our present day appreciation of thylakoid organisation we can now describe the State transitions in structural terms by which LHC diffuses laterally between appressed and non-appressed regions in response to surface phosphorylation [3, 5, 8].

The relationship between structure and function will continue to be an important facet of photosynthesis research for some years to come. An area of consideration which has yet to develop is the role of the polar lipids which, together with the intrinsic protein complexes, make up the photosynthetic membrane. In the case of higher plant thylakoids the major polar lipid components are the electroneutral galactolipids, monogalactosyldiacylglycerol (MGDG) and digalactosyldiacylglycerol (DGDG) having levels of about 50% and 25% of the total lipids, respectively [14]. A special property of MGDG which has been discussed in considerable detail in terms of its possible function in the intact membrane [30, 63] is that on isolation it forms non-bilayer structures [34]. Another important feature is that both lipids have mainly 18 carbon long acyl chains which are extremely unsaturated, a property which makes the thylakoid membrane highly fluid at normal temperatures [7]. The remaining polar lipids are phosphatidylglycerol (PG) and sulphoquinovosyldiacylglycerol (SQDG) which typically have concentrations in higher plant thylakoids of 10 and 12%, respectively, while phosphatidylcholine (PC) makes up most of the remaining 3%. Although the values given represent an average percentage composition, the relative amounts of the thylakoid polar lipids can vary with species or environmental conditions. PG and SQDG are the only lipids having a net negative charge and are characteristic of the membranes of photosynthetic organisms. Although PG is found in many different types of biological membranes, in the thylakoid a proportion of this lipid contains trans-Δ3-hexadecenoic acid [22]. This form of PG is unique to fully mature thylakoids and there have been speculations about the involvement of this particular molecular species in maintaining the organisation of pigment protein complexes within the membrane [65]. SQDG is an intriguing lipid class since it is an unusual molecule found in high concentrations only in photosynthetic membranes. Recent considerations of surface electrical properties of thylakoids and isolation of specific protein complexes are giving hints as to the localization and function of this sulpholipid and it is the implications of these new developments which we wish to discuss here.

Sulphoquinovosyldiacylglycerol: Structure, occurrence and biosynthesis

Sulphoquinovosyldiacylglycerol, commonly referred to as the "plant sulpholipid", was first discovered by Benson and co-workers in 1959 [10]. It was the availability and use of [^{35}S]-sulphate that led to its identification and allowed for its subsequent analysis. Studies on its degradation products [18] revealed that deacylation yielded sulphoquinovosylglycerol which exhibited a molecular rotation [M]^{25}D of +31 000°, characteristic of alkylα-D-glucopyranosides. In addition, its rotational shift in cupra B of −370° indicated three adjacent equatorial hydroxyl groups as in glucosides [46]. SQDG resulting from either deacylation of the lipid or from chemical synthesis [51] has been shown to have similar infra-red spectra [38] and the structure has also been confirmed by X-ray crystallography of its rubidium salt [58]. Complete degradative analysis of SQDG has now been carried out and it is described as: 1,2-diacyl-[6-sulpho-α-D-quinovopyranosyl-(1'→3)]-sn-glycerol. D-quinovose is 6-deoxy-D-glucose and the prefix sulpho- denotes a sulphonic acid group. The chemical structure of the headgroup of the lipid is shown in Figure 1. It is important to note that, in contrast to most naturally occurring organosulphur compounds, including sulphur containing lipids, where sulphur occurs in an ester (C−O−SO$_3^-$) linkage, SQDG contains a sulphonic acid group in which carbon

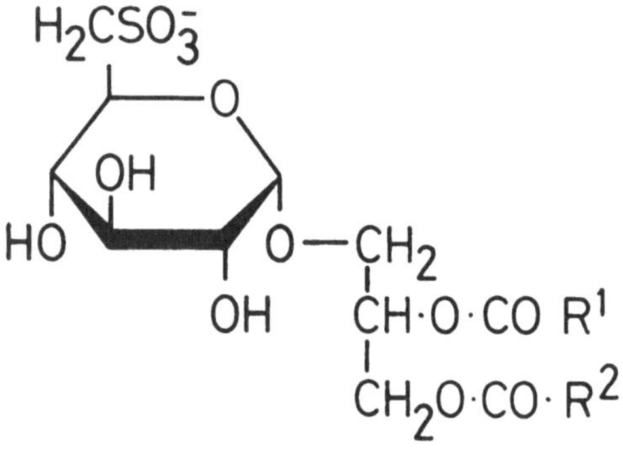

Figure 1. The chemical structure of the plant sulpholipid, SQDG (1,2-diacyl-[6-sulpho-α-D-quinovopyranosyl-(1'→3)]-sn-glycerol).

is directly bonded to sulphur as $C-SO_3^-$. Sulphonic acids of this type are chemically very stable and strongly acidic in a wide pH range.

SQDG has been reported to occur in a variety of organisms although some characterisations are only preliminary. It has thus been found to occur in phytoflagellates, cyanobacteria, green, red and brown algae, purple sulphur and non-sulphur bacteria [43] and in higher plants [41]. In higher plants it occurs in very low amounts in non-photosynthetic tissue but in higher amounts in photosynthetic tissue. In the latter it is only found in the chloroplasts [19] and in particular in the thylakoid membranes [21].

The apparent lack of a lipase specific for sulpholipid [41] has prevented the unequivocal characterization of its distribution within the thylakoids. Examination of thylakoid fragments shows SQDG to be present in both appressed and non-appressed regions, the actual amounts varying with the type of procedure involved for the preparation of the membranes [16, 35, 54]. Results obtained with detergent derived particles also point to an association of SQDG with protein complexes found in both regions of the membrane (see later section). The use of lipid antibodies [64] has indicated that polar lipids are in general most abundant at the internal rather than the outer surface. Nevertheless, the same techniques suggest that SQDG may be preferentially located at the external surface although most of it is not accessible to antibodies from either side of the membrane. In addition, hydrolase treatment of membranes (Harwood JL, personal communication) has confirmed that the majority of sulpholipid molecules are inaccessible from the aqueous media. Recently, it has been argued that the acidic lipids are located in the inner leaflet of the bilayer [69] supporting the proposal that the charged head groups could act as a proton conducting pathway [39]. Other evidence, however, tends to indicate that at least two thirds of the phosphatidylglycerol molecules are in the outer leaflet of the bilayer [30].

The positional distribution of fatty acids in the molecular species sub-fractions obtained from SQDG has not been unequivocally determined. It appears that linolenic (C18:3) and palmitic (C16:0) acids are the major constituents while linoleic (C18:2) and oleic (C18:1) acids occur in lesser amounts. It has been shown that the proportion of palmitic acid at the C-1 or C-2 position of the glycerol backbone varies with the plant type [40]. In most plants examined, however, SQDG is mainly composed of the palmitoyl/linolenoyl and palmitoyl/linoleoyl species [42, 57].

The metabolism of SQDG has attracted a considerable amount of research, but the actual biosynthetic pathway remains unclear and it is not yet established whether the chloroplast is autonomous in SQDG biosynthesis or not. The biosynthetic routes suggested have been summarized [41]. Briefly, a pathway analogous to glycolysis has been proposed and called the 'sulpho-glycolytic pathway'. This results in the formation of a nucleotide diphosphate sulphoquinovose. The precursor(s) of the pathway remain a matter of controversy and more recently [52] the very existence of the sulpho-glycolytic pathway has been questioned. The catabolism of SQDG in plant tissues also remains obscure mainly through the lack of identification of a sulpholipid specifc enzyme.

Correlations between the appearance of chlorophyll and the presence and concentration of sulpholipid in photosynthetic tissues have been found. Thus, although the initial levels of this lipid are significant, it has been shown that the amounts increase on greening of etiolated tissue or on maturation of protoplasts [45, 12, 48]. It has been suggested that SQDG may assist the orientation of chlorophyll molecules in the membrane [9] and model systems have indicated the possibility of interactions between these two types of molecules [71]. In contrast, however, it has been shown that *Chorella pyrenoidosa* [68] can be grown heterotrophically so as to completely lack sulpholipid and have normal levels of chlorophyll, but such cultures could not photosynthesise.

Most interestingly, in a series of labelling experiments on leaf sulpholipid [42] it was found that certain molecular species of the lipid were labelled faster than others. It was thus suggested that besides a structural role the sulpholipid may have a metabolic role within the membrane.

Surface electrical charge on the thylakoid membrane

Free flow particle electrophoresis measurements on isolated thylakoid membranes made by Mercer et al. [48] and more recently by Nakatani et al. [55] and Nakatani and Barber [56] suggest that the electrical charge on the outer surface is due to acidic and basic residues of exposed segments of intrinsic proteins. Variations of pH and experiments involving the use of chemical modifiers suggest that the negative charges are derived from carboxyl groups of aspartic and glutamic acid residues. At about pH 4.3

the surface is isoelectric and below this pH value it becomes positively charged. The use of 1,2-cyclohexanedione treatment indicated that most of this charge is due to the guanidino group of arginine. These conclusions gain support from other studies including treatments with carbodiimides [11] and calcium binding [62].

Thus there is no evidence to suggest that acidic lipids contribute to the electrical charge on the outer surface. Far less is known about the nature of the electrical charge on the inner lumenal surface. Using inside-out vesicles derived from the appressed lamellae of the grana, Mansfield et al. [47] concluded that the inner surface is more negatively charged than the outer but no chemical modification studies were conducted to investigate the nature of the ionisable groups involved. However, the isoelectric point of the inner surface was found to be at pH 4.0, a value higher than expected even if the acidic lipids were preferentially located in the inner leaflet of the bilayer.

For the sake of argument, if it is assumed that PG and SQDG are located entirely in either the outer or inner leaflet then they would be expected to contribute significantly to the electrical charge density on either surface. As can be seen in Figure 2 a surface pressure-area isotherm of SQDG indicates that the area occupied by its head group, prior to the collapse point of the monolayer, is about $0.48 \, nm^2$. A slightly smaller headgroup size has been obtained for PG of 0.40 to $0.42 \, nm^2$ [31]. Taking the combined level of SQDG and PG as 22% of the total lipid and assuming a random distribution, then each negative charge due to the acidic lipids would occupy $200 \, nm^2$ corresponding to a charge density of about $-8 \times 10^{-2} \, C \, m^{-2}$. This calculation has assumed an average head group area for the total lipids of about $0.60 \, nm^2$ (MGDG and DGDG have larger head groups than SL and PG) but no allowance has been made for the presence of protein. If we assume the thylakoid membrane has a lipid to protein weight ratio of 1:1 then there will be about 500 lipid molecules of molecular weight 1 KDa to each protein complex of 500 KDa. Assuming that the protein complexes carry no net charge and can be taken as cylinders having an exposed end surface area of $200 \, nm^2$ then the area per negatively charged lipid increases to approximately $360 \, nm^2$ so that the surface charge due to the acidic lipids reduces to $-4.4 \times 10^{-2} \, C \, m^{-2}$. This minimal value is only slightly higher than the values for the outer and inner surfaces measured using a range of techniques [3]. However, as already stated above, in the case of the outer surface it seems that the net negative charge is due entirely to carboxyl groups of amino acids. Therefore it can be concluded that for the outer surface SQDG or PG must be closely associated with protein in such a way that their charges are inaccessible to the external aqueous phase. It is possible, however, that the majority of the acidic lipids are located almost entirely at the inner surface although direct experimental evidence tends not to support this [30]. In fact, as stated earlier, both antibody and

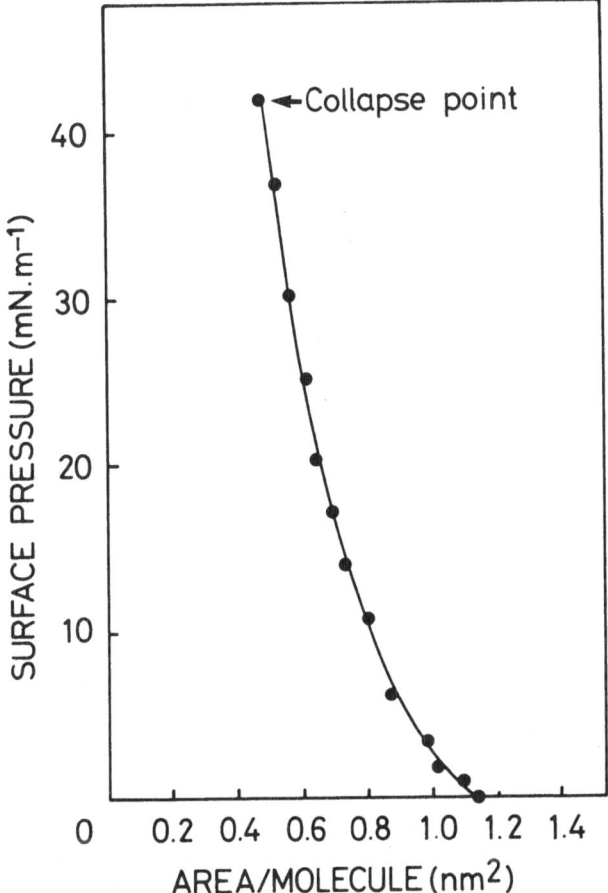

Figure 2. Force-area isotherm obtained from SQDG at 20° on a distilled water sub-phase [31].

lipase treatments, as well as the isoelectric point measurements argue against this possibility.

Alternatively, the electrical characteristics of the inner surface may, like the outer surface, also be due to amino acid residues so that the majority of SQDG and PG in the thylakoid membrane would be closely associated with proteins.

Location and function of SQDG

What evidence is there for the suggestion that acidic lipids are closely associated with proteins? For some years now Remy and his colleagues [65] have argued that PG plays an important role in the formation of oligomeric forms of LHC. However, although there have been speculations there has been no evidence concerning the specific location and function of SQDG. Nevertheless experimental data is now emerging which suggests that SQDG may play a very important role in the structure and function of at least two important protein complexes. In collaboration with Pick and Weiss we have found that SQDG is tightly bound to CF_0-CF_1 complex isolated either from spinach chloroplasts or the halotolerant alga *Dunaliella salina* [60]. The estimated level of SQDG ranged from 5 (in spinach) to 20 (in *D. salina*) lipid molecules per CF_0-CF_1 complex and the lipid could not be exchanged with phosphatidylcholine, phosphatidylserine or with glycolipids. Moreover, removal of some of this tightly bound SQDG (10 to 30%) with extensive detergent treatment was accompanied by an inhibition of ATPase activity which could not be restored by the addition of glycolipids. Clearly these results suggest that the association of SQDG with CF_0-CF_1 is very strong and is essential for the integrity of the enzyme. Interestingly the addition of excess SQDG and other acidic lipid, to isolated CF_0-CF_1 causes an inhibition of its activity [59]. SQDG is important, however, at low concentrations for reconstituting the ATP-Pi exchange of this enzyme but this is due to the ability of SQDG to decrease the proton permeability of membrane vesicles composed of thylakoid galactolipids [61]. Although less well studied than CF_0-CF_1 we also have evidence that SQDG is firmly bound to the PS2 protein complex [33]. Using Triton X-100 and sucrose density centrifugation we have isolated a PS2 complex from spinach chloroplasts which contains a small number of polypeptides and redox components. Associated with this complex are SQDG and MGDG possessing unusually saturated fatty acids. Taking 40 chlorophyll-*a* molecules per PS2 reaction centre indicates that there are about 12 molecules of SQDG associated with each isolated complex. As in the case of CF_0-CF_1, it has been shown that excess acidic lipids result in an inhibition of PS2 activity when monitored as oxygen evolution from PS2 enriched membrane fragments [36].

These interesting findings raise the question about how much of the total SQDG is closely associated with protein complexes. Above we crudely estimated that there are about 500 polar lipid molecules per complex, a number also derived by considering chlorophyll-lipid ratios [15]. Since SQDG makes up about 12% of the lipid composition then 50 sulpholipid molecules are available for interacting with each complex assuming that they are all about the same size. As yet there is no experimental evidence to suggest that there is specific association of sulpholipid with other main intrinsic complexes; the calculated pool of 50 molecules per complex would increase if this acidic lipid is preferentially associated only with PS2 and CF_0-CF_1.

Although it is now accepted that there is a lateral separation of protein complexes between the appressed and non-appressed regions of the thylakoids and that there is transmembrane asymmetry of polypeptides, a similar distribution of SQDG along or across the membrane is not established. On balance it seems that this lipid occurs in both membrane regions consistent with its interaction with PS2 (normally in the appressed region) and CF_0-CF_1 (localized exclusively in the non-appressed region).

Conclusion

Experimental evidence is emerging to support the idea that SQDG is closely associated with intrinsic complexes of the thylakoid membrane. This association is probably partly stabilized by electrostatic interactions between its negatively charged sulphonate group and positive charges within the proteins. In some cases the interactions may be very strong as suggested by the resistance of some SQDG molecules bound to CF_0-CF_1 to exchange with other acidic lipids. It is possible that SQDG has a function similar to cardiolipin in the inner mitochondrial membrane. Cardiolipin is a negatively charged lipid which is required for the catalytic activity of cytochrome oxidase [66] and CF_0-CF_1 ATP synthetase [44] and has been shown to be firmly bound to the former [67]. We suggest that SQDG probably does not form a significant part of the general lipid matrix or is involved in "protein packaging" as postulated from MGDG [32, 63]. Rather we suggest that this acidic lipid plays a more specific and intimate role in the catalytic activity of proteins and therefore does not readily exchange with other lipids in the membrane. Consistent with this proposal is that the lipid class contains a high degree of saturated acyl fatty acid chains indicative of a "boundary" rather than a "bulk" lipid. Indeed, ratio labelling experiments indicate that saturated forms of sulpholipid are important for the functional activity of the thylakoid during their illumination [41, 42]. The close association of this lipid with proteins functioning in photosynthesis is further emphasised by the fact that a wide range of photosynthetic bacteria have membranes containing lipids quite different to those of the higher plant thylakoids but still retaining a significant level of sulpholipid [43].

Acknowledgements

We wish to thank the Agricultural and Food Research Council for financial support and acknowledge our collaborative interaction with Dr. Uri Pick which has given substance to the ideas presented.

References

1. Allred DR and Staehelin LA (1985) Plant Physiol 78:199–202
2. Andersson B and Anderson JM (1980) Biochim Biophys Acta 593:427–440
3. Barber J (1982) Annu Rev Plant Physiol 33:261–295
4. Barber J (1983) Plant, Cell and Environ 6:311–322
5. Barber J (1985) In Staehelin A and Arntzen CJ eds. Enc Plant Physiol Photosynthetic Membranes. Berlin, Heidelberg, New York: Springer-Verlag in press

6. Barber J (1985) In: Barber J ed. Photosynthetic Mechanisms and the Environment Vol. 6 Topics in Photosynthesis pp 91–134. Amsterdam, Elsevier
7. Barber J, Ford RC, Mitchell RAC and Millner PA (1984) Planta 161:948–954
8. Bennett J (1983) Biochem. J. 212:1–13
9. Benson AA (1964) Annu Rev Plant Physiol 15:1–16
10. Benson AA, Daniel H and Wiser R (1959) Proc Natl Acad Sci (USA) 45:1582–1587
11. Berg S, Dodge S, Krogmann DW and Dilley RA (1974) Plant Physiol 53:619–627
12. Bolton P and Harwood JL (1978) Planta 139:267–272
13. Bonaventura C and Myers J (1969) Biochim Biophys Acta 189:366–383
14. Chapman DJ, DeFelice J and Barber J (1983) Plant Physiol 72:225–228
15. Chapman DJ, DeFelice J and Barber J (1984) In: Siegenthaler P-A and Eichenberger W eds. Structure, Function and Metabolism of Plant Lipids. pp 457–464. Amsterdam, Elsevier
16. Chapman DJ, DeFelice J and Barber J (1985) Photosynth. Res in press
17. Cramer WA, Widger WR, Herrmann RG and Trebst A (1985) Trends Biochem Sci 10:125–129
18. Daniel H, Miyano M, Mumma RO, Yagi T, Lepage M, Shibuya T and Benson AA (1961) J Am Chem Soc 83:1765–1766
19. Davies WH, Mercer EI and Goodwin TW (1965) Phytochemistry 4:741–749
20. Deisenhofer J, Epp D, Miki K and Huber R (1984) J Mol Biol 180:385–400
21. Douce R, Holtz RB and Benson AA (1973) J Biol Chem 248:7215–7222
22. Dubacq JP and Tremolieres A (1983) Physiol Végét 21:293–312
23. Duysens LNM (1952) Ph.D. Thesis, University of Utrecht
24. Duysens LNM (1954) Nature 173:692
25. Duysens LNM (1965) Arch Biol (Liège) 76:251–275
26. Duysens LNM and Amesz J (1962) Biochim Biophys Acta 64:243–260
27. Duysens LNM and Sweers HE (1963) In: Japanese Soc. Plant Physiologists, ed. Studies on Microalgae and Photosynthetic Bacteria pp 353–372. Tokyo, University of Tokyo Press
28. Duysens LNM and Talens A (1969) In: Metzner H ed. Progress in Photosynthesis Research Vol. II, 1073–1081. Tübingen
29. Duysens LNM, Huiskamp WJ, Vos JJ and van der Hart JM (1956) Biochim. Biophys Acta 19:188–190
30. Giroud C and Siegenthaler P-A (1984) In: Siegenthaler P-A and Eichenberger W eds. Structure, Function and Metabolism of Plant Lipids pp 413–416. Amsterdam, Elsevier Science Publishers BV
31. Gounaris K (1983) Stability of photosynthetic membranes of higher plant chloroplasts Ph.D. Thesis, University of London
32. Gounaris K and Barber J (1983) Trends Biochem Sci 8:378–381
33. Gounaris K and Barber J (1985) FEBS Lett. 188:68–72
34. Gounaris K, Mannock DA, Sen A, Brain APR, Williams WP and Quinn PJ (1983) Biochim Biophys Acta 732:229–242
35. Gounaris K, Sundby C, Andersson B and Barber J (1983) FEBS Lett 156:170–174
36. Gounaris K, Whitford D and Barber J (1983) FEBS Lett 163:230–234
37. Haehnel W (1984) Annu Rev Plant Physiol 35:659–693
38. Haines TH (1971) Prog Chem Fats Other Lipids 11:297–345
39. Haines TH (1983) Proc Natl Acad Sci (USA) 80:160–164
40. Harwood JL (1980) In Stumpf PK ed. The Biochemistry of Plants. Vol. 4, pp 1–55. New York, London, Academic Press Inc
41. Harwood JL (1980) In Stumpf PK ed. The Biochemistry of Plants. Vol. 4, pp 301–320, New York, London, Academic Press Inc
42. Heinz E and Harwood JL (1977) Hoppe-Seyler's Z. Physiol. Chem 358:897–908
43. Imhoff JF (1984) In Siegenthaler P-A and Eichenberg W, eds. Structure, Function and Metabolism of Plant Lipids. pp 175. Amsterdam, Elsevier Science Publishers B.V.

44. Kagawa Y, Kandrach A and Racker E (1973) J Biol Chem 248:676–684
45. Leech, RM, Rumsby MG and Thomson NW (1973) Plant Physiol 52:240–245
46. Lepage M, Daniel H and Benson AA (1961) J Am Chem Soc 83:157–159
47. Mansfield RW, Nakatani HY, Barber J, Mauro S and Lannoye R (1982) FEBS Lett 137:133–136
48. Mercer F, Hodge AJ, Hope AB and McLean JD (1955) Aust J Biol Sci 8:1–18
49. Millner PA and Barber J (1984) FEBS Lett 169:1–6
50. Mitchell P (1968) Chemiosmotic coupling and energy transduction. Bodmin UK: Glynn Res
51. Miyano M and Benson AA (1962) J Am Chem Soc 84:59–62
52. Mudd JB, Dezacks R and Smith J (1980) In Mazliak P, Benveniste P, Costes C and Doucer eds. Biogenesis and Function of Plant Lipids. pp 57–66 Amsterdam, Elsevier/North-Holland Biomedical Press
53. Murata N (1969) Biochim Biophys Acta 172:242–251
54. Murphy DJ and Woodrow IE (1983) Biochim Biophys Acta 725:104–112
55. Nakatani HY, Barber J and Forrester JA (1978) Biochim Biophys Acta 504:215–225
56. Nakatani HY and Barber J (1980) Biochim Biophys Acta 591:82–91
57. Nishihara M, Yokota K and Kito M (1980) Biochim Biophys Acta 617:12–19
58. Okaya Y (1964) Acta Crystallogr 17:1276–1282
59. Pick U, Gounaris K, Admon A and Barber J (1984) Biochim Biophys Acta 765:12–20
60. Pick U, Gounaris K, Weiss M and Barber J (1985) Biochim Biophys Acta 808:415–420
61. Pick U, Weiss M, Gounaris K and Barber J (1985) Biochim Biophys Acta Submitted
62. Prochaska LJ and Gross EL (1975) Biochim Biophys Acta 376:126–135
63. Quinn JP and Williams WP (1983) Biochim Biophys Acta 737:223–266
64. Radunz A (1980) Z Naturforsch 35c:1024–1031
65. Remy R, Tremolieres A, Duval JC, Ambard-Brettevile F and Dubacq JP (1982) FEBS Lett 137:271–275
66. Robinson NC (1983) Biochemistry 21:184–188
67. Robinson NC, Strey F and Talbert L (1980) Biochemistry 19:3656–3661
68. Sinensky M (1977) J Bacteriol 129:516–524
69. Sundby C and Larsson C (1985) Biochim Biophys Acta 813:61–67
70. Trebst AV (1974) Annu Rev Plant Physiol 25:423–458
71. Trosper T and Sauer K (1968) Biochim Biophys Acta 162:97–107
72. Whitmarsh J (1985) In: Staehelin A and Arntzen CJ eds. Enc. Plant Physiol. Photosynthetic Membranes. Berlin, Heidelberg, New York: Springer-Verlag. in press

Photosynthesis Research 9, 251–259 (1986)
© *1986 Martinus Nijhoff/Dr. W. Junk Publishers, Dordrecht.*

Photosynthesis 3.5 thousand million years ago

JOHN M. OLSON[1] and BEVERLY K. PIERSON[2]

[1] Institute of Biochemistry, Odense University, Campusvej 55, DK-5230 Odense M, Denmark
[2] Biology Department, University of Puget Sound, Tacoma, Washington 98416, USA

(Received 23 July 1985)

Key words: anoxygenic photosynthesis, cyanobacteria, microfossil, stromatolite, sulfur bacteria

Abstract. The recent discovery of stromatolites and microfossils in 3.5-Ga-old sedimentary rock formations is evidence for the existence of phototrophic prokaryotes at that time. Values of $\delta^{13}C$ for sedimentary organic carbon strongly suggest autotrophic CO_2 fixation, and the existence of large deposits of sedimentary sulfate is consistent with a photosynthesis dependent on reduced sulfur compounds for reducing power. The ancient photoautotrophs are thought to have contained only one kind of reaction center with either chlorophyll *a* or bacteriochlorophyll *a* as primary electron donor and with one or more iron-sulfur centers as secondary electron acceptors. Light-harvesting pigments might have been chlorophyll *a*, bacteriochlorophyll *a*, or possibly bacteriochlorophyll *c*.
A new proposal is made to explain how these organisms could have survived an intense UV flux at the earth's surface in the absence of an ozone layer. Photochemically produced ferric iron was abundant in sediments, and the UV-absorption of this ferric iron would have been sufficient to shield those organisms living below the water-sediment interface.

Introduction

Three and one half thousand million years (Ga) ago there appear to have been at least four types of filamentous prokaryotes living in stromatolitic communities (Table 1) at the interface of water and sediment in shallow bays or lakes [12, 16]. In the absence of substantial O_2-levels in the atmosphere, there was no ozone layer to block out the solar UV, and the communities appear to have been exposed to the full UV flux from the sun only slightly attenuated by passage through the overlying water. In previous papers [4–6] one of us (JMO) postulated that early life must have evolved deep under water to escape the deadly effects of the solar UV radiation, but it is now clear that some forms of life actually flourished in or on the sediments at the bottom of shallow bays or lakes [3, 12, 17]. How can one account for this apparent anomaly? A variety of protective mechanisms have been proposed [8]; we suggest another simpler and more likely one.

In honor of Prof. L.N.M. Duysens on the occasion of his retirement.

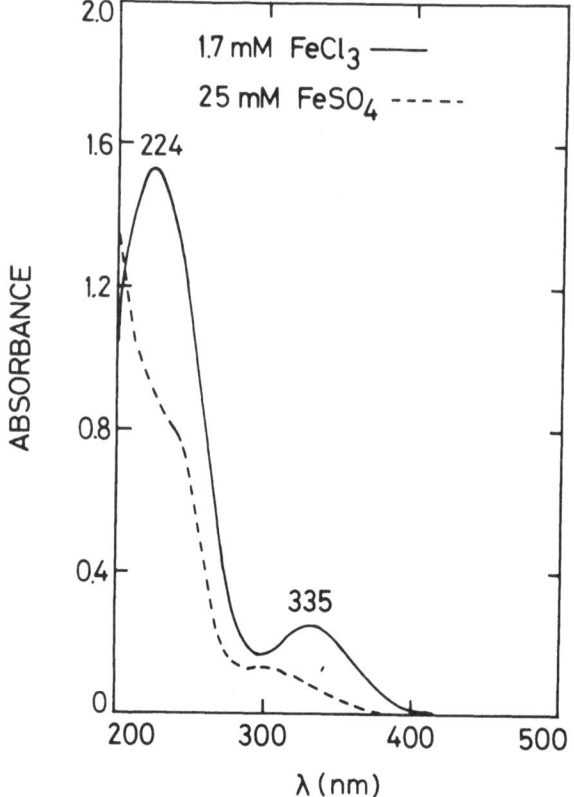

Figure 1. Absorption spectra of 1.7 mM FeCl₃ (solid line) and 25 mM FeSO₄ (dashed line) in 0.1 M HCl.

Table 1. Early record of stromatolites and microfossils [11, 16, 17]

Age (Ga)	Stromatolites	Microfossils	Geological unit and location
3.56 ± 0.03	+	+	Warrawoona Group, Australia
3.54 ± 0.03	+	+	Onverwacht Group, South Africa
3.09 ± 0.09	+	−	Insuzi Group, South Africa
2.79 ± 0.01	+	−	Uchi Greenstone Belt, Canada
2.77 ± 0.01	+	+	Fortescue Group, Australia

Advantages of a sandy environment

The key to this problem comes from a consideration of contemporary microbial communities found in beaches where the sand is mixed with organic materials and the environment is highly reducing. Such communities commonly contain cyanobacteria in the upper layers, purple bacteria in the middle layers, and green sulfur bacteria in the lower layers. These contemporary communities do not need localized protection from the sun's UV radiation, but they would be protected, if necessary, by the ferric iron present in the sand (see Figure 1). We may suppose that the prokaryotes living 3.5 Ga ago would have been protected from the sun's UV radiation, if they had lived in the sediment and if the sediment had contained sufficient ferric iron to absorb the UV radiation. The next question is whether there would have been sufficient ferric iron in the sediments where organisms that became microfossils and/or stromatolites were found.

Banded-iron formations

Since the upper continental crust contained about 10% FeO during the early Precambrian [13], it is likely that there was abundant iron in the environment of the stromatolite-forming organisms. In the absence of significant levels of oxygen in the atmosphere one might have expected that the level of ferric iron would have been vanishingly small in Precambrian sediments, but that was not the case. The oldest known sedimentary rocks (3.8 Ga ago) are the Isua Supracrustals (see Table 2) which contain banded-iron formations in which at least 30% of the iron is in the oxidized (ferric) form [15]. Other similar banded iron-formations were formed subsequently during the early Precambrian Era (up to 2.5 Ga ago). They are of sedimentary origin, containing at least 15 percent iron, and characterized by alternating thin (ca. 1 mm) layers of quartz and iron bearing sediment (Table 3).

The ferric iron was most probably precipitated from a solution of Fe^{2+} ($6.7 \leqslant pH \leqslant 8$) in the form of insoluble FeOOH. But in the absence of substantial amounts of oxygen in the atmosphere (and dissolved in the surface water) how does one account for the tremendous amount of ferric iron in the banded-iron formations? Some have proposed that oxygen-evolving photosynthesis was responsible. That explanation seems unlikely, at least for the oldest banded-iron formations that date from 3.8 Ga ago. An alternate proposal is that FeOOH would have formed by the near-UV photo-oxidation of hydrated ferrous ion according to the following reactions [1]:

1) $Fe^{2+}(aq) \rightleftharpoons Fe(OH)^+(aq) + H^+$

2) $Fe(OH)^+(aq) + H^+ \xrightarrow{h\nu} Fe(OH)^{2+}(aq) + 0.5\,H_2$

3) $Fe(OH)^{2+}(aq) \rightleftharpoons Fe(OH)_2(aq) + H^+$

4) $Fe(OH)_2(aq) \rightleftharpoons Fe(OH)_3(aq) + H^+$

5) $Fe(OH)_3(aq) \longrightarrow FeOOH(s) + H_2O$

Table 2. Geochronologic tabulation of the oldest known banded-iron formations [15]

Approximate age of deposition (Ga)	Geological unit and location
3.7–3.8	Isua Supracrustal Belt, Greenland
3.3–3.6	Sebakwian Group, Zimbabwe
3.1–3.5	Konsky Series (I), Ukraine, USSR
3.0–3.4	Imataca Complex, Venezuela
3.0–3.4	Swaziland Supergroup, South Africa
3.0–3.3	Tobacco Root and Ruby Ranges, Montana, USA

Table 3. Mineral assemblages in unaltered, relatively unmetamorphased banded-iron formations [15]. Minerals with asterisks contain ferric iron. Density values are from ref. [9].

Mineral	Chemical formula	Density $(g\,cm^{-3})$
Quartz	SiO_2	2.6
Magnetite*	$Fe^{2+}Fe_2^{3+}O_4$	5.2
Hematite*	$Fe_2^{3+}O_3$	5.3
Siderite	$Fe^{2+}CO_3$	4.0
Dolomite-ankerite	$CaMg(CO_3)_2 \cdot CaFe^{2+}(CO_3)_2$	ca. 2.9
Calcite	$CaCO_3$	2.7
Greenalite	$(Fe^{2+}, Mg)_6 Si_4 O_{10}(OH)_8$	ca. 3.0
Stilpnomelane*	$K_{0.6}(Mg, Fe^{2+}, Fe^{3+})_6 Si_4 O_{10}(OH)_2$	ca. 2.8
Minnesotaite	$(Fe^{2+}, Mg)_3 Si_4 O_{10}(OH)_2$	3.0
Riebeckite*	$Na_2 Fe_3^{2+}Fe_2^{3+}(Si_8 O_{22})(OH)_2$	3.3
Pyrite	FeS_2	5.0
Pyrrhotite	$Fe_{x-1}S_x$	ca. 4.6

Above pH 6.5 solutions of ferrous ion show a strong near-UV to visible absorption due to the high concentration of $Fe(OH)^+$ (see Figure 2) [2]. Irradiation with near-UV to visible light causes the formation of FeOOH and release of H_2 with a quantum yield between 1 and 5% [1]. The estimated rate of photoprecipitation is 150–$750\,mg\,cm^{-2}\,yr^{-1}$ in the tropics or 50–$250\,mg\,cm^{-2}\,yr^{-1}$ at the poles based on the solar output between 300 and 450 nm only [1].

These estimated rates of photoprecipitation compare favorably with the estimated rate of 10–$40\,mg\,cm^{-2}\,yr^{-1}$ for iron deposition 2.5 Ga ago in the Hamersley basin (Australia) banded-iron formation [13].

UV stopping power of iron-bearing sediment

Although no microfossils or stromatolites have been found in banded-iron formations, we have used their properties to estimate the UV stopping power of iron-bearing sediments in general. Thus we have estimated the

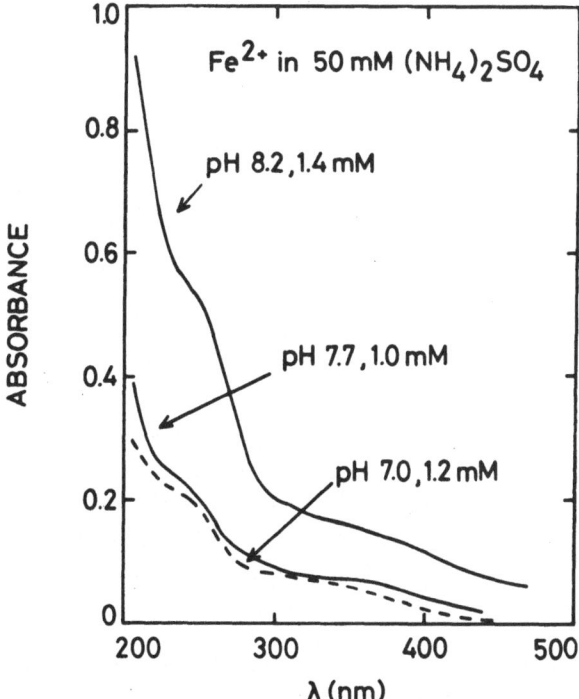

Figure 2. Absorption spectra of ferrous ion at various values of pH and in the presence of 50 mM $(NH_4)_2SO_4$. Adapted from [2].

UV absorbance at 265 nm for the banded-iron formations described in Table 3 [15]. The 2-mm bilayer of iron-bearing sediment and quartz contains 10% Fe^{2+} and 5% Fe^{3+}. The iron-bearing sediment is assumed to have a dry density of approximately 2.4 g cm^{-3}, and the quartz layer 1.6 g cm^{-3} (the value for contemporary sand from the west coast of Jutland). Light-scattering by the sediment particles is neglected for simplicity, and the total absorbance at 265 nm is estimated to be approx. 150 from the values of $\epsilon_{265}(Fe^{2+}) = 12 \text{ cm}^{-1} \mu M^{-1}$ and $\epsilon_{265}(Fe^{3+}) = 380 \text{ cm}^{-1} \mu M^{-1}$ obtained from Figure 1. For most purposes the contribution of Fe^{2+} can be neglected, and the absorbance of a 1-mm layer of mixed quartz and iron-bearing sediment will be about $(150/2)$ $(5\% \text{ } Fe^{3+})^{-1} = 15$ $(1\% \text{ } Fe^{3+})^{-1}$. We suppose that the sediments associated with 3.5-Ga organisms and stromatolites contained more than 0.1% Fe^{3+}, and therefore the absorbance of a 1-mm layer of sediment would have been greater than 1.5 at 265 nm.

What kind of photosynthesis?

There is a continuous record of biological CO_2 fixation which goes back at least 3.5 Ga ago [10]. This record has been preserved in the isotopic composition of sedimentary organic carbon (kerogen) as compared to the isotopic composition of carbonate carbon. The relative excess or depletion of ^{13}C in a sample, defined as

$$\delta^{13}C = \left[\frac{(^{13}C/^{12}C) \text{ sample}}{(^{13}C/^{12}C) \text{ standard}} - 1 \right] \times 10^3$$

is referred to the Peedee belemnite standard (PDB). Values of $\delta^{13}C$ for contemporary cyanobacteria span the range, -10 to $-30\%_{00}$, PDB, and for photosynthetic bacteria the range is -20 to $-35\%_{00}$, PDB. The geological record shows a continuous signal of $-27 \pm 7\%_{00}$, PDB, for organic carbon and a continuous signal of about $0.4 \pm 2.6\%_{00}$, PDB, for carbonate carbon. The $\delta^{13}C$ record strongly implies that autotrophic photosynthesis was occurring at least 3.5 Ga ago.

Contemporary photoautotrophs obtain reducing power for CO_2 fixation primarily from water (cyanobacteria) or reduced sulfur compounds (cyanobacteria, purple sulfur bacteria, and green sulfur bacteria). The existence of water-splitting cyanobacteria in the Gunflint microflora about 2 Ga ago is inferred from the striking morphological similarities between several microfossils and contemporary cyanobacteria [11], but to the authors' opinion claims for oxygen-evolving photosynthesis before 2.5 Ga ago are open to question. Neither microfossil morphology nor chemical/geological evidence persuade us of the existence of photosynthetic water-splitting at very early times. On the other hand the existence of sedimentary sulfate evaporites 3.5 Ga ago makes it reasonable to propose that autotrophic photosynthesizers may have utilized reduced sulfur compounds in the hydrosphere, thus contributing sulfate to the deposits found in the Onverwacht and Fig Tree Groups (Africa) and in the Warrawoona Group (Australia) [3, 15].

What kind of photosynthesizers?

Of the four major types of filamentous prokaryotes found as microfossils in the Warrawoona Group, we may suppose that at least some were photoautotrophs carrying out a sulfur-based CO_2 fixation. We may also suppose that these ancient photosynthesizers were similar in certain fundamental aspects to contemporary organisms living under similar conditions. For example many cyanobacteria can live in the presence of H_2S (maximum concentrations between 0.3 and 5 mM) and can convert CO_2 to cell material without evolving any oxygen and with photosystem (PS)-2 completely inoperative [7]. In this kind of photosynthesis H_2S is converted to elemental sulfur which is expelled from the cell. (Cyanobacteria are unable to oxidize the sulfur to sulfate). The PS-1 reaction center which carries out the photochemistry contains chlorophyll (Chl) a as the primary electron donor (P700) and two iron-sulfur centers as secondary electron acceptors.

Carbon dioxide fixation by the reductive pentose phosphate cycle requires 2 NADPH and 3 ATP. The path of linear electron transport is as follows:

$$H_2S \rightarrow PC/Cyt\, c \rightarrow Chl\, a \quad (P700) \xrightarrow{h\nu} (Fe.S)_1 \rightarrow (Fe.S)_2 \rightarrow ferredoxin \rightarrow NADP.$$ The main light-harvesting pigments in addition to Chl a are allophycocyanin, phycocyanin, and sometimes phycoerythrin.

Other contemporary examples are the green sulfur bacteria which also fix CO_2 with a single reaction center similar to the PS-1 reaction center of the cyanobacteria. In the former reaction center the primary electron donor is bacteriochlorophyll (BChl) a instead of Chl a, and the path of electrons is $H_2S \rightarrow Cyt\, c \rightarrow BChl\, a \quad (P840) \xrightarrow{h\nu} (Fe.S) \rightarrow ferredoxin$. Other forms of reduced sulfur (S^0, $S_2O_3^=$ and $SO_3^=$) are also used. The green sulfur bacteria do not use the reductive pentose phosphate cycle for CO_2 fixation, but instead use the reductive tricarboxylic acid cycle which requires reduced ferredoxin as well as NADH, NADPH, and ATP. The main light-harvesting pigment is BChl c or a similar pigment.

The purple sulfur bacteria form the final class of organisms for comparison. These bacteria contain BChl a or BChl b, and their reaction centers contain quinones as secondary electron acceptors instead of iron-sulfur centers. CO_2 fixation is via the enzymes of the reductive pentose phosphate cycle as in the cyanobacteria, but the reduction of NAD(P) requires reverse electron flow from the quinone pool in the cyclic electron transport chain. The path of electron flow is

$$H_2S \rightarrow Cyt\, c \rightarrow BChl\, a \ (P870) \xrightarrow{h\nu} Q_1 \rightarrow Q_2 \rightarrow Q_{pool} \rightarrow NAD(P).$$

Other forms of reduced sulfur are also used. The main light-harvesting pigment is BChl a or b.

The organisms living 3.5 Ga ago could have contained either Chl a or BChl a or both. With at least four different morphological types of cells present in the fossil record one can easily imagine the existence of physiological diversity among these putative primitive phototrophs. Some could have been Chl a organisms and others BChl a organisms. The spectral distribution of light would have been suitable for both kinds of chlorophyll, and in a stromatolitic community there might have been an advantage in having a mixture of light-harvesting pigments. The phycobilins (especially phycoerythrin) seem to have been especially suited for capturing blue-green light deep under water. BChl c and related pigments, on the other hand, appear to be suited for either blue light (ca. 440 nm) found deep under water or far-red light (700–800 nm) found at the surface. We would hazard a guess that BChl c and related pigments might have existed 3.5 Ga ago, but that the phycobilins (especially phycoerythrin) evolved later in deep-water organisms.

The reaction centers of the ancient organisms were probably of the PS-1 type with iron-sulfur centers as secondary electron acceptors and either

Chl *a* or BChl *a* as the primary electron donor. These reactions centers could reduce ferredoxin directly and could utilize either the reductive tricarboxylic acid cycle or the reductive pentose phosphate cycle for CO_2 fixation. The PS-2 type reaction centers with quinones as secondary acceptors probably had not yet evolved [6].

Sulfide may well served as the most common reductant for autotrophic photosynthesis for hundreds of millions of years before the sulfide concentration in the hydrosphere fell below the minimum level required, perhaps about 0.1 mM [7]. During this earliest phase of autotrophic photosynthesis, large amounts of elemental sulfur should have accumulated. There might be some evidence for this accumulation in the geological record unless the record has been too badly damaged by metamorphosis and oxidation. After the average sulfide level dropped below about 0.1 mM, photosynthetic organisms would have been restricted to specialized environments where the sulfide level remained high or would have evolved to oxidize elemental sulfur to sulfite. During this second phase, some mineral sulfites should have accumulated and been preserved in very early sedimentary rocks. Finally as the supply of elemental sulfur was depleted, photoautotrophs would have been forced to utilize sulfite as the source of reducing power with the release of sulfate. It appears that by 3.5 Ga ago photosynthesis was already in the third phase of oxidizing sulfur compounds all the way to sulfate. Eventually "sulfate-evolving" photosynthesis was largely replaced by oxygen-evolving photosynthesis in which two types of Chl *a*-containing reaction centers (PS-1 and PS-2) cooperated to extract electrons from water and deliver them to NAD(P). The evolution of the PS-2 reaction center and the water-splitting enzyme is not likely to have occurred in one mighty act of creation, but probably required many intermediate stages of development [4, 5] before oxygen evolving photosynthesis emerged sometime between 3 and 2 Ga ago.

References

1. Braterman PS, Cairns-Smith A and Sloper RW (1983) Nature 303, 163–164
2. Ehrenfreud M and Leibenguth J-L (1970) Bull. Soc. Chim. Fr. 2494–2505
3. Groves DI, Dunlop JSR and Buick R (1981) Sci Am. (Oct.) 56–65
4. Olson JM (1970) Science 168, 438–446
5. Olson JM (1978) Evolutionary Biology 11, 1–37
6. Olson JM (1981) BioSystems 14, 89–94
7. Padan E (1979) Annu. Rev. Plant Physiol 30, 27–40
8. Rambler MB and Margulis L (1980) Science 210, 638–640
9. Roberts WL, Rapp, Jr. GR and Weber J (1974) Encyclopedia of minerals New York: Van Nostrand Reinhold
10. Schidlowski M, Hayes JM and Kaplan IR (1983) In Schopf JW, ed. Earth's earliest biosphere, pp. 149–186. Princeton NJ: Princeton Univ. Press
11. Schopf JW (1974) Evolutionary Biology 7, 1–43
12. Schopf JW and Walter MR (1983) In Schopf JW, ed. Earth's earliest biosphere, pp. 214–239. Princeton, NJ: Princeton Univ. Press

13. Trendall AF and Blockley JG (1970) Bull. Geol. Surv. West Aust. 119
14. Veizer J (1983) In Schopf JW, ed. Earth's earliest biosphere, pp. 240–259. Princeton, NJ: Princeton Univ. Press
15. Walker JCG, Klein C, Schidlowski M, Schopf JW, Stevenson DJ and Walter MR (1983) In Schopf JW, ed. Earth's earliest biosphere, pp. 260–290. Princeton, NJ: Princeton Univ. Press
16. Walsh MM and Lowe DR (1985) Nature 314, 530–532
17. Walter MR (1983) In Schopf JW, ed. Earth's earliest biosphere, pp. 187–213. Princeton, NJ: Princeton Univ. Press

Photosynthesis Research 9, 261–272 (1986)
© 1986 Martinus Nijhoff/Dr. W. Junk Publishers, Dordrecht.

Detection of rapid induction kinetics with a new type of high-frequency modulated chlorophyll fluorometer

U. SCHREIBER

Lehrstuhl Botanik I, Universität Würzburg, Mittlerer Dallenbergweg 64,
D-8700 Würzburg, FRG

(*Received 7 August 1985*)

Key words: chlorophyll fluorescence, fluorometer, fluorescence quenching, Kautsky effect, photosynthesis, photosystem II

Abstract. A newly developed modulation fluorometer is described which operates with $1\,\mu sec$ light pulses from a light-emitting diode (LED) at 100 KHz. Special amplification circuits assure a highly selective recording of pulse fluorescence signals against a vast background of non-modulated light. The system tolerates ratios of up to $1:10^7$ between measuring light and actinic light. Thus it is possible to measure the "dark fluorescence yield" and record the kinetics of light-induced changes. A high time resolution allows the recording of the rapid relaxation kinetic following a saturating single turnover flash. Examples of system performance are given. It is shown that following a flash the reoxidation kinetics of photosystem II acceptors are slowed down not only by the inhibitor DCMU, but by a number of other treatments as well. From a light intensity dependency of the induction kinetics the existence of two saturated intermediate levels (I_1 and I_2) is apparent, which indicates the removal of three distinct types of fluorescence quenching in the overall fluorescence rise from F_0 to F_{max}.

Abbreviations

Q_A and Q_B, consecutive electron acceptors of photosystem II; PS II, photosystem II; P 680, reaction center chlorophyll of photosystem II; F_0, minimum fluorescence yield following dark adaptation; F_{max}, maximum fluorescence yield; DCMU, 3-(3,4-dichlorophenyl)-1,1-dimethyl-urea; DCCD, N,N'-dicyclohexylcarbodiimide; PQ, plastoquinone; DAD, diaminodurene.

Introduction

Chlorophyll fluorescence has found numerous applications as a sensitive indicator of photosynthetic reactions (for reviews, see refs. [21, 25, 27, 29]). It was Duysens and Sweers [12] who introduced the concept of the fluorescence quencher Q controlling fluorescence yield. These authors recognized that variable fluorescence originates primarily from the pigments serving photosystem II (PS II) and that fluorescence quenching is closely correlated with the availability of a PS II acceptor molecule Q_A in the oxidized

---

Dedicated to Prof. L.N.M. Duysens on the occasion of his retirement.

state. The Q-concept has been extremely helpful for the interpretation of fluorescence changes during the past 20 years. However, Duysens and Sweers [12] also clearly demonstrated a second type of fluorescence quenching which does not depend on oxidized Q_A. Recently, there has been renewed intensive interest in the differentiation between the two main types of fluorescence quenching, the photochemical quenching (Q-quenching) and the non-photochemical quenching [5, 6, 20, 28].

The discoveries made by Duysens and Sweers largely depended on the development of a special type of modulated fluorescence measuring system, which made it possible to study the effects of light preferentially absorbed by pigment systems I or II on the fluorescence yield. The fluorescence excitation beam was chopped at 50 Hz by means of a disc mounted on a synchronous motor, and only the 50 Hz component of the signal was amplified by a phase and frequency sensitive amplifier. Hence, when continuous actinic light of various spectral compositions was applied, no direct signal due to stray light components or fluorescence of this illumination was detected. The modulated signal recorded only the indirect effect of this light on the redox state of the quencher Q_A.

In recent years, substantial progress has been made in the characterization of the detailed reaction mechanisms in the vicinity of PS II (for reviews, see refs. [1, 8, 40, 42]). In particular, the existence of a primary electron acceptor between the reaction center chlorophyll P 680 and Q_A has been recognized [19, 38]; P 680$^+$ was found to cause fluorescence quenching [7, 13] and a two-electron gate mechanism at the secondary PS II acceptor Q_B was discovered [3, 39].

So far, fundamental research on photosynthetic primary reactions, mostly carried out by biophysicists, has had little impact on the applied photosynthesis studies of plant physiologists. This has been mostly due to the requirements of very specialized experimental equipment to analyse the rapid primary electron transport steps. It was the intention of the present work to design a fluorometer which is sufficiently flexible to serve the sophisticated requirements for analysing the primary reactions and to be used as a tool in plant physiological studies.

This paper describes a new type of modulated chlorophyll fluorometer, which may be considered a modern version of the fluorometer introduced by Duysens and Sweers [12], which allows detection of the rapid fluorescence changes reflecting PS II primary reactions. Outstanding properties of the new fluorometer are a high time resolution, extreme insensitivity to non-modulated signals, great compactness and easy operation. The new measuring system is based on a high frequency light emitting diode (LED), substituting for a mechanically chopped excitation light source, and a photodiode replacing the photomultiplier. In some examples of experimental applications the performance of the new fluorometer is demonstrated in particular with respect to the resolution of rapid fluorescence kinetics. It will be shown that the system is capable of resolving the fluorescence decay kinetics corresponding to the reoxidation of the primary PS II acceptor Q_A by the secondary acceptor Q_B. From the light intensity dependence of the light-on induction kinetics the existence of three distinct components in variable fluorescence from F_0 to F_{max} is demonstrated.

Materials and Methods

Spinach (*Spinacia oleracea* L., Yates Hybrid 102) was grown in the greenhouse at day/night cycles of 13/11 h. Intact spinach chloroplasts were isolated following standard procedures [15, 16]. Intactness of the outer membrane was estimated by the ferricyanide method [14]. Usually, about 80% of the chloroplasts had intact envelopes. Class D chloroplasts were obtained from intact chloroplasts by mild osmotic shock and isotonic resuspension, as described previously [34].

Chlorophyll fluorescence was measured with a newly developed "pulse-amplitude-modulation" fluorometer, features of which are detailed in the following section. This fluorometer is the prototype of an instrument which has become commercially available (PAM 101, 102, 103 chlorophyll fluorometer, H. Walz, Effeltrich, Germany). The fluorometer is equipped with four-armed fiber-optics connecting a suspension cuvette with an LED emitter, a photodiode detector, a Xenon flash lamp (EG & G FX 6A, flash duration about 5 μsec) and a source for continuous actinic light (150 W halogen lamp, Osram Xenophot HLX). Actinic light intensity was varied by use of neutral density filters (Schott NG series). Flash-triggering as well as opening and closing of electro-magnetic shutters (Compur electronic-m) were controlled by trigger and timer circuits provided by the fluorometer control unit. Full shutter opening occurred within 0.8 msec. Kinetic traces were recorded on a Digital Storage Oscilloscope (Nicolet Explorer III) from where they were photographed.

The measuring system

Most chlorophyll fluorometers employed in past photosynthesis studies make use of the same light both for fluorescence excitation and for driving photosynthetic reactions (actinic light) (for a technical review, see ref. [33]). With such fluorometers, separation between incident light and fluorescence depends on optical filters only — any light which can pass the red cut-off filter in front of the photodetector must be eliminated from the incident light (normally by a blue filter). Another feature of such fluorometers is the linear dependence of the fluorescence signal on incident light intensity, i.e. any change in actinic light intensity will cause corresponding fluorescence changes, irrespective of any true changes in fluorescence yield (fluorescence intensity/light intensity).

In modulation fluorometers, as e.g. introduced by Duysens and Sweers

[12], a modulated fluorescence measuring beam is used in addition to the non-modulated actinic light. As the amplifier system selects the modulated signal, actinic illumination can be varied within wide ranges without corresponding signal changes. With constant measuring light intensity the signal reflects relative fluorescence yield, which in green plants may vary by a factor of up to 5, depending on the redox state of the quencher Q_A and the extent of non-photochemical fluorescence quenching.

Conventional measuring systems for modulated fluorescence employ a mechanically chopped measuring beam and a lock-in amplifier for phase and frequency sensitive signal amplification. Such systems are normally limited with respect to frequency, response time, sensitivity and selectivity. To monitor the rapid fluorescence relaxation kinetics following a single turnover flash a time resolution of less than 150 μsec is required and over-saturation of the amplifier system must be avoided. Furthermore, the intensity of the measuring light should be low enough as not to have any actinic effect on the sample. In the following, a new type of modulated fluorescence measuring system is described which combines selectivity with speed and sensitivity.

Figure 1 shows a schematic diagram of the new measuring system. A master pulse generator controls the current pulses which drive emission from a light emitting diode (LED). The LED light pulses, after passing an optical short-pass filter ($\lambda < 680$ nm), will excite pulsed fluorescence in a sample. An optical long-pass filter ($\lambda > 700$ nm) rejects all stray excitation light and lets the long wavelength fluorescence pass to a photodiode detector, together with any long wavelength component of stray actinic light and of fluorescence excited by such light. Separation of the latter signals from the pulsed fluorescence signal is achieved in two steps, first, by an AC-coupled pulse amplifier and second, by a selective amplifier which is synchronized with the master pulse generator which also controls the LED driver.

To reach a high amplitude pulse signal, without inducing actinic effects by the measuring beam, the LED pulse was chosen to be only 1 μsec wide, at frequencies of 1.6 KHz or 100 KHz. Currents of up to 150 mA during single pulses were allowed for the given LED (Stanley, USBR 2000). A PIN photodiode (Hamamatsu, S 1723) with 6 V negative bias was sufficiently fast to resolve the single pulse signals. The combination of short-pass (Balzers DT Cyan) and long-pass (1 mm Schott RG 9) filters effectively blocked stray measuring light to reach the detector. Highly selective amplification of modulated fluorescence is favored by the special character of the pulse signal, which is extremely short and rapid.

Figure 2 shows the signal at the output of the AC-coupled pulse amplifier, which is determined by the loading (positive peak) and unloading (negative peak) of the coupling capacitor. By appropriate electronic circuits it is possible to amplify the difference between the "loading peak" and the

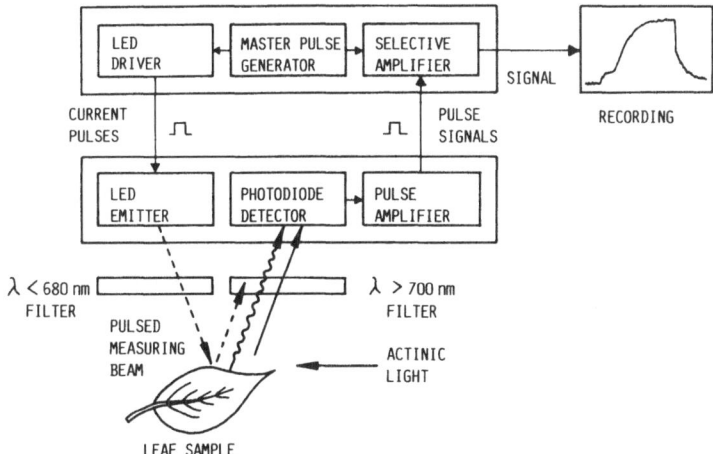

Figure 1. Schematic diagram of the pulse modulation chlorophyll fluorometer. A master pulse generator controls LED pulse emission and selective amplification of the pulse fluorescence signals. Optical short-pass and long-pass filters between the LED, sample and photodiode detector assure that no stray LED light can reach the detector. The modulated fluorescence yield is modified by the action of non-modulated actinic light, any direct signal of which is rejected by the combination of an AC-coupled pulse amplifier and the selective amplifier. In practice, the various light paths are connected via flexible, multibranched fiberoptics.

"unloading peak", which is unaffected by any overlapping signals, provided these do not show significant changes within the $2.5\,\mu sec$ between the two sampling periods. Rapid electronic switches (CMOS) which are controlled by the same pulse generator which controls the LED driver, in combination with suitable decimal timers, assure that signal amplitudes are sampled at the correct times with respect to the LED pulse.

Due to the very large linearity range of the photodiode and to the highly selective amplification system, ratios of up to $1:10^7$ between measuring light and actinic light are tolerated by the measuring system. This feature makes it possible to record fluorescence at low measuring light intensity (e.g. $10\,mW/m^2$) and still drive photosynthetic reactions at saturating intensities of unfiltered white light (e.g. $1000\,W/m^2$).

When applying saturating single turnover flashes (about $10^6\,W/m^2$) even a dynamic range of $1:10^7$ is not sufficient to prevent amplifier overloading, which would require about $300\,\mu sec$ to recover, spoiling resolution of the rapid relaxation kinetics. To avoid such overloading, a gating circuit was designed which short-circuits the photodiode output and prevents signal sampling at the selective amplifier for the duration of the flash.

For the recording of rapid kinetics a high modulation frequency is essential. On the other hand, high frequency pulse illumination may have an actinic effect on the sample. To allow high frequency measurements without significant preillumination effects, a special circuit was developed

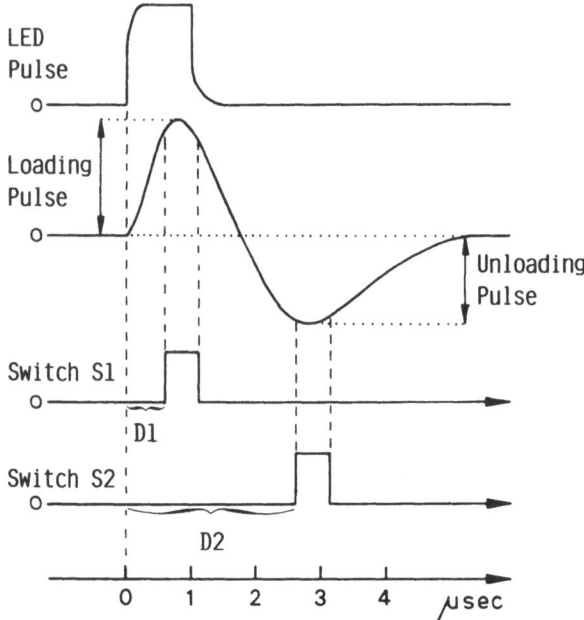

Figure 2. Time correlation between an LED measuring pulse, the signal at the output of the pulse amplifier and the sample-hold switches of the selective amplifier. The square LED pulse produces a corresponding fluorescence pulse which is transformed by the photodiode detector and the AC-coupled preamplifier into a characteristic signal, consisting of a loading pulse and an unloading pulse. The selective amplifier contains logical elements which enable it to sample the signal amplitudes at the peaks of the loading and unloading pulses and to store these amplitudes in capacitors at the two entrances of a differential amplifier until the next pulse signal arrives. The switches S_1 and S_2 are operated in synchrony with the LED pulses and are appropriately delayed, D1 and D2, to coincide with the peaks of the pulse signals.

by which modulation frequency is switched from 1.6 KHz to 100 KHz 3 msec before triggering of a flash or of continuous actinic illumination.

System performance and experimental results

Rapid decay kinetics following a flash. Figure 3 shows oscilloscope traces of the fluorescence relaxation kinetics following a saturating single turnover xenon flash in spinach leaves (A, B) and isolated chloroplasts (C, D). In Figure 3A the penetration of the PS II inhibitor DCMU into a spinach leaf is demonstrated. With increasing penetration times there was a small increase in F_0, a marked increase in the maximal yield induced by the flash and progressive retardation of the decay kinetics. These curves reflect the well-known effect of DCMU on the PS II acceptor complex: The inhibitor competes with PQ for the binding site at the B-protein [23, 24, 41] and thus blocks Q_A^- reoxidation by Q_B. Due to the 130 μsec gating period, the very

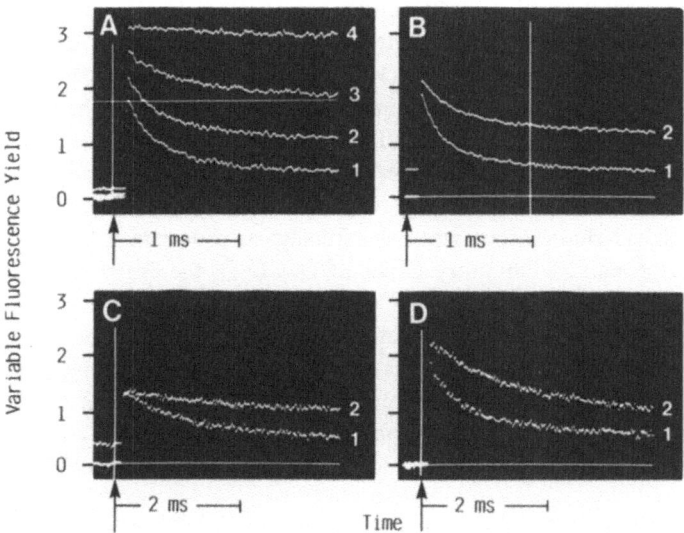

Figure 3. Fluorescence dark decay kinetics following application of a single turnover flash. Modification of the decay kinetics in spinach leaves (A, B) and class D chloroplasts (C, D) by various treatments. (A) Effect of 10^{-5} DCMU, penetrating via the upper epidermis of the leaf from which the cuticle was partially removed by gentle rubbing with carborundum powder; (1) control, (2) 5 min, (3) 10 min, (4) 20 min penetration time. (B) Effect of short heat-pretreatment; (1) control; (2) leaf exposed for 5 min in 46 °C hot water, then rapidly cooled back to 25 °C. (C) Effect of ATP-hydrolase inducing reverse electron flow. Following light activation, intact chloroplasts were osmotically shocked, isotonically resuspended and 0.5 mM ATP was added; (1) before addition of DAD, (2) 30 sec following addition of 0.2 mM of reduced DAD, which serves as donor for reverse electron flow. (D) Effect of 3×10^{-4} M DCCD on class D chloroplasts; (1) control, (2) after 5 min incubation of the inhibitor. In all experiments: The measuring beam of 50 mW/m² (integrated intensity) was switched on 1 sec before flash triggering. 3 msec before the flash pulse, frequency was increased from 1.6 to 100 kHz, with a corresponding increase of integrated measuring beam intensity. No signal is recorded during a 130 μsec gating period after the flash. Temperature: 25 °C with leaves (A, B); 12 °C with chloroplasts (C, D). One unit of variable fluorescence corresponds to yield at F_0.

first part of the decay kinetics is not recorded. However, from detailed previous work by Duysens and co-workers [10, 13] and other investigators [26, 45] it is known that after a saturating flash there is a biphasic fluorescence rise, correlated with the re-reduction of P^+_{680}, followed by a plateau, which lasts for about 100 μsec before the first decay phase sets in. Hence, the initial fluorescence values should be close to the true maximal amplitudes induced by the flash. An increase of flash-induced fluorescence by DCMU has been described before [37].

Other treatments which slow down the Q^-_A reoxidation kinetics are characterized by the data of Figures 3B, C, D. Figure 3B shows the effect of a short heat-treatment of a leaf. Phenomenologically this effect is similar to that of DCMU-treatment, but without the pronounced increase of flash-induced fluorescence. Obviously, the heat treatment affects not only the PS II donor site [44] but the acceptor complex as well. In Figure 3C the effect of an active ATP-hydrolase in spinach chloroplasts is demonstrated. It has been shown before that under similar conditions ATP-hydrolysis can induce reverse electron flow and corresponding fluorescence stimulation [30, 32, 34]. It is apparent that the reverse electron pressure prevented rapid Q_A^- re-oxidation. Contrary to the action of DCMU, no increase of flash-induced fluorescence was seen. Finally, as shown in Figure 3D, there was also a marked inhibitory effect of DCCD on the decay rate. This substance is known to block proton channels [2, 36] in the CF_0 part of the ATP-ase complex, as well as in other membrane bound protein complexes. An inhibitory action of DCCD at the PS II acceptor side has been observed before [31]. Hence, one might consider the possibility of a proton channel serving Q_B^- protonation which may be essential for rapid Q_A^--reoxidation. So far, not much is known on protonation reactions at the PS II acceptor side.

These data demonstrate that the new modulation fluorometer is sufficiently fast and sensitive to record the rapid relaxation kinetics of PS II. In the past, these kinetics have been accessible only by point to point measurements, using weak detection flashes after variable delays with respect to an actinic flash [4, 10, 17, 26, 45]. The ease, with which such kinetics can now be measured, should stimulate its use as a tool to characterise the state of the PS II acceptor complex. This protein complex, which is known to be the site of PS II herbicide action [23, 24, 41] may also play a key role in the detrimental effects of environmental stress (see e.g. ref. [22]), which are expressed at physiological and biochemical levels.

Light intensity dependence of induction kinetics

An important practical advantage of a modulation fluorometer is the possibility of recording fluorescence yield over a wide range of intensities of actinic light using unchanging measuring beam intensity and amplifier gain. Figure 4 shows fluorescence induction kinetics of a spinach leaf at actinic intensities varied by a factor of 600. Whereas at moderate intensities the well-known induction pattern, O–I–P, is displayed (Figure 4A), at high light intensities the rise kinetics became considerably more complex and showed (Figure 4B, C, D) two characteristic intermediary levels, here called I_1 and I_2. With increasing intensities the I_1-level was raised, until at about $2000 \, W/m^2$ it reached a saturation level, which amounted to 50–60% of F_{max}. At very high intensity there was a pronounced plateau or even a dip following I_1. The rise from I_1 to I_2 showed two steps which, depending on the leaf sample and the state of dark-adaptation, could be more or less pronounced. As the I_1-level, the I_2-level also approached a saturation value at very high light intensity. The last phase from I_2 to P, which was distinctly slower than the preceding phases, amounted to 10–15% of total variable fluorescence.

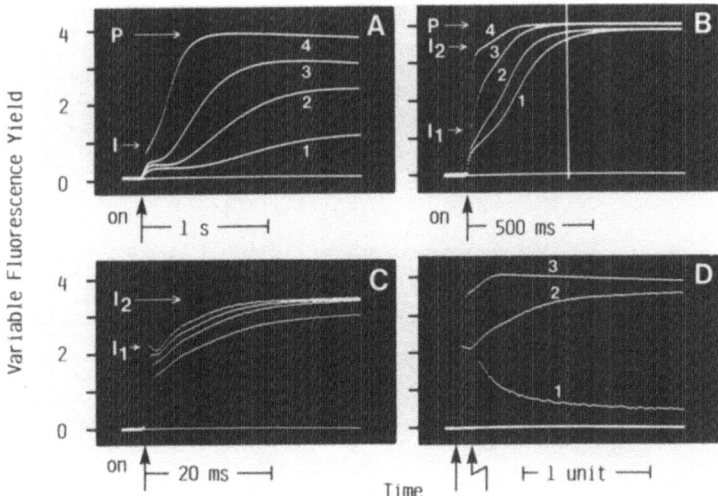

Figure 4. Fluorescence induction kinetics of spinach leaves at different actinic light intensities. (A–C) Increasing actinic intensities; recordings at different time scales. (A) (1) $5 \, W/m^2$, (2) $10 \, W/m^2$, (3) $20 \, W/m^2$, (4) $40 \, W/m^2$. (B) (1) $40 \, W/m^2$, (2) $100 \, W/m^2$, (3) $250 \, W/m^2$, (4) $500 \, W/m^2$. (C) curves from bottom to top, 500, 1000, 2000, $3000 \, W/m^2$. (D) Comparative recordings of fluorescence increases induced by saturating continuous light ($3000 \, W/m^2$) and a saturating single turnover flash. (1) Flash illumination, 1 msec/time unit; (2) continuous illumination, 20 msec/time unit; (3) 40 msec/time unit. For other conditions, see Fig. 3.

A saturating single-turnover flash induced a fluorescence increase to a level which closely corresponded to the saturated I_1-level (Figure 4D). This feature is further elaborated by the data of Figure 5. By lowering the temperature to $0\,°C$, the I_1-level was increased, with a corresponding diminution of the I_1-I_2-phase (Figure 5A, C). At the same time also the flash-induced fluorescence rise was increased, and due to a slowing-down of the dark decay rate, any underestimation of flash-induced fluorescence (by the $130\,\mu sec$ gating period) would have been smaller at the low temperature. When a flash was triggered at a moment during induction which corresponded to the I_1-level, this did not cause any appreciable fluorescence rise beyond the I_1-level, neither at $25\,°C$ nor at $0\,°C$ (Figure 5B, D).

These data suggest that variable fluorescence is composed of three parts with distinctly different properties, namely the 0-I_1, I_1-I_2 and I_2-P phases. The 0-I_1 rise probably corresponds to the "photochemical phase", first described by Delosme [9] who was investigating fluorescence induction at extreme light intensities in Chlorella. Delosme distinguished this "photochemical phase", corresponding to the reduction of the quencher Q_A [12, 18], from a slower "thermal phase", which he attributed to the removal of a "non-photochemical quencher R", closely related to the secondary PS II acceptor pool. Vernotte et al. [43] demonstrated fluorescence quenching

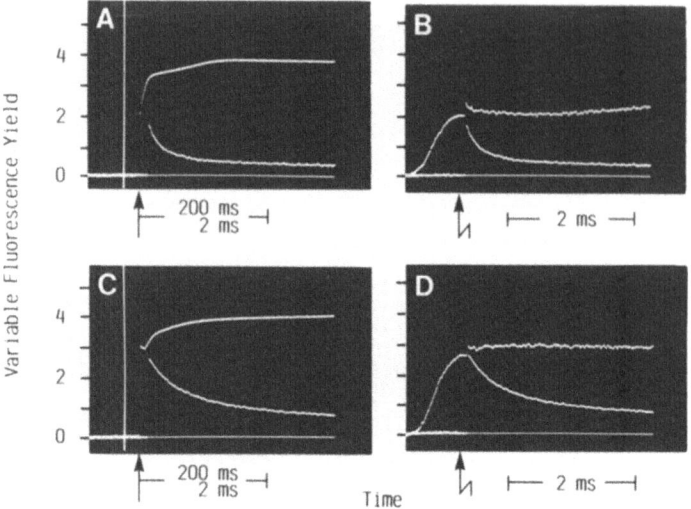

Figure 5. Effect of temperature lowering on fluorescence increases induced by saturating continuous light $(3000 \, W/m^2)$ or a saturating single turnover flash. Spinach
leaves were in contact with an aluminum block at $25\,°C$ (A, B) or $0\,°C$ (C, D). In (B)
and (D) a saturating flash was triggered 1 msec following admission of continuous
light, when the fluorescence rise has reached the saturated I_1-level. The two time scales
in (A) and (C) apply to the curves with continuous illumination (200 msec) and flash
illumination (2 msec). Other condition as in Figs. 3 and 4.

by the oxidized PQ-pool. While the extent of this quenching (about 10%)
is too small to account for the $I_1 I_2$ phase, it could be responsible for the
I_2-P phase.

Recently, there has been renewed interest in the differentiation between
photochemical and non-photochemical quenching [5, 6, 11, 20, 28, 35],
which is essential for practical applications of chlorophyll fluorescence
methods in plant physiology [29]. Interpretation of fluorescence changes
requires information on the relative contributions of the different quenching
components to the overall changes. In practice, saturating light pulses are
applied for complete reduction of PS II acceptors, with the assumption of
complete removal of photochemical quenching. The remaining quenching has
been considered to represent "non-photochemical quenching".

The above data on fluorescence changes at high light intensities demonstrate that it is not possible to reach fluorescence saturation by short light
pulses, however intense, as the I_1-level cannot be surpassed. When longer
pulses are applied, inducing a saturated fluorescence response (I_1-I_2-P phases)
quenching which is likely to be of "non-photochemical" nature is also removed.

These considerations show that differentiation between photochemical
and non-photochemical quenching by the saturation pulse method [5, 6,
11, 20, 28, 35] requires well defined saturation pulses, with respect to

pulse length and intensity. In principle the use of saturating single turnover flashes may provide the clearest results if true photochemical quenching is to be determined. On the other hand, with the use of more extended pulses, inducing the maximal P-level, photochemical as well as non-photochemical quenching appears to be eliminated, with the exception of the so-called "energy-quenching". In this latter way the differentiation is between "redox quenching" and "energy quenching", rather than between photochemical and non-photochemical quenching. As the 0-I_1 rise, also the I_1-I_2 and I_2-P rises are related to the reduction of the electron transport chain between the two photosystems. Hence, application of saturating "long flashes" is suitable for determination of the energy status of the thylakoid membrane, giving important information for the plant physiologist on the photosynthetic performance of leaves. Further investigations will have to clarify the detailed differences in information obtained with flashes inducing saturation at I_1, I_2 or P.

In conclusion, the newly developed pulse modulation fluorometer has been shown to be capable of measuring the rapid reoxidation of Q_A^- by Q_B, and to be unaffected even by very high actinic light intensities which were shown to be essential for complete removal of "redox quenching". By these features in combination with its compact design, the pulse modulation fluorometer opens new ways in basic photosynthesis research as well as in applied plant physiological studies.

Acknowledgements

I wish to thank Ulrich Heber and Wolfgang Bilger for stimulating discussions and encouragement. Ulrich Schliwa is thanked for excellent electronic engineering.

References

1. Amesz J and Duysens LNM (1977) In primary processes of photosynthesis (Barber J, ed.) pp. 149–185 Amsterdam: Elsevier
2. Azzi A, Casey RP and Nalecz MJ (1984) Biochim Biophys Acta 768:149–185
3. Bouges-Bocquet B (1973) Biochim Biophys Acta 314:250–256
4. Bowes JM and Crofts AR (1980) Biochim Biophys Acta 590:373–384
5. Bradbury M and Baker NR (1981) Biochim Biophys Acta 63:542–551
6. Bradbury M and Baker NR (1984) Biochim Biophys Acta 765:275–281
7. Butler WL (1972) Proc Nat Acad Sci US 69:3420–3422
8. Crofts AR and Wraight CA (1983) Biochim Biophys Acta 726:149–185
9. Delosme R (1967) Biochim Biophys Acta 143:108–128
10. Den Haan GA, Gorter de Vries H and Duysens LNM (1976) Biochim Biophys Acta 430:265–281
11. Dietz K-J, Schreiber U and Heber U (1985) Planta, in press
12. Duysens LNM and Sweers HE (1963) In studies on microalgae and photosynthetic bacteria, pp. 353–372. Tokyo: University of Tokyo Press
13. Duysens LNM, den Haan GA and van Best JA (1975) In Proc 3rd Int Congr Photosynth (Avron M, ed.) Vol 1, pp. 1–12. Amsterdam: Elsevier

14. Heber U and Santarius KA (1970) Z Naturforsch 25b:718–728
15. Heber U (1973) Biochim Biophys Acta 305:140–152
16. Jensen RG and Bassham JA (1966) Proc Nat Acad Sci US 56:1095–1101
17. Joliot A (1974) In Proc 3rd Congr Photosynth Res, Rehovot (Avron M, ed.) Vol 1, pp. 315–322, Elsevier: Amsterdam
18. Kautsky H, Appel W and Amann (1960) Biochem Z 332:277–292
19. Klimov VV, Klevanik AV, Shuvalov VA and Krasnovskii AA (1977) FEBS Lett 82:183–186
20. Krause GH, Briantais JM and Vernotte C (1982) Biochim Biophys Acta 679:116–124
21. Krause GH and Weis E (1984) Photosynth Res 5:139–157
22. Kyle DJ, Ohad I and Arntzen CJ (1984) Proc Nat Acad Sci US 81:4070–4074
23. Laasch H, Schreiber U and Urbach W (1983) FEBS Lett 159:275–279
24. Lavergne J (1982) Biochim Biophys Acta 682:345–353
25. Lavorel J and Etienne AL (1977) In primary processes of photosynthesis (Barber J, ed.) pp. 203–268. Amsterdam: Elsevier
26. Mauzerall D (1972) Proc Nat Acad Sci US 69:1358–1362
27. Papageorgiou G (1975) In bioenergetics of photosynthesis (ed. Govindjee) pp. 319–371. New York: Academic Press
28. Quick WP and Horton P (1984) Proc R Soc Lond B 220:371–382
29. Renger G and Schreiber U (1985) In light emissions by plants and bacteria (Govindjee, Amesz J and Fork DC, eds.) New York: Academic Press, in press
30. Rienits KG, Hardt H and Avron M (1974) Eur J Biochem 43:291–298
31. Sane PV, Johanningmeier U and Trebst A (1979) FEBS Lett 108:136–140
32. Schreiber U and Avron M (1979) Biochim Biophys Acta 546:436–447
33. Schreiber U (1983) Photosynth Res 4:361–373
34. Schreiber U (1984) Biochim Biophys Acta 767:70–79
35. Schreiber U, Bilger W and Schliwa U (1985) Photosynth Res, in press
36. Solioz M (1984) Trends Biochem Sci 9:309–312
37. Van Best (1977) Doctoral Thesis, State University of Leiden, the Netherlands
38. Van Best JA and Duysens LNM (1977) Biochim Biophys Acta 459:187–206
39. Velthuys B and Amesz J (1974) Biochim Biophys Acta 333:85–94
40. Velthuys BR (1980) Ann Rev Plant Physiol 31:545–567
41. Velthuys BR (1981) FEBS Lett 126:277–281
42. Vermaas WFJ and Govindjee (1981) Photochem Photobiol 34:775–793
43. Vernotte C, Etienne AL and Briantais J-M (1979) Biochim Biophys Acta 545:519–527
44. Yamashita T and Butler WL (1968) Plant Physiol 43:2037–2040
45. Zankel Kl (1973) Biochim Biophys Acta 325:138–148

Photosynthesis Research 9, 273–283 (1986)
© *1986 Martinus Nijhoff/Dr. W. Junk Publishers, Dordrecht.*

A Monte Carlo method for the simulation of kinetic models

J. LAVOREL

ARBC, CEN Cadarache 13108, St. Paul-lez-Durance, France

(*Received 24 September 1985*)

Key words: Monte Carlo, kinetics, simulation, model

Abstract. The purpose of this note is to illustrate the feasibility of simulating kinetic systems, such as commonly encountered in photosynthesis research, using the Monte Carlo (MC) method. In this approach, chemical events are considered at the molecular level where they occur randomly and the macroscopic kinetic evolution results from averaging a large number of such events. Their repeated simulation is easily accomplished using digital computing. It is shown that the MC approach is well suited to the capabilities and resources of modern microcomputers. A software package is briefly described and discussed, allowing a simple programming of any kinetic model system and its resolution. The execution is reasonably fast and accurate; it is not subject to such instabilities as found with the conventional analytical approach.

Abbreviations

MC, Monte Carlo; RN, random number; PSU, photosynthetic unit

Introduction

Much of the research effort devoted to photosynthesis — more generally to biology — is now concerned with the kinetic studies of molecular events and their modeling. The latter activity implies the ability to generate a quantitative time-dependent representation of the target phenomenon and the optimal fitting of its constitutive parameters with respect to experimental data. I see at least two reasons why modeling is taking an increasing importance as a research tool. Firstly, as we are nearing a complete description of every corner in photosynthetic arcana, the urge is felt more intensely than ever to achieve a quantitative — and not only qualitative — explanation of phenomena. Secondly, as digital computing has become an affordable technique in research laboratories, computer literacy tends to be the rule rather than the exception among scientists.

This note describes the state of a project of kinetic simulation by Monte Carlo (MC). I dedicate it to Lou Duysens who, years ago, led the way to

---

Dedicated to Prof. L.N.M. Duysens on the occasion of his retirement.

using digital computing in modern photosynthesis research. Specifically, my purpose was to inquire whether implementing the MC approach on a typical 8-16 bit personal computer was at all realistic in terms of accuracy and rapidity. As the MC method is not yet well known in our field (but see [3, 5]), I shall briefly explain it, after comparing it with the conventional technique of numerical integration. A description of the structure and functioning of a software system and a discussion of random number generators follow. This software is illustrated in solving the classical Michaelis-Menten enzyme catalysis. In conclusion, a proposal for a dedicated MC machine for improved speed performance is mentioned.

Principle of the Monte Carlo method

As is well known, a kinetic system needs be only of moderate complexity to be intractable by human means. In such cases, a solution for the integration of the set of differential rate equations governing the kinetic system is not available in analytical or convenient form. To handle such a situation, one usually resorts to numerical integration. Let us assume that the set of equations contains the following one:

$$d(C)/dt = k(A)(B) \tag{1}$$

which expresses, by the law of mass action, that the rate of formation of the product C depends on the concentration of substrates A and B (rate constant k). Numerical integration essentially replaces differentials by finite differences. That is, the time t is incremented by finite steps Dt and, from an initial set of conditions on concentrations, the system is allowed to evolve by small increments. For instance, assuming that concentrations $(A)_n$, $(B)_n$ and $(C)_n$ are known at step n, the situation concerning C at step $n + 1$ shall be:

$$(C)_{n+1} = (C)_n + D(C) \tag{2a}$$

where $D(C) = k(A)_n(B)_n Dt.$ \hfill (2b)

Since for practical reasons the step size Dt cannot be made infinitely small, the process given by Eqn (2b) is only approximate, the result being that numerical instabilities — even divergence — of the solution may occur in an unpredictable manner. Powerful methods (Runge–Kutta, predictor-corrector [4]) have been devised to tackle this problem by automatically adjusting the step size Dt optimally. But their complexity entails a serious computing overhead.

The Monte Carlo approach is stochastic. In essence, it mimicks repeatedly an event of probability p and, by the law of large numbers, the average frequency of success equals p to any desirable accuracy. This scheme ideally applies to chemical kinetics. For instance, in a first order reaction ($A \rightarrow B$), the rate equation may be written as:

$$-d(A)/(A) = kdt. \tag{3}$$

The left member is the fraction of A molecules having reacted during dt or, in probabilistic language, the probability of success. It is seen that the probability of the event: $A \rightarrow B$ for any member of the A population is kdt, i.e. it only depends on the time interval dt, it does not depend on the past history, nor on the size of the population. This simple scheme defines a Poisson process [1]; consequently, the average fraction of unreacted A at time t is $\exp(-kt)$, which is also found as the solution of the deterministic, macroscopic Eqn (3). The direct simulation of the above stochastic process is easily translated into a general algorithm:

1. Select at random a member of the total population $A + B$;
2. if the selected item belongs to A, decide with probability kdt to transform it into a member of B;
3. repeat.

(Adaptation to a bimolecular process is obvious: in steps 1 and 2 selections must be performed and the success in step 2 is conditional on both selections being successful). How the above random selection or decision is made refers to a particular probability distribution which is known as the uniform distribution. It can be realized with the classical urn model containing N identical balls numbered 1 to N; a ball is drawn with replacement. It is seen that the probability of drawing a number n $(1 \leqslant n \leqslant N)$, i.e. the ball n, is simply 1/N and that, by the law of additive probabilities of exclusive events, the probability of drawing a number in the interval 1 to m $(1 \leqslant m \leqslant N)$, i.e. any ball with number $\leqslant m$, is m/N. The first outcome is a random selection, the second a random decision on an event of probability m/N. In digital machines, this idealized urn model is implemented rather imperfectly by software or hardware devices known as random number (RN) generators.

Step 1 in the above algorithm may be conveniently simplified. Actually, as given, it amounts to a topological selection, in as much as it preserves the information pertaining to the spatial organization of the population. This information is not needed in the present context (it is on the contrary essential, for instance, when modeling the light-induced evolution of connected PSU's [3] or the wandering of excitons in antenna's [5]). An equivalent, simpler alternative is the 'analog' selection, whereby step 1 above reads as:

1'. Take a RN in the cardinal $A + B$; if RN $<$ cardinal A, a number of A has been selected.

The analog selection has the added advantage over the topological one of running faster and requiring much less memory space.

Material

The software has been implemented on GOUPIL-3 (SMT) running under FLEX (TSC), with a 8-16 microprocessor MC6809 (MOTOROLA). The binary module was written using FLEX's editor and assembler; the present version is less than 500 bytes. The BASIC driver was written in SBASIC (SMT, similar to TSC's XBASIC); the present source is about 1.5 K. Minimal graphic functions are most useful for visualizing a simulation in progress; the accompanying figures are direct screen printouts with a 256*256 pixel resolution.

The only specific hardware requirement concerns the RN generator; as explained below, it has been implemented using a programmable timer MC6840 (MOTOROLA) as a free running 16 bit counter at the 2 MHz system frequency. If this or equivalent hardware is not available, an all-BASIC software could be written (with RND function mandatory) at the cost of a much slower execution.

Software and hardware

The MC processor

This is comprised of a BASIC driver and a binary module, the latter essentially performing the above described algorithm at full speed repeatedly. This is of high practical importance because the MC method relies on the law of large numbers, implying many repetitions of every elementary operation. The translation of the evolution algorithm is straightforward; however, great care was taken to define appropriate structures representing the MC object in memory. The design was intended to be general enough so that any kinetic model, with any complexity, could in principle be set up without having to rewrite any portion of the program (see pre-processor below). This is in opposition to the 'one problem – one program' approach, thus allowing the user to enter – if needed modify in the course of execution – his or her model in a special, but easily mastered high level language.

Two structures, STOECH and ESPECE, stand as the internal MC representation of the kinetic model. STOECH is a linked list holding the information related to each reaction (or partial reaction: a chemical equilibrium has 2 partial reactions, forward and backward). Each element of the list is comprised of the following fields: (1) numerical representation of the rate constant, (2) pointer(s) to the substrate(s) and (3) pointer(s) to the product(s). ESPECE is a sequential list, whose elements are pointed to in STOECH and correspond to the different chemical species; each element contains the 'capacity', a scaling constant (playing a role in the analog selection similar to N in the urn model explained above) and the numerical representations of initial, current and updated concentrations. The distinction between the latter two quantities arises as follows. During a cycle of operation of the MC processor, reactions are treated sequentially and the corresponding concentrations ('current') are modified or not according to the stochastic decision taken in relation to the fixed, 'updated' concentrations. Now a cycle corresponds to one unit of evolution of all the reactions in parallel (and not in sequence); in order to correct for the sequential, arbitrary ordering of reactions, 'updated' is assigned the value of 'current' only at the end of a cycle.

The kinetics pre-processor

This is the specialized language interfacing the user with the MC processor. It is presently in development. Two functions are performed:

(a) Mapping the kinetic model onto the MC structures. This is the interfacing language properly. The program accepts strings such as:

$$\text{``}2H_2O \rightarrow 4H^+ + O_2 + 4e\text{''} \tag{4}$$

and parses them extracting the substrate(s) S_i and product(s) P_i (separated by the arrow symbol); the species names are stored after checking against multiple definitions. The name is an unlimited alphanumeric string; however, if it starts with a numeral, the latter is interpreted as a stoichiometric coefficient (e.g. in Eqn (4), "$4H^+$" is analyzed as the species "H^+" with coefficient 4, whereas "O_2" is the species "O_2" with default coefficient 1). At this stage, a stoichiometric matrix is progressively being constructed, one line per reaction and one column per species. For instance, assuming H_2O, H^+, O_2 and e were given the index numbers 0, 1, 2 and 3, Eqn (4) is translated into the following line:

$$
\begin{array}{lcccccc}
\text{column:} & 0 & 1 & 2 & 3 & 4 & \dots \\
\hline
\text{line:} & 1 & -4 & -1 & -4 & 0 & \dots
\end{array}
\tag{5}
$$

The convention is obvious: positive/negative/zero for S_i/P_i/irrelevant species. Next, the program prompts for the rate constant, implied to be given in a standard (molarity, second) or at least coherent system of units. The procedure is terminated upon entering the null string (Carriage Return). Finally, the program prompts for the capacities and initial concentrations for all species previously defined.

(b) Translating the phenomenological quantities (concentrations, rate constants) into integer constants. As the MC processor is designed to perform only simple 16/8 bit operations (addition, subtraction, comparison), this translation involves some approximations (16 bits is worth 5 decimal digits). An error is flagged if the 16 bit range is exceeded. In fact, nothing forbids to extend this range, at the expense, however, of speed performance. The parameters, once translated, are passed to the appropriate locations in STOECH and ESPECE.

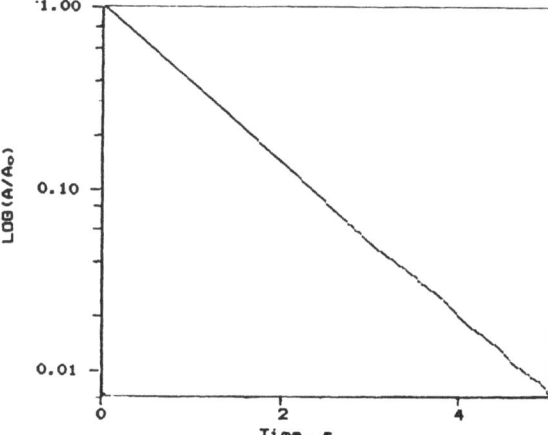

Figure 1. Exponential test for the RN generator used in this study. The first-order reaction: $A \rightarrow B$ was simulated with initial concentration = 1535 arbitrary units (au), rate constant $k = 1\ s^{-1}$; step size on time axis = $0.6514 \cdot 10^{-3}$ s; average of 20 runs (each run is 7650 RN drawings). The last point (at $t = 4.983$ s) is (A) = 11.2 au, whereas the theoretical expected value is 10.5 au.

RN generators

Having a good generator is most critical for MC processing. This means that the generator must be fast and statistically correct for the intended purpose, two qualities often incompatible. The first quality is obviously needed in view of the repetitive nature of the MC method. However, I was rather pleased to observe that this requirement is actually not as stringent as it would seem a priori, as quite a small number of repetitions are sufficient in practice in order to get a decent simulation (see Figure 2). Nevertheless, a fast generator remains highly desirable, the more so the more complex is the model and the larger is the total amplitude of numerical constants. Two classes of generators are known: software and hardware. The software type (e.g. the RND function in BASIC) is inherently slow and was therefore avoided. After numerous trials, I have come to a simple hardware scheme consisting of a 16 bit free-running counter. An analogy is a fast spinning disk where angular positions are read at intervals; randomness tends to result due to the above quantity being modulo 2*PI provided the period of reading is irregular and, on the average, much longer than the rotation period. At any rate, a generator must be tested for correctness. A large amount of work has been devoted to this topic [2]. Randomness may be tested in a number of ways that are not always equivalent (a generator may pass one test successfully and not another). For the purpose of MC kinetics simulation, it is therefore necessary to check the generator carefully for a number of properties defining useful correctness. Two properties seem intuitively required here: no 'hole' in the distribution and no correlation between successive drawings. Usually, once a property and its probability distribution have been defined, a generator is tested using the standard χ-squared technique [2]. I have used an equivalent, much quicker method; I call it the 'exponential' test. It simply amounts to run a first-order reaction on the MC processor using the generator under test and to visualize the substrate decay in a semi-log plot, which should be a straight line (over the full 16 bit range). Such a test is depicted in Figure 1. During initial testing, this generator displayed a definite departure from theory in the range of small numbers. The effect was traced back to a 'hole' effect. It has been largely corrected by software: when the concentration integer number falls below a given threshold (here set at 16), it is up-scaled along with the associated capacity number; corresponding down-scaling is performed when the BASIC driver reads the concentration.

Results

The Michaelis-Menten scheme of enzyme catalysis has been simulated for illustrative purposes. The kinetic model would be fed into the pre-processor as follows:

$$E + S \rightarrow ES$$

$$k = \underline{1.E6}\ M^{-1}\ s^{-1}$$

$$ES \rightarrow E + S$$

$$k = \underline{35.4}\ s^{-1}$$

$$ES \rightarrow E\ (+ P)$$

$$k = \underline{63}\ s^{-1}$$

E: $C_0 = \underline{3.9\,\mu M}$ $C_i = \underline{3.9\,\mu M}$

S: $C_0 = \underline{1\,mM}$ $C_i = \underline{1\,mM}$

ES: $C_0 = \underline{3.9\,\mu M}$ $C_i = \underline{0}$

(k, C_0 and C_i stand for rate constant, capacity and initial concentration, respectively; underlined items are entered by the user; note that the third reaction does not actually mention the product P: this is allowed).

Figure 2 shows the time course of (S) and (E) for various numbers of runs. The induction phase of the enzymatic reaction is particularly apparent, as is the pseudo 0-order of the decay of S (Figure 2F). Notice that even a single run (Figures 2A and D) yields quite an acceptable representation of the

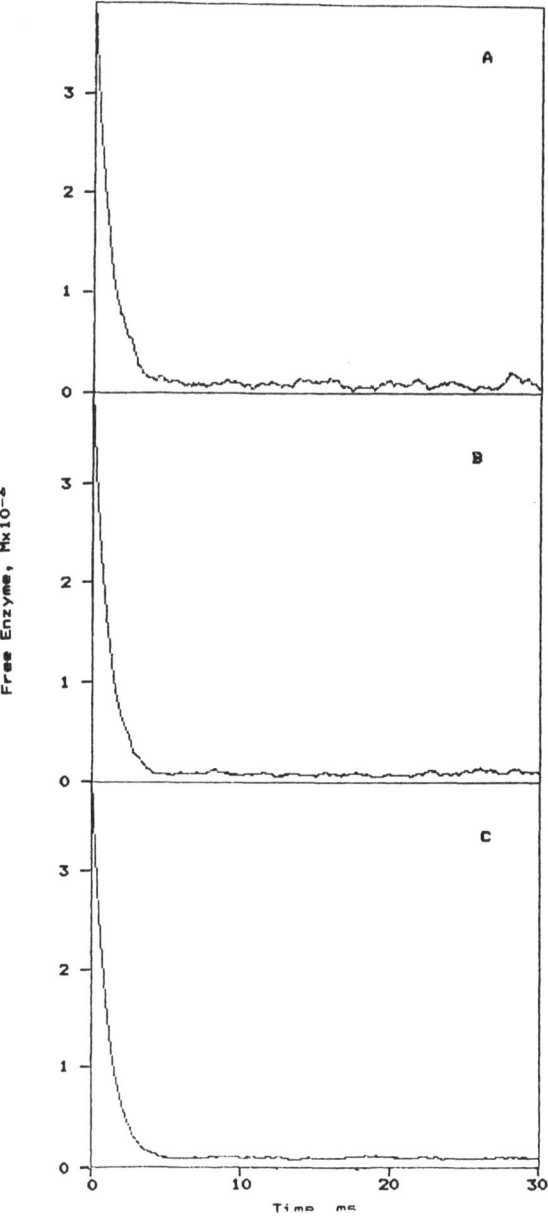

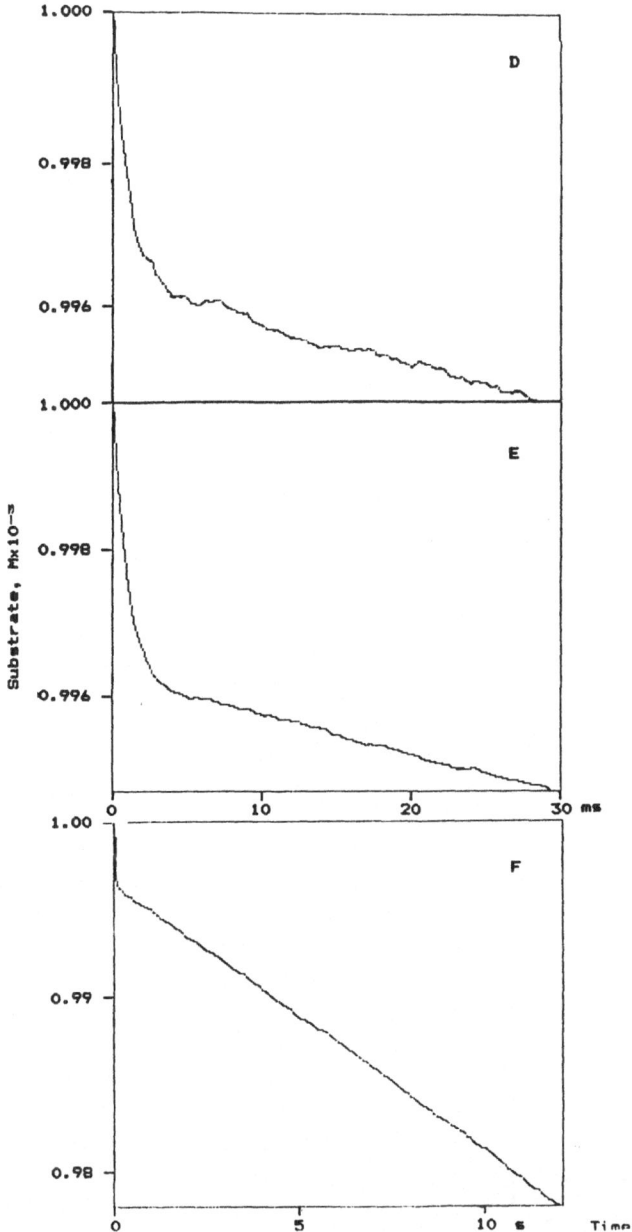

Figure 2. Simulation of the Michaelis-Menten scheme of enzymatic kinetics (see parameters in text). Step size on time axis $= 3.92\,\mu s$. (A), (B), (C): time course of free enzyme E concentration for 1, 4 and 16 runs, respectively. (D), (E): time course of substrate S concentration for 1 and 4 runs. (F): as (D) with step size on the time axis $= 52.26\,\mu s$.

kinetic system. As a run takes only little time (a few seconds, including the graphic display), the ability of quickly visualizing a model with a set of parameters, to modify them conveniently – and even the model itself – can be very helpful in the early stage of model formulation. It should be pointed out that the noise appearing in Figure 2 is a direct reflection of molecular fluctuations experienced by the kinetic system (and wholly un-related to the instabilities seen in numerical integration). This is a new source of information, of theoretical and practical value (see [3]). At any rate, I find it esthetically pleasant, as it adds a touch of realism to the simulation.

Conclusion

The project in its present state demonstrates several positive features of the MC approach: it is simple to program and to run, it may be organized into a general software package of very modest size, accepting any kind of kinetic model (and permitting run-time modifications), it is not subject to numerical instabilities and is reasonably accurate and fast. I have shown also that this approach is well adapted to the capabilities of typical modern microcom-puters.

At least two directions are apparent for improvement. The first is con-cerned with the quality of the RN generator. If the principle of the free-running counter is kept, then it seems obvious that the faster it shall be the better its statistical correctness. The second major improvement is dealing with the inherent slowness of MC processing: if a low noise level is required, averaging is necessary (and the signal-to-noise ratio only goes as the square root of the number of runs). Also, as the complexity of the target model increases, the cycle gets slower. One possibility would be to concentrate the MC processing in hardware, ideally using a VLSI component, similar to the arithmetic co-processors now commercially available. But this is probably out of reach for economical reasons. Meanwhile, a more modest – yet ef-fective – solution might be tried. One may think of a dedicated machine built of discrete active components. Indeed, all functions (logical and arith-metic) in the MC processing can obviously be realized using standard com-mercial integrated circuits. Whatever the design, it is clear that a parallel architecture – e.g. 'one reaction – one discrete MC processing unit' – would naturally impose itself, which could be an important factor for achieving a high execution speed.

I shall be glad to communicate a detailed version of the described soft-ware to interested colleagues for their personal use. The cost would have simply to cover secretarial expenses.

References

1. Feller W (1957) An Introduction to probability theory and its applications, Vol 1, pp 142–154. New York: John Wiley

2. Knuth DE (1969) The art of computer programming, Vol 2: Seminumerical Algorithms, pp 1–160. London: Addison-Wesley
3. Lavorel J (1973) Physiol Végét 11:681–720
4. Pennington RH (1970) Introductory computer methods and numerical analysis, pp 471–491. London: MacMillan
5. Sebban P and Barbet JC (1985) Photobiochem Photobiophys 9:167–176